Joseph Jacobelli
Asia's Energy Revolution

Joseph Jacobelli

Asia's Energy Revolution

China's Role and New Opportunities as Markets
Transform and Digitalise

DE GRUYTER

The contents of this book are solely of a general nature and shall not be construed as or relied upon in any circumstances as professional, targeted financial, or investment advice or be considered to form part of any offer for sale, subscription, solicitation, or invitation to buy or subscribe for any securities or financial products. The author may have owned in the past or may currently own or may own in the future one or more securities and other financial instruments mentioned in this book. You should obtain the assistance of a professional for advice on investments, tax, the law, or other professional matters before reliance is placed upon any content in this book. The author has made best effort to ensure that the information provided in this book is accurate as of November 2020, however the author does not give an explicit or implied assurance or guarantee in respect of the accuracy, reliability or completeness of the information.

ISBN 978-3-11-126293-2
e-ISBN (PDF) 978-3-11-069985-2
e-ISBN (EPUB) 978-3-11-069990-6

Library of Congress Control Number: 2021933273

Bibliographic information published by the Deutsche Nationalbibliothek
The Deutsche Nationalbibliothek lists this publication in the Deutsche Nationalbibliografie; detailed bibliographic data are available on the internet at http://dnb.dnb.de.

© 2023 Walter de Gruyter GmbH, Berlin/Boston
This volume is text- and page-identical with the hardback published in 2021.
Cover image: pictures on right side: China Longyuan Power Group Corporation Limited/
    left side: Sefa Kart/iStock/Getty Images Plus
Typesetting: Integra Software Services Pvt. Ltd.
Printing and binding: CPI books GmbH, Leck

www.degruyter.com

Dedicated to

Caterina, Jean-Luc, Luca, and Shirley.
Thank you for all of your encouragement,
for all of your patience, and
for all of your love.
                              – Joseph

Acknowledgements

My humble and sincere gratitude goes towards the great number of people who helped me in this project in a great variety of ways and areas, including brainstorming, securing data, and fact checking. From inception to manuscript completion, Douglas Wong spent many hours in helping by exchanging ideas and providing so much value-added and constructive criticism. A big thanks also goes to a variety of people without whom this book would not have been possible. I would like to note that the help and opinions provided are purely their own and do not directly or indirectly represent the view of their current or previous employers. For the first chapter, on Asia's economy and energy, Justin Jimenez put me on the right track. Four friends most kindly contributed their thoughts on Asia's economy and energy future, namely Manu Bhaskaran, Richard Jerram, Mark Matthews, and John Seel. For the second and third chapters that discuss the changing energy landscape in Asia and analyse the dual movements towards clean energy and digitalisation, I have provided a great number of real-world insights or case studies that should help the reader in better visualising some of the developments. For this I thank Aden's Fulvio Bartolucci, Asia-Pacific Energy Research Centre's David Wogan, ENN Holdings' Grace Wei, Gradient's Giorgio Fortunato, InterContinental Energy's Alex Tancock, Longyuan Power's Yuan Wang, Macquarie's Charles Yonts, Mayer Brown's Rupert Burrows, Mott MacDonald's Duncan Barker, Swire Properties' Priscilla Li, Synergis' Paul Snelgrove, WM Motor's Shaun Garrard and Rupert Mitchell, and WoodMackenzie's Mark Hutchinson. For the fourth and final chapter, on financing growth, I am most grateful for all of the value-added help and comments from CPI Global's Donovan Escalante and June Choi, HKEX's Grace Hui, independent consultant Sam Hilton, Ms Yuan Sun, Pinsent Mason's Paul Haswell, Schulte Research's Paul Schulte, the UN's Anthony Miller, and WWF-Hong Kong's Jean-Marc Champagne. I also offer my gratitude to the many friends who offered unlimited encouragement including David Bae, Thomas Jastrzab, David Morgan, Jan Napiorkowski, Keith Punzalam, and John Yeap. Last but not least, I would like to give my thanks to the-team-behind-the-scenes at my publisher, De Gruyter, including Jaya Dalal and Stefan Giesen.

Giuseppe (Joseph) Jacobelli
Hong Kong S.A.R.
15 November 2020

Foreword

Asia's record of spectacular growth has boosted the region's living standards, and this achievement is admired and envied by developing economies around the world.[1] While many attribute this miraculous economic development to the region's young and growing population, it is evident that 'demographics does not always define national destiny'. A quick glance at the countries in Africa and the Middle East shows that many continue to be mired in poverty despite their young population and abundant natural resources.

The 'secret sauce' behind Asia's success is the ability to harness raw young talent with appropriate "tools" to take advantage of emerging opportunities. For Asia, the opportunities arose from expanding global trade, and the search for low-cost highly efficient manufacturing centers. Essential factors for attracting a broad spectrum of manufacturers into the region were abundant supplies of reliable clean energy, supported by an infrastructure shaped around a flexible, preferably digital, backbone. Going forward, the demands of 21st century sustainable production will require as much as three to four times the current level of electricity generation by 2050. Despite the region's wide range of energy production and transmission options (as outlined in Chapter 2 of this book) there will be daunting logistical and technological challenges in sorting through emerging options for transforming the landscape to meet the energy needs of the next century (Chapter3).

The immediate challenge to ensure the smart development of Asia's energy resources centers on finding and committing the enormous amount of capital required to build the energy-related resources demanded by the emerging digital economy. It will often take years to develop energy options before they become operational; a characteristic that is abhorrent to short-sighted investors in global capital markets who focus on quarterly profits. However, financing green, sustainable, and socially inclusive projects (Chapter 4) does have appeal to the growing cult of 'ESG' investors. Certainly, the size of the financing requirements (likely in the range of trillions of dollars) will require tapping the global capital markets; no single nation or region will have the required savings.

Private–public partnerships that engage multilateral development banks will be vital for attracting the requisite trillions of dollars from global investors who are looking for safe yet high-return investments. Bank loans and 'vanilla' instruments (e.g., bonds and equity) commonly used to raise funds will not be adequate for Asia's many upcoming cross-border and long-dated infrastructure projects. The challenge for global investment bankers will be to develop specialised financial products with varying tenors, risk exposures, and credit guarantees to appeal to a wide range of global investors.

---

[1] The views expressed here are my own and do not necessarily reflect those of the Milken Institute.

Joseph Jacobelli has surveyed a broad range of issues essential for understanding and navigating the complex business and financial nuances associated with choosing among Asia's many energy options. This volume will become an essential reference for all who seek a deeper understanding of how investors and policy makers will be making the choices that will reshape Asia's future.

Dr. William Lee
Chief Economist-Milken Institute

Preface

With much fanfare, albeit online, on 15 November 2020 a group of leaders representing 15 Asian economies inaugurated the Regional Comprehensive Economic Partnership (RCEP), the world's largest free-trade agreement creating a formidable trading block, representing almost one-third of the world's population and the global economy. It was a rare example of some form of Asian unity given the region's vast geographical, political, and cultural differences. The agreement was ten years in the making and still needs a lot of fine tuning. But it has put yet another global spotlight on the region. Most experts are forecasting that Asia will enjoy some of the fastest gross domestic product (GDP) growth rates in the world over the next few years while the population will grow from the current 4.5 billion plus. Now that there is an increasing number of economies getting on the decarbonisation band wagon globally, the region's energy complex will increasingly attract global attention.

The Rationale for the Book

Having 'lived' the region's energy developments for the past 30 years, in early 2019 I started to get excited about the gradual formation of a huge makeover of the energy landscape. I saw the advent of a quiet, massive revolution. Quiet because change is occurring at different speeds and depths in individual electricity markets. Massive because the region is already the biggest energy consumer in the world, and it will get even bigger in the coming three decades. I feel that these transformations in Asia will impact the world as a whole. Also, I believe that the growth and transformation of the region's energy complex offers an incredible amount of investment and business opportunities for market incumbents as well as for new entrants. These thoughts led me to a journey that involved hundreds of hours in researching and writing this book. I also confess that another incentive in writing this book is that I had missed the opportunity to share a vision back in 2010. I had then started on a book project about why clean energy would become mainstream in the very short term. Work priorities meant that I put the project aside. I got back to it three years later. By then clean energy had already become mainstream. In fact, today power from wind or solar farms is no longer treated as some form of exotic separate source of energy. It was a fortuitous event that I switched careers in 2019 from analysing the developments in the Asian electric power industry to directly working in it, thus becoming a market participant.

Since starting to research and write this book in early 2020, two considerable events impacting Asia's energy development occurred. One is the COVID-19 pandemic. The other is key Asian economies announcing their net-zero targets. Both have proven important for an acceleration in clean energy generation additions. The expected massive build-up of solar, wind, and other clean energies capacities in the region has also promoted interest in investing in digitalisation. This is because new digital technologies and solutions are key to optimising energy supply, including helping grid networks to cope with the intermittent output from solar and wind farms, that is, no sun or wind, no solar or wind power.

The Objectives of the Book

The goal of my work is to provide some basic tools or intelligence for corporates, financial institutions, investors, and others to understand the energy landscape and then to identify new business or investment possibilities. The growth that the region is set to witness over the next 30 years from the energy mix transitioning to become green and sustainable and from digitalisation, offers almost limitless opportunities. I aimed at using plain language even though energy and digitalisation can get quite complex and technical. I also resisted giving laundry lists of facts, technologies, and trends. Rather, I focused on key factors while providing some colour, including contributions from other on-the-ground experts.

I mentioned that this book will look at change in the next three decades. Thirty years may sound like an eternity to some. In the world of energy, it is a relatively short time. Generation facilities need one or more years in the planning. Construction can be one to two years for a solar farm and as much as six or more years for a nuclear power plant. Then they will operate for 20 to 40 years. This is the reason why in the energy industry 30 years is not an eternity.

Change is occurring because of two short-term trends and two long-term ones. These trends all point to Asia's energy sector being the new promised land for business growth, for assets' investment, and related opportunities. The vast possibilities are not limited to investing into projects such as solar power plants. There are opportunities for equipment suppliers and professional services firms such as lawyers, business consultants, and engineering firms as well as financial institutions – and, of course, there are plentiful prospects for firms involved in digital technologies and solutions.

The two short-term trends are consumption growth and more open markets. The existing gigantic, and rapidly growing, energy requirements in the region are set to boom further. Most experts expect energy consumption to grow several folds in the region in the coming decades. Another trend is the push by an increasing number of jurisdictions in the region to get the market to drive the price of energy rather than having some kind of regulatory body lead the game. This will help the energy markets become much more effective and more efficient.

In the next 30 years, a very dramatic shift in the energy mix will be witnessed in most key economies, including the region's largest: China, India, Indonesia, and Japan. Electricity, the chief source of primary energy consumption, will no longer principally come from burning polluting thermal coal. The type of resources will be much more varied, and the despatching or delivery of those sources will see a breath-taking shift partly because of the entry in the markets of newer technologies, such as various forms of energy storage. Another long-term trend is the advent of digital technologies and solutions spurred by incredible improvements that have helped processing power to be significantly faster and cheaper. Artificial intelligence (AI), blockchain, and the Internet of Things (IoT) are being rapidly adopted by the energy industry; these and other digital solutions are explained in detail in the 'New Opportunities Through the Digital Door' section of Chapter 3. The technological digital solutions will drive the optimisation of the demand and the distribution of energy, making the sector much more efficient.

Now, none of the elements that I have mentioned are novel for those of us who 'live' the energy industry in the region. However, one must look at all of these elements in the aggregate; thinking of it as just one gigantic revolution. A multidisciplinary approach must be very widely implemented by all actors involved in the sector. In my view this is the rational way forward. Many of the largest Asia energy markets are still at various stages of development and some of the more mature ones are actually going through a metamorphosis in the construct of their domestic energy systems.

Finally, apart from trying to help in the identification of business and investment opportunities, I also hope to have addressed some misconceptions. For example, the misconceptions that non-local companies cannot succeed in the local energy businesses, that they cannot earn attractive returns on investment in such (perceived) closed economies as China and Japan, or that Chinese firms are not strong at technological innovation. Such perceptions are simply not true, and I will reference examples of companies in the energy sector that did succeed and examples of Chinese firms that are incredibly innovative. Also, I should point out that there are two important areas that I purposely did not tackle, politics and cybersecurity. This book strived to be apolitical and as such I did not address the geopolitical ramifications of the massive changes that we are witnessing in the Asian energy markets. It also did not address the cyber-security dimension of the massive digitalisation of the sector given that there is already an enormous amount of expert literature on the subject.

It is my sincere hope that this book will bring you a great amount of food for thought and will provide you with sufficient material to identify how to take advantage of the enormous numbers of opportunities offered by Asia's energy revolution.

Contents

Acknowledgements —— VII

Foreword —— IX

Preface —— XI

Chapter 1
The World's Centre of Economic Growth and Energy Transition —— 1
1.1   Why Asia? Why Energy? —— 1
1.1.1   Continuation of Asia's Stupendous Economic Growth —— 1
1.1.2   More Growth Will Need More Energy —— 5
1.2   More Views on Asia's Economic Growth and its Future —— 6
1.2.1   Question One: Growth Potential? —— 7
1.2.2   Question Two: What are the Growth Elements? —— 10
1.2.3   Question Three: What are the Policies Critical for Growth? —— 15
1.2.4   Question Four: What are the Risks to Economic Growth? —— 17
         Notes —— 20

Chapter 2
Asia's Dramatically Changing Energy Landscape —— 21
2.1   The Energy Consumption is Unique and is Rising —— 21
2.1.1   Plain Vanilla Look at the Growth Upside —— 22
2.1.2   Higher Demand from Shift to Electric Vehicles —— 28
2.1.3   Energy Efficiency Could be Growth Containment Factor —— 37
2.1.3.1 Saving Money: The Regulation and the Implementation —— 37
2.1.3.2 Energy Savings Case Studies in the Region —— 38
2.1.3.3 Energy Mega-Users: Buildings —— 41
2.1.3.4 Making a Business Out of Saving Energy —— 44
2.1.3.5 Increasingly Brighter Lighting Prospects —— 48
2.2   Huge Regulatory Shifts Transforming the Energy Industry —— 51
2.2.1   Australia – Deregulation Lessons Brutally Learned —— 56
2.2.2   Japan – The Land of the Slowly Rising Competition —— 62
2.2.3   Singapore – Progressive but Paced, Highly Controlled Reform —— 69
2.2.4   China – a Giant's Long Build Up, Accelerated Adoption —— 75
2.3   Changing Nature of Industry Players —— 83
2.3.1   UK: Precursor of Change BG and its Complete Metamorphosis —— 84
2.3.2   Australia's Major Energy Retailers: Similar Yet Different Models —— 86
2.3.3   New Zealand's Retailers' Valuable Experience —— 89
2.3.4   Japan's Tokyo Gas, KDDI: Cautiously Progressive Approach —— 91

2.3.5	Singapore's Transition to Newborn Utilities	96
2.3.6	China's Private and State-Owned Fast Changers	97
	Notes	103

Chapter 3
Twin Transformations: New Fuel Mix and New Tech —— 106

3.1	Impact of Changing Energy Fuel Sources and Price Trends	109
3.1.1	Fuel Mix Metamorphosis to Green and Clean from Brown and Dirty	110
3.1.1.1	Australia – Vast Energy Resources, a Curse in Disguise?	113
3.1.1.2	East Asia – Road to Less Energy Dependence	118
3.1.1.3	China – Dash to Clean Energy, Pressures from Pollution and Emissions	123
3.1.2	Asian Clean Energy Prices' Fall is Cliff-Hanger for Coal, Gas, Oil	144
3.2	New Opportunities through the Digital Door	150
3.2.1	New Digital Tech and the New Energy World	152
3.2.1.1	Getting One's Head Around Digital Energy Tech: How it Works	152
3.2.1.2	New Digital Energy Tech, Many New Benefits	159
3.2.2	The Real Life of Digital Energy Tech: Some Use Cases	164
3.2.2.1	South Korea Has a Surprising Tech Innovator	164
3.2.2.2	Asia's Little Big Digital Energy Technologies Giant	168
3.2.2.3	The New Breed of New Energy Digitalisation Market Entrants	169
3.2.3	China's Fast-Track Energy Tech Digitalisation	177
3.2.3.1	Augmenting Reality of Digital Leadership	177
3.2.3.2	China's Smartening of Energy and its Earth-shaking Near-Term Future	181
	Notes	190

Chapter 4
Financing the Growth —— 198

4.1	Financing Energy Digitalisation Tech: Rich and Well-Proven Channels	198
4.2	Alphabet Soup and Drivers of Green Capital Growth	202
4.2.1	The Alphabet Soup: What's in a Definition?	203
4.2.2	The Drivers of Green Capital Growth	206
4.3	Plain Vanilla Financing Turns Green	211
4.3.1	Corporates Building Green Equity Credentials	211
4.3.2	Bond Issuers Learning from Chameleons	221
4.3.3	Green Bank Borrowing	228
4.4	Future Greening of Finance and Financial Instruments	238
4.4.1	Higher Adoption and Capital Flows	238

4.4.2	New Tools and Paths to Green Liquidity —— **241**	
4.4.3	Revaluating New Valuations —— **248**	
4.4.4	Crystal Gazing into the Green Future —— **251**	
	Notes —— **253**	

Chapter 5
Conclusions and Suggestions —— 259

List of Figures —— 267

List of Tables —— 271

List of Abbreviations —— 273

About the Author —— 275

Index —— 277

Chapter 1
The World's Centre of Economic Growth and Energy Transition

1.1 Why Asia? Why Energy?

The energy sector is one of the largest industrial sectors and one of the most vital in the global economy. That of Asia is already the biggest in the world by a wide margin. In the next 30 years, it is slated to climb even further. Its size and growth will create an enormous amount of investment and new business opportunities for companies directly or indirectly involved in the sector – for energy companies, energy equipment manufacturers, professional services providers such as consultancies and law firms, and brand new innovative businesses that do not even exist yet. The objective of this first chapter is to offer a macro and quantitative evaluation of Asia's economic and energy positions in the world so as to provide context to the discussion of the way the region's energy sector is poised to evolve in the next 30 years or so.

1.1.1 Continuation of Asia's Stupendous Economic Growth

The world's centre of economic gravity has been shifting to Asia for many years and will continue to do so over the next few decades. In this book, the Asia region is defined as comprising the whole of Asia and Asia Pacific, including Australia, New Zealand, and the Indian subcontinent. The region accounts today for more than a third of global output, based on data from the World Bank, and is home to some of the world's fastest growing economies. Crucially, a sustained strong pace of growth is not possible if the continent does not have enough energy to power the growth. Also, it is economically and politically important that the energy must be affordable and that its supply must be reliable and secure. These subjects are discussed in detail in the next chapter of this book.

There are tons of books and studies available that tell the narrative of Asia's growth. Looking at tables with lots of numbers and data-heavy charts may not be the most exciting thing in the world, yet it is absolutely crucial to look at a few data points so that one may be able to better gauge the extent of the potential business and investment opportunities, including their size and their growth propositions.

Asia now accounts for more than 60 per cent of the world's population, based on numbers from the United Nations' Department of Economic and Social Affairs (Figure 1.1). The region added a massive 1.35 billion people or so from 1990 to 2018. By comparison, Africa comprises about 17 per cent of the total, the Americas roughly 13 per cent, and Europe approximately 10 per cent. Between 2018 and 2050, Asia is expected to add another huge number of people. A number close to three-quarters of a billion, based on the department's mid-range growth forecast.

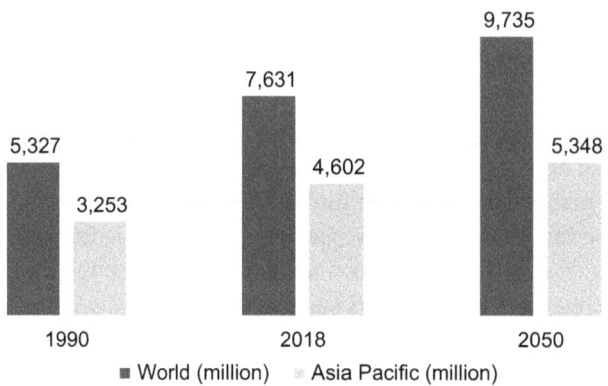

Figure 1.1: Asia and World Population Forecasts.
Source: United Nations, Department of Economic and Social Affairs, Population Division (2019). World Population Prospects – Population Division – United Nations. [online] Un.org. Available at: https://population.un.org/wpp/Download/Probabilistic/Population/ [Accessed 4 September 2020].

At the same time, the demographics of the region will also dramatically change during the next 30 years, and beyond. The emerging economies in the region such as Indonesia, the Philippines, or Sri Lanka will continue to have relatively younger, fast-growing populations. Some other jurisdictions such as Australia, China, and Japan are facing fast-ageing populations. These dynamics mean that the region's share of the global population will decline 5.4 percentage points – or about one-tenth – during the period. Still, it will manage to account for a massive 54.9 per cent of the global total.[1]

Apart from being home to more than half of the world's population, Asia also accounts for an incredibly large swathe of total global output. It comprised about 34 per cent of global output as of 2018, up from less than 29 per cent at the beginning of the decade. The other regions accounted for far less (Figure 1.2).

The sharp 52 basis points increase between 2010 and 2018 was due to rapid economic growth. GDP change in East Asia and the Pacific averaged 4.6 per cent between 2000 and 2018, while that of South Asia averaged 6.5 per cent, according to data compiled by the World Bank Group.[2] That compares with 2.8 per cent average growth for the world as a whole. Only that of Sub-Saharan Africa came close to the growth rates seen in Asia, with average growth of 4.7 per cent during the period, albeit its global economic footprint is tiny (Figure 1.3). Its average annual per capita GDP growth rate has also been impressive. For each of the past three decades, developing Asia's rates of GDP per capita were 4.9 per cent (1990–1999), 6.2 per cent (2000–2009), and 5.5 per cent (2010–2018) while that of developed

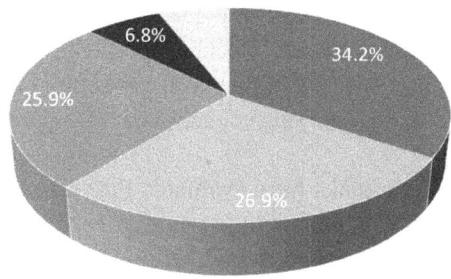

- Asia
- North America
- Middle East & Africa
- Europe and Central Asia
- Latin America and Caribbean

Figure 1.2: Global GDP Share Breakdown by Region.
Source: The World Bank (2019). Gross Domestic Product Ranking Table, World Development Indicators, The World Bank. [online] Worldbank.org. Available at: https://datacatalog.worldbank.org/dataset/gdp-ranking [Accessed 4 September 2020].

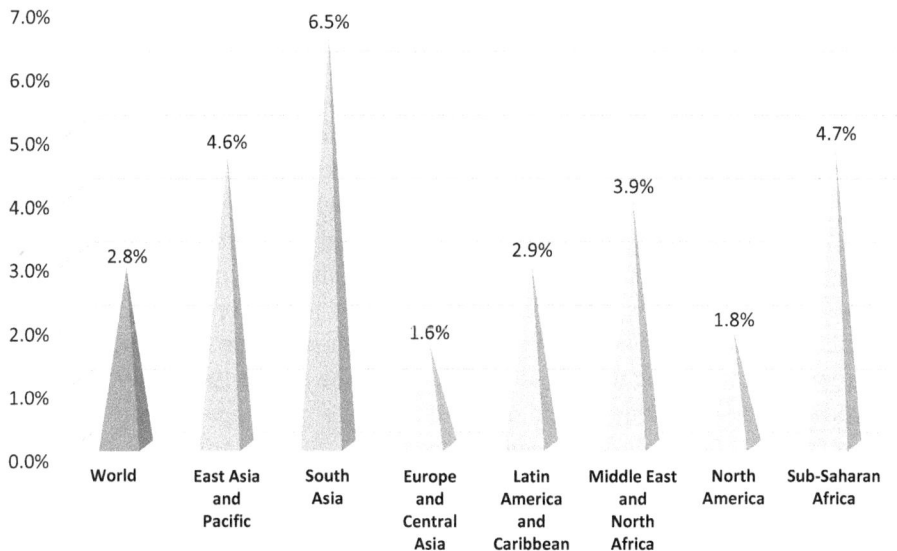

Figure 1.3: World Average Annual GDP Growth 2000–2018.
Source: The World Bank (2020). 4.1 World Development Indicators: Growth of Output, World Development Indicators, The World Bank. [online] wdi.worldbank.org. Available at: http://wdi.worldbank.org/table/4.1 [Accessed 4 September 2020].

Asian economies were 1.3 per cent, 0.7 per cent, and 1.5 per cent respectively, based on data from the Asian Development Bank.[3] Looking specifically at the last decade, from 2010 through 2018 (Figure 1.3), the higher GDP per capita rates were driven by China (7.3 per cent), East Asia (6.4 per cent), South Asia (5.8 per cent), and Southeast Asia (4.1 per cent).[4]

A very large number of studies and books have evaluated the reasons behind Asia's economic growth success, and it is almost impossible to comprehensively summarise. Still, one of the views that may be worth highlighting is from the Asian Development Bank. The regional multi-lateral institution recently published a book evaluating the region's growth in the past 50 years. The authors argue that the achievements were chiefly due to more optimal policies and more optimal institutions. They believe that

> [t]hese policies and institutions helped develop and nurture market economies and a vibrant private sector which, in turn, led to sustained technological adoption and innovation. This process benefited from governments' (i) pragmatism in making policy choices, including the practice of testing or piloting major policy changes before full-scale implementation; (ii) ability to learn lessons from its own and others' achievements and mistakes; and (iii) decisiveness in introducing (sometimes drastic) reforms when needed.[5]

These statements may appear to some readers a little generic, but the Asian Development Bank's book backs up the thesis with over 600 pages. The point is that what seems to have worked in Asia, perhaps strangely so, is the way governments have set policies, built a foundation, and then gone on to execute on these policies. Asia's economies differ a great deal from one another in a multiple of ways, be it cultural, demographic, or political. Let's take Thailand as an example. The kingdom has seen its fair share of political upheavals, natural disasters, and economic earthquakes in recent decades. These include in the past 100 years or so 13 successful and nine unsuccessful coup d'états, and in just the past 25 years, three economic recessions (not including the recession induced by the COVID-19 pandemic) as well as a dozen minor to major natural disasters. Still the nation managed to post one of the fastest economic growth rates in the region. Some institutions, including the energy-related ones, managed to relatively retain policy transparency and consistency.

In the next three decades, the order of the main engines of the region's GDP growth will change I believe. China will soon become the world's largest economy, but its speed of growth will be slower than that of the previous three decades. The economic policy of Chinese authorities shifted its focus away from a manufacturing-intensive one, in favour of a service-based economy in recent years. Apart from China, the East Asia markets are becoming increasingly mature and their populations are ageing, too. To be clear, the pace of growth of the Chinese and East Asian economies is still healthy albeit the pace is slower.

Simultaneously, the region's growth push will come from steadily rising consumption and productivity gains in South Asia and Southeast Asia while the economies of China and East Asia will reach a near-mature to mature status.

Prior to the COVID-19 global economic slowdown, leading public and commercial institutions had highly positive long-term outlooks for the region in general. For example, global professional services provider PricewaterhouseCoopers (PwC)[6] forecasted that not only would China replace the US as the world's largest economy but also that India would claim the number two spot and that Indonesia would rise to number four by 2050, after the US. While the current economic malaise will slow their ascent, their long-term absolute growth trajectories will not change.

1.1.2 More Growth Will Need More Energy

A rising population and steady GDP growth mean that more energy is necessary given that the bulk of the economies in the Asia region are not developed ones. The more mature energy markets, including Australia, Hong Kong, Japan, New Zealand, and Singapore, today have economies that have already shifted away from having a dominant manufacturing sector and are more service industry-oriented or commercial services-focused, sectors that require significantly less energy consumption than manufacturing. Meanwhile, the less developed economies in the region with vibrant and growing industrial activity, including India, Indonesia, the Philippines, Sri Lanka, and Thailand, will need a lot more energy to satisfy the growing appetite from their industrial sectors. The great need for more energy over the next three decades is evaluated in detail in the next chapters but at this juncture it is simply worth highlighting that the region is already the world's largest user of the world's energy resources.

In terms of energy usage, just like the size of its population and that of its economic output, the region is already an incomparable giant. The Asia region (Figure 1.4) accounted for a share of 44 per cent of the world's primary energy in 2019, equivalent to 2.2 times that of North America, which had a 20 per cent share, and 3.1 times that of Europe, which had a 14 per cent share, based on data from the widely used BP Statistical Review of World Energy. On the electric power generation side of things, the region's share was even larger, standing at 47 per cent of the world's total, 2.3 times more than that of North America's share and 3.2 times more than that of Europe.[7] Still, energy resources are the lifeline to economic growth. They are essential for the powering of the economic engine.

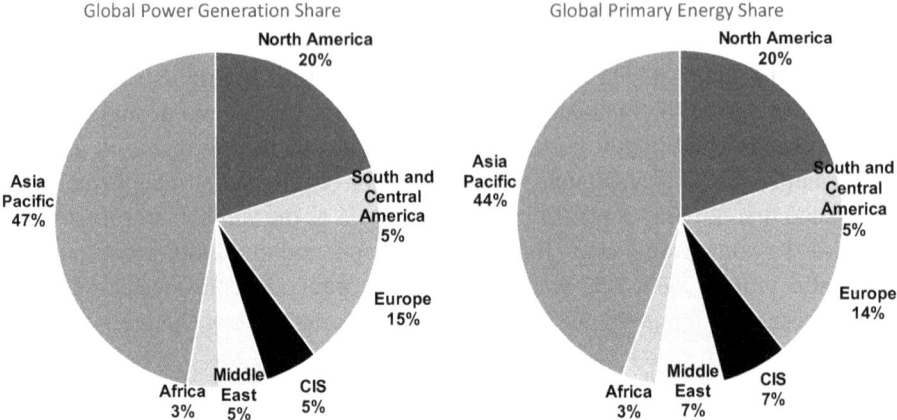

Figure 1.4: Global Energy and Power Output Share in 2019.
Source: bp p.l.c. (2020). Statistical Review of World Energy. [online] BP Global. Available at: https://www.bp.com/en/global/corporate/energy-economics/statistical-review-of-world-energy.html [Accessed 4 September 2020].

1.2 More Views on Asia's Economic Growth and its Future

The future growth of the Asian economies is fundamental to the discussions in the rest of the chapters. To reinforce this basic premise, this author called on four friends who are highly experienced and renowned economists with three to four decades each in macro-economic analysis and investment advice experience, chiefly focused on the Asian region. They each offer broad and in-depth views on the region's economic growth over the next 30 years. Four questions were posed to each of them. They included whether Asia will continue to grow at a pace higher than that of North America and Western Europe, what are likely to be the most exciting elements of the growth, and the policies absolutely imperative for the growth. They were also asked to identify the key risks to the growth and discuss if the region was well positioned to address these risks.

The first expert is Manu Bhaskaran. Based in Singapore, Manu has worked for the government of Singapore, for a global investment bank, and as the director of Asian economic research at a strategic advisory firm working with financial institutions, government agencies, and corporations. The second expert is Dr Richard Jerram. Richard is another leading economist who worked for a consultancy and as a lecturer in his early career. He subsequently worked for global investment banks in Singapore and Tokyo for more than two decades. The third expert is Mark Matthews. Mark has held senior positions managing research and equity sales at several leading global financial institutions in the past 30 years. He has lived and worked in Asia for most of his life, including Hong Kong, Taiwan, and Thailand, and is now based in Singapore. Last but definitely not least is John Seel. John is yet another highly experienced hand in Asia's economics who has worked for some of the world's largest global investment banks, including many years in Hong Kong, Shanghai, and Seoul. Needless to say, the views expressed by Manu, Richard, Mark, and John are their own personal views and do not represent in any way the views of the institutions they work for now or have worked for in the past.

1.2.1 Question One: Growth Potential?

The first question asked by this author pertained to consensus seemingly believing that the 'Asian Economic Century' is far from over. And that many experts remain highly positive towards Asia's economic growth over the next few decades relative to North America or Western Europe. The drivers highlighted by Manu included demographics in much of emerging Asia, infrastructure needs, improving business environments, and the intra-Asia trade prospects. Richard underlined institutions, especially the progressive improvement in governance, as well as education as two drivers. Mark emphasised that the drivers will differ by each of Asia's three geographical clusters. John stressed the complex interplay of global forces, domestic policy decisions, and demographics.

Manu Bhaskaran

There are good reasons to expect most Asian economies to sustain relatively high growth compared to their peers for the next two decades. Although some countries in the region are ageing, overall demographic factors remain supportive of growth, as does the continued flow of workers from rural to urban areas. Shifting workers from relatively low productivity activities in farming to higher value work in the cities and towns raises growth. The growth of Asia's middle class will continue. Consumer spending will thus be a powerful driver of growth.

An infrastructure boom that has just begun in recent years will be another powerful force for good growth. Infrastructure spending will raise growth through higher investment in construction activities in the near term. In the longer term, better roads, railways, mass transit, ports, airports, and power will also be enhancing competitiveness and relieve a constraint on the region's growth potential.

The business environment will be further boosted by improving 'ease of doing business', as governments all over the region are cutting red tape and deregulating. Over time, this should spur a higher investment rate and so a higher pace of economic growth.

While protectionism could well be a headwind to export growth, there is a countervailing force in the form of increasing regional integration, which will help countries in Asia such as in the ASEAN region reap synergies from increased integration. Asian countries such as Singapore and Vietnam have also been adept at negotiating free trade agreements with all the major trading blocs and nations, so helping to protect themselves to some extent from that protectionism.

Richard Jerram

Economists with a focus on Asia often view its continued rise as inevitable. Over the past 50 years, one country after another in the region has followed the path blazed by Japan, sucking workers out of the low-productivity agricultural sector, into export-led manufacturing and then moving up the value-added curve. However, this disguises the fact that outside Asia (unless you discovered oil), almost no country has gone from poor to rich over the same period. Catch up to the West is the exception, not the norm, so one can reasonably ask whether it will continue.

Asia's success can broadly be explained by three factors: institutions, education, and demographics. Of these, demographics looks like the biggest challenge. Most countries are past their 'Lewis Point' where they can grow simply by taking surplus workers out of agriculture and moving them into manufacturing. More worryingly, populations are ageing rapidly. Japan's woes have made headlines for many years, but China's working-age population is set to shrink by about one-sixth over the next 30 years. Many other countries face similar issues. This might not necessarily seem like a problem if one were concerned with per capita incomes, but an ageing population is also a less dynamic one and it introduces questions about the sustainability of public finances.

More positively, institutions are mostly strong and education standards are high. Measures of state capacity and effectiveness show continued improvement. This is evident in surveys such as the World Bank's Ease of Doing Business rankings or Transparency International's Corruption Perception Index. This steady improvement in governance helps to explain why the major South East Asian economies did not see growth rates decelerate in the wake of the 2008–2009 Global Financial Crisis, despite the more difficult global environment. They were able to generate stronger domestic demand by pursuing economic reform. Standards are still relatively low – they are emerging markets, after all – so there is still plenty of room for improvement.

Extrapolation can be a dangerous game, but the potential for continued catch-up still exists, with income levels well below those in the West. Moreover, the track record and current institutions suggest that continued progress should be possible, even if demographics mean that relative growth rates until 2050 are not as wide as they have been over the past three decades.

Mark Matthews

There are many reasons to be positive on the long-term economic growth of the economies in the Asia region in general. This growth will be in three main clusters: China, the Indian subcontinent, and Southeast Asia. We are unlikely to see much growth in North Asia's former tiger economies of the 1980s, South Korea and Taiwan, and also see little or no growth in Japan. With the China and India mega clusters, we have approximately 1.4 billion people in each today. This will obviously change radically over the next 50 years. Data modelling indicates the Chinese population will decline towards the 1 billion mark by the end of the century. India's though will do the exact opposite, climbing probably to 1.7 billion or above. They are on totally different trajectories. China is more of the qualitative growth while India's is more of quantitative growth given the robust demographics while China has the technology. So, they will both grow but in very different ways. For India, without a doubt, the Internet and technology will be a big part of that at the very least over the next 10 years. The double-digit growth in India and in China over the next five years or so for sure will be in the Internet space. These are the two big growth engines thanks to the transition to more of a technologically driven economy in China. China is far ahead of India and probably the entire world in its successful efforts to upgrade all of its basic infrastructure including airports, energy, rail, roads, and telecommunications over the past 20 years or so. Now China over the next five years will be focusing on new economy infrastructure. This includes electric vehicles charging stations, inner city rail for the mega cities such as the Greater Bay Area, 5G, and the IoT among the many areas. This will provide China with a technological sophistication that is relatively unprecedented. As data is the new oil, like some people like to say, this technological supremacy may continue to generate some geopolitical dislocations.

John Seel

Much of Asia has experienced a remarkable economic transformation over the past half century. The region's growth was driven by the interplay of global forces, domestic policy decisions, and demographics. Globalisation brought immense benefits to Asian economies, which became deeply imbedded in global supply chains and financial networks. This brought massive inflows of investment capital and technology that upgraded Asia's industrial base and its international competitiveness. Asia was able to take advantage of globalisation in large part because of appropriate policies that encouraged investment and manufacturing-driven growth. Although Asian governments were hardly laissez-faire in their development policies, they mostly allowed key industries to remain in private hands. This improved the efficiency and flexibility of their economies, especially during their early high-growth period.

They maintained stable monetary and fiscal environments, which helped keep inflation low. Modest and stable inflation in turn reduced political pressures for price controls, avoiding the distortions in economic incentives that they have generated in other regions. Resource allocation was relatively efficient, and consistently high spending on infrastructure reduced bottlenecks and encouraged the development of manufacturing as well as service industries. Underlying Asia's growth was a positive demographic wave, with a large working-age population, improving levels of education, and a massive increase in urbanisation.

In the 21st century, many of these growth factors are under pressure. Globalisation has stalled, under pressure from slower world growth, the Sino–US trade tensions, and populist economic policies, and there are signs that it may go into reverse. There is less room for Asian governments to use infrastructure investment to offset reduced external demand; major Asian economies already have efficient infrastructure, so the return from spending on more would be limited (with the exception of less developed economies and others such as the Philippines that have underinvested in the past). Similarly, there is not much space to expand manufacturing capacity in an environment of excess capacity for many products including steel, automobiles, consumer electronics, and most light manufactured products. Underlying these pressures is the ebbing of Asia's demographic dividend, as their populations age and labour forces begin to contract.

1.2.2 Question Two: What are the Growth Elements?

Ensuing from the first question, the second enquired about the actual elements necessary for the drivers of Asia's economic growth in the next three decades. Manu talks about some of the multi-dimensional improvements that are likely to be witnessed including internal consumption and urbanisation. Richard discusses the impact of the potential shift away from globalisation and how to avoid the 'middle income trap'. Mark brought up growth from a low base of some regions and the impact of new economy businesses. John identified several elements including the openness to the adoption of new tech.

Manu Bhaskaran

What Asia is experiencing is not just economic growth; it is economic development that is multi-dimensional, going beyond simple quantitative outcomes. What Asia is witnessing is much more exciting; it is transformation in many areas. Companies will become more professionalised in what they do, adopting best practices from their developed counterparts. Innovation in processes and products will become more important as a source of economic growth. But for that to happen, there will have to be more and better universities and research institutes. Products available to consumers will become more sophisticated to appeal to a more educated and wealthier consumer, which will be the case in goods as well as services. Government services and policy making will also become more modern in keeping with the advances in society.

Certainly, this means that consumer spending will become more and more important so that the growth in each economy is not so hugely dependent on exports as in the past. And much of the economic dynamism will be concentrated in new urban agglomerations that will emerge over time, such as the major ones that China is investing heavily in.

Richard Jerram
Three elements of Asia's economic growth are set to preoccupy economists over the next three decades. First, how will the growth dynamic be affected by the shift away from globalisation? The rise of Asia has been in parallel to flows of foreign investment and the building of supply chains around the region. That process had already been staggering under the weight of rising trade barriers – which predates the emergence of President Trump – and looks set to intensify in the wake of the disruption from COVID-19. The success of the ASEAN economies over the past decade, through better domestic management, gives room for optimism. A region powered by its own internal dynamics would be more resilient and better placed to withstand whatever difficulties the developed world will face.

Second, can the low- to middle-income economies – which is still most of them – avoid the middle-income trap? Admittedly, there is some debate about whether the middle-income trap exists, but more helpfully we can argue that political structures and institutions that support the early stages of development can become a barrier as the economy grows more sophisticated. Taiwan and South Korea are encouraging examples of economies where a relatively smooth transition to competitive democracy and rising prosperity is possible, but it is not clear China can make a similar shift. A successful adjustment would create hundreds of millions of middle-class consumers and allow the development of currently unimaginable products and services.

Third, can India find a path to continued development? Structurally, India is so different from South East and East Asia that it can seem artificial to try to put it into the 'Asian Miracle' box. It has not followed the export-led manufacturing model and now it is probably too late. Moreover, education standards are a barrier to exploiting its demographic potential. Recent years have been reasonably successful, but due either to easy fixes of obvious problems, or at the cost of a build-up of financial sector imbalances, sustainability looks doubtful. However, if India can find some way of maintaining high growth rates, the sheer size of its population means that it will become an economic powerhouse to balance China's influence in the region.

Mark Matthews

Post COVID-19, China, India, and Southeast Asia, with its 700 million plus people, should all produce annual economic growth in the range of 5 per cent broadly speaking. The Asian century is far from over. The dark horse of course could be India given that its economy is coming from a very low base, especially relative to that of China. One of the reasons we should continue to see healthy economic growth out of China is the fact that in more recent quarters the nation has had to go on its own, turning inward in an 'inner loop'; a new term now more emphasised than Belt and Road strategy. China wants to focus on driving domestic consumption and weening itself from external exports. This may actually allow India to attract foreign investment and thus improve its basic infrastructure. Another dimension is the possibility that we are witnessing the creation of a quadrilateral grouping comprising Australia, India, Japan, and the US. There is little doubt the growth rates will remain higher in Asia than the rest of the world for at least the next decade or two.

So, for India and the whole of Asia the most exciting aspect of the growth is in the area of the new economy. This is because the technology or digital aspects will pretty much impact all sectors; from consumer staples to insurance there are so many endless possibilities. The fact that so many sectors can be digitalised is the reason why these Internet companies are such giants in terms of market capitalisation. Their offering is endlessly scalable in a way, for example, a traditional bank is not. Actually, technology is not a particularly relevant tern anymore. Different sectors have embraced or are in the process of embracing technology. Of course, it does not mean that every company in a sector will be able to do it because it does require thinking outside of the box and being nimble, which for many companies is difficult to do. So, there will be lots of disruption and lots of new entrants. Just like there has been around the world. One can recall what Amazon was like many years ago compared to today. There will be some incumbents of course that will not only survive but will prosper, such as Walmart, which through its own innovation it prevented being overwhelmed by Amazon and the way that a Macy's or a Sears was. In Asia, it will be like that, too. Clearly one advantage that Asia has, apart from North Asia, is its demographics relative to say Europe. So, in essence, for Asia the two big drivers for growth over the coming decades are the young population and the embracing of new technologies.

John Seel

Asia retains key advantages that should help its economies grow at levels above the global average going forward. First, Asian economies are notable for their openness to technological innovation, allowing new technologies to be developed and expanded to the mass market quickly. Examples include the rapid growth of online industries in such areas as finance, gaming, and communications in a number of regional economies, as well as their rapid adoption of online communications and AI to enhance productivity in more traditional industries. The rapid adoption of mobile payment systems and fintech platforms in countries such as China, Malaysia, and South Korea have enhanced productivity in what had been a relatively traditional and inefficient sector of their economies. Asia's ability to adapt technology and commercialise it successfully in the mass market has largely offset the relative decline of electronics manufacturing as a regional growth driver. It has also differentiated Asia from other emerging regions such as Russia or Latin America, which, despite pockets of technological skill, have not commercialised their technology to the same extent.

A related strength is Asia's relatively high level of educational attainment. The development of Asia's education systems has been a key enabler of its move up the industrial ladder from light manufacturing to more complex manufacturing to high technology and services. According to the Asian Development Bank, the average years of schooling for Asian workers aged 20–24 increased from 3.5 in 1960 to 8.9 in 2010, and some countries have achieved near-universal access to tertiary education.[8] Not only are average years of education high by comparison to other developing regions, but the content of education is often relevant to the needs of the economy. North Asian economies in particular have benefited from the widespread availability of technological education. Their consistently high demand for overseas education has also given them a core of skilled professionals and managers who understand global trends, technology, and organisations. This has been particularly important for China and South Korea, helping them to adapt to an increasingly globalised economy. A better educated workforce is more flexible and innovative, and also more able to take on high value-added roles in service industries. This advantage should help such economies adapt to future changes in growth drivers and the trade environment.

Another potential source of future growth for Asia is continued urbanisation and industrialisation. The movement of hundreds of millions of workers from manual work in the countryside to higher-paying jobs in factories and cities created massive improvements in productivity, as well as a sustained upswing in domestic savings and consumption. This process is still in its early stages in many South and Southeast Asian economies, where farming remains overly labour-intensive and inefficient. There is great potential for macroeconomic growth if these countries can encourage and manage a smooth transition from primary to secondary and tertiary industries. This will require careful, foresighted planning of cities, infrastructure, and industrial development zones, as well as a balance of policies between encouraging investment and protecting the interests of domestic workers.

Asia's growth going forward should continue to benefit from its competitiveness in manufacturing. Even though there are political pressures in some developed economies to shorten supply chains and bring manufacturing production back onshore, Asia is likely to remain a key base for global manufacturing. Manufacturing industries generally develop clusters of related producers and service providers, which contribute inputs and support services that each manufacturer needs to be competitive. Clustering also gives manufacturers access to experienced workers and managers, who otherwise would have to be trained at higher cost. Such networks are difficult and time consuming to reproduce and cannot be created by political fiat. In addition, the ageing and shrinking labour force in many developed countries will keep labour costs high, making it difficult for labour-intensive industries to operate competitively. As a result, Asia's dominance of many fields of manufacturing is likely to continue.

Another area of growth potential for Asia is in services. For example, many Asian countries still lag in the quantity and quality of healthcare, particularly in specific sectors such as rural populations or the elderly. With ageing populations and declining family sizes in many countries, Asia has a growing need for nursing homes and specialised elder care, an important sector of health care in most developed countries. Similarly, most Asian countries have room to develop their insurance and social welfare sectors to support access to quality healthcare services. Financial services are another potential growth driver. Asian countries have been relatively open to new financial technologies, such as online payment services, but this innovation often exists side by side with a traditional banking sector that is hobbled by financial repression and inconsistent risk management. China and India are both examples of this relationship. In China, for example, excessive state control over bank interest rates and inadequate prudential regulation of the equity market have driven a string of speculative bubbles, as Chinese savers in search of higher returns flock into other assets such as real estate, shadow banking products, and peer-to-peer lending. Meanwhile, Indian banks have been hobbled by excessive state intervention and inadequate risk management. There is tremendous potential for growth inside the financial sector and beyond if each country adopts a regulatory framework that allows more market signals and minimises arbitrary government interference, while focusing on prudent risk control.

1.2.3 Question Three: What are the Policies Critical for Growth?

The third of the questions revolved around government policies and, in particular, identifying the key policies that need to be in place for Asia's policy makers to ensure continued economic growth in the long term. And, whether these policy makers should do more of the same or should do something different. Manu looked at socio-economic and business ecosystem related policies. For Richard it is economic governance and the private sector related ones. For Mark it is the public–private interplay related ones. And, for John it is very much the localised policies in the different economies.

Manu Bhaskaran

Governments have to improve the socio-economic and business eco-systems in many ways. They can remove the impediments to business such as bad infrastructure, poorly designed labour laws, corruption, inefficient tax regimes, unhelpful bureaucrats, and bloated state enterprises and banks. They can massively invest in human capital – improve early intervention programmes so that even children of the poor have good nutrition and a good start to their education, build better schools with well-incentivised teachers who actually turn up to teach, improve universities and vastly expand vocational training. And they can expand in a sustainable way social safety nets and programmes to reduce inequality. That will make the social changes that accompany high economic growth and productive but dislocating transformations more sustainable politically.

Richard Jerram

The basic policy challenge is to continue to improve economic governance to enable the private sector to prosper. If anything, this has become more urgent as relations with the rest of the world become less supportive. Globalisation has already stalled over the past decade and the risk is that it goes into reverse in coming years.

A look at the World Bank's Ease of Doing Business survey shows that the micro details matter, as much as any broader macro top-down policy drive. How easy it is to set up a company? Find a loan? Pay taxes? How secure are property rights? How good is infrastructure? How bad is corruption? The good news – if one can call it that – is that most countries in the region lag well behind best practices in the developed world, so there is still plenty of room for improvement. The benefits from reform are evident in rising living standards and this generates a tailwind for further progress.

However, there is a related challenge for China. The tools that the communist party needs in order to maintain power, such as control of information flows and the legal system, will increasingly be barriers to building a more complex economic system. Understandably, political priorities dominate economic ones, but, in turn, delivering rising prosperity is important for political legitimacy. China needs to find a way to liberalise markets in a way that does not threaten Party dominance, but unfortunately the world has few – if any – successful examples.

That is not to say that Anglo-American capitalism is necessarily the target. However, there is a need to recognise that markets give important signals about resource allocation – not just for products, but also for labour and capital. There is a role for the state in providing public goods and managing externalities, and environmental issues could be at the forefront of this debate in coming decades, but governments need to learn to put greater trust in markets.

Mark Matthews

One can look at the US experience. It is perhaps the most obvious one to look at as a precedent. It was the first country to embrace the Internet. Most of the major new economy companies in the world are American from Silicon Valley. Back in the 1950s and 1960s there was a very strong government collaboration with the private sector. Primarily government in the form of the military. If one looks at companies like HP and Intel, the major contracts that they secured when they first got started was with the military. They were strongly supported by the US government and also indirectly there was close cooperation and collaboration with leading US educational institutions, including MIT and Stanford University. This was also important to develop what Silicon Valley is today. But that was very much at the beginning. Since then the development has been primarily led by private companies with relatively minimal government involvement. What the degree of government involvement should be with Asia's digitalisation shift is difficult to say. It could happen organically if one just lets the private sector develop it.

In a lot of Asian countries, such as India, for example, there is very little government directive or support to develop the new economy. It is happening anyway because young people are interested and involved, and there is capital that can be invested in the new economy companies. It cannot really be bad if governments are willing to support and are engaged and help in building the innovative processes.

John Seel

It is inaccurate and misleading to speak of Asia as a whole, because there are still huge differences in development between Asian economies. On one end of the scale, Japan is a slow-growth developed country, while countries like Laos, Cambodia, and Myanmar are still in the early phases of their economic transition. This economic diversity has been an important driver of Asian growth since the 1970s, when Japanese investment in lower-wage economies such as South Korea and Taiwan contributed to a 'flying goose' pattern of regional development. Although Japan no longer drives this flow of regional investment, it will remain an important potential driver of Asian growth. Higher costs and shrinking labour pools will continue to encourage manufacturing industries to move out of Asia's more developed economies. With their proximity in geography and time zones and their relatively convenient access to regional supply chains, South and Southeast Asia are well positioned to attract this investment. Development of regional transport infrastructure, including projects related to China's One Belt/ One Road initiative, will support this expansion of the region's economic synergies. This expansion of intra-regional synergies will become even more important for Asia if barriers to globalisation reduce trade and investment flows between Asia and the rest of the world.

One of the key challenges for high-growth economies is to maintain their economic momentum once they reach a moderate level of per capita income. It is not easy for economies to avoid plateauing and to keep their economic momentum alive. This is known as the 'middle income trap', and it has affected Asia as well as other regions of the world. One of the key factors creating this trap is a slowdown in productivity growth, and the best way to avoid it is to continue pursuing structural reform. The difficulty of moving beyond the 'middle income trap' is that the needed reforms are often politically or socially sensitive, such as encouraging flexibility of the labour market, enhancing economic participation by women, or streamlining the legal system. Pushing ahead with such reforms often takes political courage, as well as an ability to explain to their importance and benefits. With productivity under pressure from ageing populations, especially in North Asia, Singapore, and Malaysia, such reforms will become even more vital for the region's pace of growth.

1.2.4 Question Four: What are the Risks to Economic Growth?

The last question this author sought opinions on is about the key risks to Asia's economic growth in the next three decades and how these risks may be managed. Manu identifies potential political instability and possible regional conflicts. Richard sees risks faced by the region as being similar to those faced by the rest of the world, such as the move away from globalisation. Mark and John both also recognise potential political tensions as one of the key risks.

Manu Bhaskaran

Political stability is key. Within countries there will be many, many dislocations because high growth is accompanied by disruption, population movements, new social mores that clash with older and more conservative ones, and so on. There will be new social forces that emerge, such as a more activist and demanding middle class. Vested interests may have gained too much power and will need to be contained. The work will be cut out for political systems that will have to make many adjustments.

There will also be political risks in the region. China is a rising power with an expanding security and economic agenda that may not always be aligned with the interests of its neighbours. Ways will have to be found to resolve the differences as they emerge. China's rise will not sit well with the presence of the US in the region – there will be tensions. A new regional security architecture will be needed to accommodate all these tensions – absent which there could be major political clashes that could derail Asia's growth promise.

High growth has to be managed as it is often accompanied by external or financial or other imbalances. Monetary and fiscal policy frameworks have to be modernised so that inflation and the external accounts do not become sources of instability as they have become on so many occasions. Financial sectors will grow rapidly – but as history has shown, without proper supervision and a good regulatory framework, growth could be interrupted too often by financial crises. A strong and independent central bank is usually quite important.

As economic growth proceeds, labour costs as well as other costs will rise. The determinants of a country's competitiveness will change and businesses and policy makers as well as workers will need to learn how to adjust flexibly and successfully to these changing competitiveness conditions.

Richard Jerram

Most of the risks to growth in Asia are similar to those in the rest of the world. One example is responding to climate change without imposing an excessive burden on economic growth. Another is how to tackle the ageing population, when income levels are not as high as in the developed world, especially as the demographic transition is happening more rapidly. It is hard to see why Asia should have an advantage in addressing these challenges.

De-globalisation is the risk that most obviously applies to Asia more then to elsewhere, if recent trends continue. Perhaps it will even accelerate in the wake of COVID-19, which has raised questions about the viability of extended supply chains or the wisdom of being dependent on imports for vital products, especially related to healthcare. This is a particular problem for the poorest countries in the region as it will be a barrier to stepping on to the development 'escalator' – it might be that Vietnam is the last beneficiary of the process. In this case, Asia will have to look inwards for growth and to some extent this is already happening.

Regional trading arrangements, which currently offer only a fairly superficial level of integration, will become more important. However, this is set to be a very slow process with many countries still viewing trade as a zero-sum game, rather than seeing the mutual benefits of Ricardian comparative advantage.

Mark Matthews
The most obvious risk is politics. If it is one of the hottest regions in the world, probably after the Middle East. Geopolitical issues in the region could derail and there could be a genuine risk of war. This would obviously hugely impact the long-term growth prospects of the region.

John Seel
Asia clearly faces a more challenging and less open global economic environment, undermining the global trade expansion that has been a key driver of regional growth. Despite this unfavourable external environment, Asian economies still have a number of ways to support future growth. One of these is to develop intra-regional trade, which has already served as a stabilising influence and has dampened the impact of changes in demand from the rest of the world. This regional integration takes advantage of differences in resource allocation, development stage, and labour cost between Asian economies. A key risk here is that political tensions, for example over conflicting claims to the South China Sea, may spill over into countermeasures affecting trade or investment. Asian policymakers will need wisdom and political cover to insulate their economic policies from tensions in their country's external relations.

In a more volatile world, it will be important for the stability of Asian economic growth to enhance the importance of domestic demand as a growth driver. Domestic demand was generally not very important for most Asian economies during their early, export-led phases of development, ranking behind external demand and investment as a growth engine. This was partly because household incomes were low, but also because governments promoted saving over consumption and kept their currencies weak to promote export competitiveness. Although the contribution of domestic demand has risen since the regional economic crisis of the late 1990s, it remains below the developed-economy average. There are good reasons for this, including a higher level of household savings, but many Asian countries also have room to encourage more domestic spending by households as well as the government. Improvements in the provision of healthcare and services for the elderly could help in this regard, both by increasing public health care spending and by reducing pressure on households to save for medical treatment and retirement. Urbanisation should also contribute to higher consumption, as governments develop urban infrastructure and as households have more opportunities to spend on services.

In a more challenging world, Asia's growth strategy and development path must evolve. Asia has faced challenging times before, and each time the region has emerged stronger and more resilient. There is no guarantee that this will continue, and it is certainly possible to construct negative scenarios; for example, China's increasingly assertive stance towards the South China Sea, Hong Kong, and Taiwan could lead to increased political tensions, greater economic autarky, or even military conflict. Finding a path to continued economic growth and prosperity will depend on the wisdom of Asia's policymakers and its people.

Notes

1 United Nations, Department of Economic and Social Affairs, Population Division (2019). World Population Prospects – Population Division – United Nations. [online] Un.org. Available at: https://population.un.org/wpp/Download/Probabilistic/Population/ [Accessed 4 September 2020].
2 The World Bank (2020). 4.1 World Development Indicators: Growth of Output, World Development Indicators, The World Bank. [online] wdi.worldbank.org. Available at: http://wdi.worldbank.org/table/4.1 [Accessed 4 September 2020].
3 Asian Development Bank (2020). Asia's Journey to Prosperity: Policy, Market, and Technology over 50 Years. [online] Manila, Philippines: Asian Development Bank. Available at: https://www.adb.org/publications/asias-journey-to-prosperity [Accessed 4 September 2020].
4 Asian Development Bank (2020). Asia's Journey to Prosperity: Policy, Market, and Technology over 50 Years. [online] Manila, Philippines: Asian Development Bank. Available at: https://www.adb.org/publications/asias-journey-to-prosperity [Accessed 4 September 2020].
5 Asian Development Bank (2020). Asia's Journey to Prosperity: Policy, Market, and Technology over 50 Years. p. 8 [online] Manila, Philippines: Asian Development Bank. Available at: https://www.adb.org/publications/asias-journey-to-prosperity [Accessed 4 September 2020].
6 PricewaterhouseCoopers (2017). The World in 2050. [online] PwC. Available at: http://www.pwc.com/world2050 [Accessed 4 September 2020].
7 bp p.l.c. (2020). Statistical Review of World Energy. [online] BP Global. Available at: https://www.bp.com/en/global/corporate/energy-economics/statistical-review-of-world-energy.html [Accessed 4 September 2020].
8 Asian Development Bank (2020). Asia's journey to prosperity: policy, market, and technology over 50 years. [online] Manila, Philippines: Asian Development Bank, pg. 187. Available at: https://www.adb.org/publications/asias-journey-to-prosperity [Accessed 4 September 2020].

Chapter 2
Asia's Dramatically Changing Energy Landscape

In the next five to ten years – or the 'short-term' in the context of energy – there are at least two significant trends in terms of energy market changes in the Asia region. It's a unique combination of rising consumption and gargantuan regulatory reform. The sharp rise in overall energy usage and drastic changes in the policies regulating energy companies and markets are driving the sector to sharp short-term changes and major longer-term transformations.

2.1 The Energy Consumption is Unique and is Rising

There are at least two reasons as to why energy demand is set to grow exponentially in the region in the next three decades.

I will first analyse the growth potential by using a simple and easy approach, namely assessing per capita electricity consumption. This basic approach will in no way result in a highly accurate answer as to how much energy the continent will need but it will provide a highly realistic sense of the realm of growth. Second, I shall turn to one important consumption driver, namely electric vehicles (EVs). I strongly adhere to a school of thought that believes that not only will EVs progressively replace the combustion engine but also that they will play a pivotal role in the future of the way energy is used. They will become tools for emission reductions, will be an important source of electric power demand, and will help to manage our energy systems given their energy storage capability.

The other side of the coin is the factor, or factors, which could potentially curtail the exponential growth in consumption. There are some obvious risks to growth. These include a massive economic slowdown post-COVID-19 or some unexpected large-scale natural calamity. Another simple curtailment factor is maturing markets. Some Asian energy markets, such as those of China, Malaysia, Thailand, and South Korea, will progressively mature thanks to rising living standards and ageing populations. The already mature markets, including Australia, Hong Kong, Japan, New Zealand, and Singapore, will see little to no energy growth. Another risk that could curtail the massive increase is energy efficiency and conservation. It is this last factor which I will examine in detail.

2.1.1 Plain Vanilla Look at the Growth Upside

In this book the word 'energy' is frequently employed. Within the energy sector our primary focus is electricity, which happens to be the major form of primary energy usage. Looking at per capita electricity consumption will help to roughly gauge what amount of energy would be required in the future. The per capita measure is by no means a highly precise one because there are simply too many variables that skew the results. Still this approach does provide quite a reasonable and realistic picture. The following pages will present some fundamental data sets and explain how some conclusions as to how much energy the region will be needing can be derived.

The measure of per capita electricity consumption is very commonly used in the world of energy research. A large number of academic papers have been written on the subject. One of the many interesting studies is by researchers at Spain's Universidad del País Vasco (the University of the Basque Country). Their article titled 'The Energy Requirements of a Developed World: Energy for Sustainable Development' highlighted that

> [a] number of authors have investigated the relation between the degree of development of a country and its energy use . . . Most studies have found strong correlations between energy use and living standards at lower energy use levels ('developing countries'), and decoupling at higher levels ('developed countries'). This can be seen when comparing the relation between the per capita total primary energy demand (TPED) and the human development index (HDI).[1]

In other words, in a less developed country, the rate of energy consumption growth is typically the same or even higher than the rate of GDP growth. This means a multiple of one or higher. In a more developed economy, energy consumption rises at a sharply lower rate than GDP growth; often the ratio can be just 0.5 times. These equations generally work in my experience.

Let us look at an example from Japan, which is one of the largest electric power markets in the world and a mature, developed economy. The nation's largest electric power company, Tokyo Electric Power Company Holdings (TEPCO), published some specific statistics about GDP elasticity, looking at a period of more than 20 years, from fiscal year 1993 (April 1992 through March 1993), until fiscal year 2015 (April 2014 through March 2015), which are shown in Table 2.1. The utility historically served the Tokyo metropolitan area, Japan's most populous region. The data the utility put together shows the rate of growth in electricity sales by the company relative to the national GDP as well as the final energy consumption rate to GDP. In recent years, GDP growth has generally been much higher than the rate of electricity growth. For example, in 2013 GDP rose 2 per cent versus negative 0.9 per cent for electricity consumption. In 2014, the data was negative 0.9 per cent versus negative 3.6 per cent, and in 2015 it was 0.8 per cent versus negative 3.6 per cent. This compares to an inverted ratio just two decades earlier. In 1993 GDP was negative 0.5 per cent versus 0.7 per cent for electricity consumption. In 1994 the data was negative 1.5 per cent

Table 2.1: Average Rates of Increase in GDP, Final Energy Consumption, Electricity Sales, and Peak Demand.

Fiscal Year	GDP(A)	TEPCO Electricity Sales (B)	Final Energy Consumption (C)	GDP Elasticity (B/A)	GDP Elasticity (C/A)	Peak Demand
	(% change from the previous year)	(% change from the previous year)	(% change from the previous year)			(% change from the previous year)
1993	(0.5)	0.7	1.1	n.a.	n.a.	(7.2)
1994	1.5	7.4	3.6	5.0	2.4	14.7
1995	2.7	2.2	3.2	0.8	1.2	1.8
1996	2.7	1.2	1.6	0.5	0.6	1.3
1997	0.1	3.1	0.6	21.0	4.2	(2.4)
1998	(1.5)	0.6	(1.2)	n.a.	n.a.	2.1
1999	0.5	2.7	2.5	5.0	4.6	0.1
2000	2.0	2.3	1.1	1.2	0.6	n.a.
2001	(0.4)	(1.8)	(1.3)	n.a.	n.a.	8.5
2002	1.1	2.3	1.5	2.1	1.3	(1.7)
2003	2.3	(2.1)	(0.9)	n.a.	n.a.	(9.2)
2004	1.5	3.9	1.3	2.7	0.9	7.2
2005	1.9	0.7	(0.4)	0.4	n.a.	(2.2)
2006	1.8	(0.4)	0.3	n.a.	0.2	(3.4)
2007	1.8	3.4	(1.7)	1.9	n.a.	5.9
2008	(3.7)	(2.8)	(7.0)	n.a.	n.a.	(0.9)
2009	(2.0)	(3.0)	(1.9)	n.a.	n.a.	(10.5)
2010	3.5	4.7	4.3	1.4	1.2	10.1
2011	0.4	(8.6)	(2.7)	n.a.	n.a.	(17.2)
2012	0.9	0.3	(1.3)	0.3	n.a.	2.3
2013	2.0	(0.9)	(0.8)	n.a.	n.a.	0.3
2014	(0.9)	(3.6)	(3.2)	n.a.	n.a.	(2.2)
2015	0.8	(3.9)	n.a.	n.a.	n.a.	(0.5)

Source: Tokyo Electric Power (n.d.). Power Demand | TEPCO Illustrated | TEPCO Illustrated. [online] www.tepco.co.jp. Available at: https://www.tepco.co.jp/en/corpinfo/illustrated/power-demand/1240162_6436.html [Accessed 21 May 2020].

versus 7.4 per cent while in fiscal year 1995 it was almost at par, namely 2.7 per cent versus negative 2.2 per cent. This trend can be found in many other mature economies ranging from Australia to France.

So, what exactly is wrong with using per capita electricity consumption to measure possible energy requirements? Nothing really. There are some reasons as to why this measure should be regarded as a broad benchmark and not a precise one. The first of the reasons is that energy consumption patterns differ power market by power market. All energy economists agree that demand rises at a much faster pace when the economy is undergoing rapid industrialisation. Demand then progressively slows down as the economy becomes industrialised or near-industrialised. In Asia, one may expect a quicker pace in Cambodia versus Vietnam, which itself would be faster than Thailand's, which in turn would be quicker than Australia or Japan, for example.

Another reason lies with the makeup of the consumption. It will be different even for countries at similar stages of development. In Hong Kong, residential consumers use about a fifth of the total energy and the services and commercial sectors almost 45 per cent. In Singapore, another city-state economy, households accounted for a little under 15 per cent of the total, over five percentage points less than Hong Kong, while the services and commercial sectors consumed about 44 per cent, just one percentage point less than Hong Kong's.

Finally, it is difficult for the calculation to fully consider different energy efficiency and energy conservation efforts in the markets one is comparing. The International Energy Agency (IEA) publishes a regular study on energy efficiency. In the 2018 edition, which looks at data up to the end of 2017, it argued that energy efficacy policy driven by the government has already had a significant impact on global energy demand in the past few decades. It concluded that the

> impact of efficiency policies has been significant over the last decades. Globally, efficiency gains since 2000 prevented 12 per cent more energy use than would have otherwise been the case in 2017. Energy efficiency is a major driver for uncoupling energy consumption from economic development.[2]

But energy efficiency and conservation efforts vary greatly from one economy to another.

Using the electricity per capita approach will permit the formation of a rough, yet clear picture of the enormous magnitude of electric power generation capacity potentially needed in the Asia region.

Table 2.2 lists the power generated and the estimated population of key electric power markets in the region in 2018. It then derives the amount of electricity generated in the power market per capita. Power consumption could have also been used but different sources would have had to have been used, which will not allow for an apples to apples comparison. The generation data is from the annual BP statistics – a really great product that is updated annually and is openly available. The population data is from another reliable source, the Population Division of the Department of

Table 2.2: Power Generation and Population Size of Key Power Markets in Asia in 2018.

Power Market	Terawatt-Hours	Population	Kilowatt-Hours Per Capita
South Korea	594.3	51,172	11,613
Taiwan	273.6	23,726	11,531
Australia	261.4	24,898	10,499
New Zealand	44.3	4,743	9,340
Singapore	52.9	5,758	9,189
Japan	1,051.6	127,202	8,267
Malaysia	168.4	31,528	5,342
China	7,111.8	1,427,648	4,981
Thailand	177.6	69,428	2,559
Vietnam	212.9	95,546	2,228
India	1,561.1	1,352,642	1,154
Indonesia	267.3	267,671	999
Philippines	99.8	106,651	935
Sri Lanka	15.5	21,229	729
Pakistan	140.6	2,12,228	663
Bangladesh	79.1	161,377	490
Developed	**2,278.1**	**237,499**	**9,592**
Developing	**9,834.2**	**3,745,948**	**2,625**

Source: Global Change Data Lab (n.d.). World Population Growth. [online] Our World in Data. Available at: https://ourworldindata.org/search?q=world+population+growth [Accessed 6 June 2020]. BP (2017). Statistical Review of World Energy | Home | BP. [online] BP Global. Available at: https://www.bp.com/en/global/corporate/energy-economics/statistical-review-of-world-energy.html [Accessed 6 June 2020]. 'Kilowatt-Hours Per Capita' calculation is by the author.

Economic and Social Affairs of the United Nations. Based on these two sets of data one can easily derive that the average amount for the region's developed power markets, comprising more than 237 million people, was almost 9,600 kilowatt-hours of generation per capita. The lowest was Japan given the high level of power usage by commercial and residential users relative to regional peers. Hong Kong was omitted because the sizeable portion of power used in Hong Kong is actually generated across the border by the Guangdong Daya Bay Nuclear Power Station and transmitted to Hong Kong. The amount by key developing power markets in the region, comprising more than 3.7 billion people, was about 2,278 kilowatt-hours of generation per capita in 2018.

In Figure 2.1 key developing markets were selected. They include Bangladesh, China, India, Indonesia, Malaysia, Pakistan, the Philippines, Sri Lanka, Thailand, and Vietnam. The amount of electricity used and the population in 2018 were used as a base. Four different scenarios conclude that by 2050 or so developing Asia will require from 3.1 times to 4.2 times more electricity.

The first scenario assumed no population growth and used Japan's 2018 average kilowatt-hours per capita. The second assumed populations to grow to the level estimated by United Nations. The third assumed no population growth and used the developing markets average kilowatt-hours per capita in 2018 as a base. The last scenario assumed populations to grow and used the same base as the third scenario. It may sound overly complicated but actually it is a very straightforward calculation. In terms of electricity volume, the biggest jumps would be with India where the usage would rise to 11,183 – 15,723 terawatt-hours from 1,561 in 2018. China's would increase to 11,803 – 13,452 from 7,112 in 2018. And Indonesia's would be raised to 2,213 – 3,174 terawatt-hours from 267 in 2018. These three power markets would account for 74 per cent to 78 per cent of the additional electric power required.

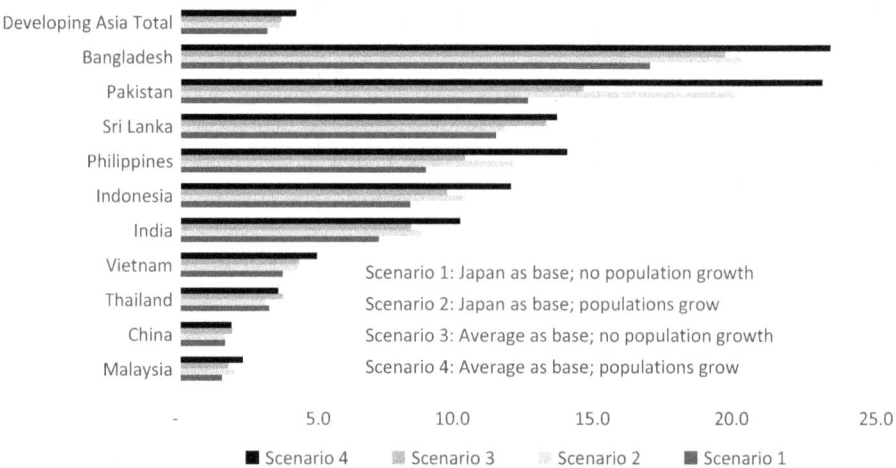

Figure 2.1: Potential Power Generation Multiple Growth in Developing Asia 2050 versus 2018 (x). Source: Global Change Data Lab (n.d.). World Population Growth. [online] Our World in Data. Available at: https://ourworldindata.org/search?q=world+population+growth [Accessed 6 June 2020]. BP (2017). Statistical Review of World Energy | Home | BP. [online] BP Global. Available at: https://www.bp.com/en/global/corporate/energy-economics/statistical-review-of-world-energy.html [Accessed 6 June 2020]. Estimates by author.

The implications are enormous when one considers that the bulk of the region will need 3.1 times to 4.2 times more power generation plants and probably a similar amount of electricity distribution grids. Hundreds of billions of dollars in capital investments will be needed to meet the power consumption required. Much of the investments may come from public sources but an enormous amount of private capital will also be needed. This will translate into a gigantic amount of new business and investments opportunities.

As indicated, the approach above is rudimentary. Its purpose was to give a flavour as to the realm of potential growth that will be witnessed in the region in coming years in terms of new power generation and transmission capacity. Forecasters have highly complex models that take into consideration dozens if not hundreds of different variables affecting electric power generation and consumption. Accuracy, forecasting experts will tell you, is extremely hard to attain. This is true not just for long-term forecasting, such as the 30-year period that was used in the number crunching above. It is also true for one-week forward or one-month forward estimates. Of course, this does not stop the experts forecasting. Given that electricity is a key element to developed and developing economies, the economic planners need to have a solid understanding of how much electric power infrastructure is likely to be necessary for an economy in the short- and long-term.

China had reached power consumption of more than 7.22 thousand terawatt-hours in 2019 with an expectation of an increase of 3 per cent to 5 per cent for 2020, which would translate into 7.44 to 7.55 thousand terawatt-hours, with the lower end more likely given the economic impact of consumption and production slowdown due to the coronavirus pandemic 2019 (or COVID-19). Interestingly, if one were to scan the forecasts by a variety of Chinese domestic think tanks and institutions, the long-term estimate is remarkably close to the 7.44 thousand terawatt-hours. In 2011 and 2012, the State Grid Energy Research Institute and the Huaneng Technical Economics Institute estimated the 2020 number to be 7.4 to 7.7 thousand terawatt-hours. Between 2013 and 2017, other think tanks and institutions estimated anywhere between 6.8 thousand terawatt-hours at the lower end to 8.0 thousand terawatt-hours at the high end and based on my calculations, averaged 7.18 thousand terawatt-hours (Table 2.3).

For 2030, the long-term estimates put consumption at a range of 8.5 to 11 thousand terawatt-hours, averaging at 10 thousand terawatt-hours, implying a compound annual growth rate of about 2.9 per cent. For 2050, the estimates put demand at 10 to 15 thousand terawatt-hours, averaging at 13 thousand terawatt-hours, implying a compound annual growth rate of about 2.9 per cent. Interestingly, these forecasts, which are derived from complex, sophisticated modelling, are quite close to my per capita-based calculation exercise discussed together with Figure 2.1.

Based on my experience I find these estimates quite realistic. Especially that for 2050. The 1 to 2 per cent rate of increase is indeed a rate that one can see with many developed economies ranging from Western Europe to Japan.

Table 2.3: China Electricity Consumption Estimates by Various Domestic Institutions.

Year	Institution	2020	2030	2050
			Terawatt-Hours	
2011	State Grid Energy Research Institute	7,400–7,700	9,500–9,800	14,300
2011	China Electricity Council	n.a.	11,000	10,000
2012	Huaneng Economic Technology Institute	7,400	10,800	13,200
2013	State Power Planning Research Centre	7,000–8,000	10,000–11,000	12,000–15,000
2015	Jingxu Wu	7,400	9,500	n.a.
2016	National Energy Administration	6,800–7,200	n.a.	n.a.
2016	China Energy Research Council	6,800	8,500	n.a.
2017	State Power Planning Research Centre	6,800–7,200	n.a.	n.a.
2017	State Grid Energy Research Institute	7,000–7,600	9,100–11,000	n.a.

Source: China Huaneng (2018). New Era Medium and Long Term Power Demand Forecast in China. [online] news.bjx.com.cn. Available at: http://news.bjx.com.cn/html/20180611/904814.shtml [Accessed 9 June 2020].

2.1.2 Higher Demand from Shift to Electric Vehicles

An important growth element of future energy demand in Asia is electric mobility. Whether for public transport, like buses, or for private transportation, the number of battery electric vehicles (BEVs) and plug-in hybrid electric vehicles (PHEVs) is set to grow exponentially around the world in general and in the Asia region in particular in the coming decades. The electric mobility growth momentum will continue to be chiefly driven by China, today's undisputed global leader, with India and many Southeast Asian nations not too far behind. Electric vehicles (EVs) will become a significant source of additional power demand. Actually, electricity is their fuel.

PHEVs are, many believe, more like a stop-gap measure as the vehicle charging infrastructure for BEVs is progressively built out. Ultimately, the chief type of EVs will be the BEVs. The transition to EVs away from internal combustion engine (ICE) vehicles is being chiefly driven by governments, and it is their policies and regulations that are pushing the massive rapid EV adoption. Of course, governments will continue to have a close collaborative partnership with the private sector, be it environmental organisations, research institutes, or car manufacturers. Governments' motivation is most typically emissions abatement, which is itself triggered by sociopolitical pressures. But, for some markets, there may also be an underlying energy security agenda, specifically the objective to cut the economy's reliance on oil imports for most or all gasoline requirements. A fuel that not only is highly polluting but also whose price has a propensity to be highly volatile. And, when its price is high, it sharply raises the national commodities import burden. Volatile oil prices are an especially serious concern for those island or stand-alone economies that have no or very limited domestic energy resources, such as Japan, Taiwan, South Korea, or Sri Lanka in the Asia region. Or, even larger economies, including China and India, which are also heavily reliant on oil imports as they do not have sufficient resources domestically.

When discussing oil price volatility, it is not a one day jump in the price that hurts economies. It is when the oil price sharply rises and lingers at that level in a tight bound range for a few weeks or a few months. The historical performance of the Cushing, Oklahoma West Texas Intermediate (Cushing, OK WTI) crude oil futures, one of the major global oil price benchmarks, clearly indicates the amount of volatility that the oil markets have endured. From a simplified perspective, the average annual price over the past 35 years or so can be considered. The price stood at $16 per barrel to $25 per barrel in 1986 to 1992, at a range of $17 to $22 in 1993 to 1997, and a low of $14 in 1998. It subsequently rose from an average of $19 to a peak of $100 in 2008. It weakened a little in 2009–2010 to $23–49 but then lingered at $93–98 in 2011–2014. This was followed by a fall to $43–51 in 2015–2017. The following year it suddenly spiked to an average of $65 only to fall to an average of $57 in 2019. In April 2020, as it has been well publicised, the prices hit an all-time low of negative $37; albeit the average for 2020, through 20 July, was about positive $37. There is little reason not to believe that the wild ride will not continue in the next three to five years (Figure 2.2). Albeit this author strongly adheres to the school of thought that concerns over emissions, geopolitical security, and other matters will mean that the usage of oil products, including gasoline, will experience a permanent downturn – not a simple cyclical one – together with coal.

The sharp uptake in EVs will be impacting the energy sector in at least three different ways. First, it will result in an absolute fall in the volume of oil, which is refined into gasoline for ICE vehicles. Second, growth in the total number of EVs will mean a rise in electricity demand and consumption which will increase electricity's share in primary energy. Third, EVs will very much be helping in the management of energy systems given EVs can be deployed as energy storage systems. The discussion in this section will solely focus on the demand part of the equation.

The amount of electricity that may be required by new EVs puts further upside on the amount of electricity that the Asian continent will need in the coming three decades and beyond. The EV global fleet reached more than 7.2 million in 2019, according to the IEA. It estimated in 2020 that by 2030 the number should reach from 140 to 245 million vehicles. The lower number is based on publicly announced policy ambitions whereas the higher number is based on a global campaign commitment for EVs to reach 30 per cent of total vehicles sold. The agency also forecasted that EVs would consume between 550 and 1,000 terawatt-hours in 2030 versus just 80 terawatt-hours in 2019.[3] This author expects the number will likely be at the higher end, probably close to or higher than the 245 million vehicles projected by the IEA in

2020. As such, their consumption will exceed 1,000 terawatt-hours, broadly equivalent to the power demand in Japan today. This bullish view arises from the fact that the global movement towards banning the sale of vehicles that use gasoline gas has been gaining deep and broad momentum.

Figure 2.2: Europe Brent and Cushing, OK WTI Spot Prices Freight-On-Board.
Source: U.S. EIA (n.d.). Cushing, OK WTI Spot Price FOB (Dollars per Barrel). [online] www.eia.gov. Available at: https://www.eia.gov/dnav/pet/hist/RWTCD.htm [Accessed 2 November 2020].

Other forecasts are more bullish than those of the IEA. Consultancy and research firm BloombergNEF argues that factors leading to rapid changes include the forecasted fall in the cost of BEV coupled with technology improvement as well as some countries banning the sale of ICE vehicles by 2030 to 2040, among other factors, will boost the numbers of BEVs. Cost-wise, BEVs should reach parity with ICE vehicles on average by the mid-2020s – from 2022 to 2030 depending on the size of the vehicle market and the geography. It expects BEVs will consume about 1,964 terawatt-hours, representing 5.2 per cent of the total electricity consumed in the world by 2040. Specifically, passenger BEVs will use about 1,290 terawatt-hours, commercial ones about 389 terawatt-hours, electricity-powered buses 216 terawatt-hours, and electric two-wheelers 69 terawatt-hours with China and European economies dominating new purchases; three-quarters by 2030.[4] Another study by accounting and advisory firm Deloitte concluded that EVs will account for at least 48 per cent of all new vehicles sales in China by 2030, 42 per cent in Europe and 27 per cent in the US.[5] In other words, about one in three vehicles sold around the world in 2030 will be EVs according to this Deloitte study.

It is worth mentioning that during the first global peak of COVID-19 in the second quarter of 2020, many speculated that the mass adoption of BEVs would accelerate because populations, especially urban dwellers, noticed the sharp fall in pollution as so many economies were in lockdown. One of many studies was published by Nature Climate Change, which found that against the mean of 2019 levels, carbon dioxide emissions in early April 2020 fell by about 17 per cent (a range of minus 11 per cent to minus 25 per cent) thanks to changes in surface transportation. Also, it said that emissions were actually 26 per cent lower on average.[6] Admittedly, it is likely that as the world goes back to the business-as-usual mode, in the very short-term, many people are likely to forget the 'good air' and prioritise convenience.

When looking at EV targets by Asian governments it is clear that there was limited commitment as of the middle of 2020. It is certain that in the coming few years, these governments will produce hard targets. As it stands, only China, Japan, Pakistan, South Korea, and Sri Lanka have made some clear and concrete commitments. Some other jurisdictions have some EVs sales targets, but these are in terms of absolute numbers, including Indonesia and Thailand. It is worth noting that China, India, and Japan are among the founding members of the EV30@30 campaign. The campaign is aimed at the country reaching a share of 30 per cent of new car sales for EVs by 2030. Other founding member nations include Canada, Finland, France, Mexico, the Netherlands, Norway, and Sweden.

China wants to see 25 per cent of total new car sales in the form of new energy vehicles by 2025. Given its overall EV market leadership in the world, Chinese planners are most likely to update and raise the objectives in the very near future. In India, the government had flirted with the idea of achieving a 100 per cent electric vehicle market by 2030. Experts and observers alike, such as Indian think-tank The Energy and Resources Institute (TERI), feel that it is not an achievable target and that the earlier target of having EVs accounting for 30 per cent of new vehicle sales by 2030 is more realistic. Japan seemingly has the clearest phase-in targets, as it wants to have various forms of EVs to account for 100 per cent of sales by 2050. Surprisingly, Sri Lanka, one of the poorest economies in Asia, targets 100 per cent electric or hybrid vehicles stock by 2040 (Table 2.4).

Table 2.4: National Electric Car Deployment Targets in Some Key Asian Markets.

	2025	2030	2035	2040	2050
	Relative to Car Sales	Relative to Vehicle Stock		Full ICE phase out or 100% EV target	
China	25% NEVs (PHEV, BEV, FCEV)				
India		30–100%			
Indonesia	2,200 EVs				
Japan		30–40% HEV; 20–30% BEV / PHEV; 3% FCEV			HEV, PHEV, BEV, FCEV 100% sales
Malaysia		100 000 EVs			
New Zealand	64,000 EVs (2021)				
Pakistan		30% EV		90% EV	
Singapore				Phase out of all ICE	
South Korea	430,000 BEVs 67,000 FCEVs (2022)	33% BEV, FCEV			
Sri Lanka				100% EV or hybrid vehicle stock	
Thailand			1.2 million EVs (2036)		

Sources: International Energy Agency (2020a). Global EV Outlook 2020: Entering the Decade of Electric Drive? [online] www.iea.org. France: IEA Publications – International Energy Agency. Available at: https://webstore.iea.org/download/direct/3007 [Accessed 11 Jan. 2021]; Reuters Staff ed., (2020). Singapore Aims to Phase out Petrol and Diesel Vehicles by 2040. Reuters. [online] 18 February. Available at: https://www.reuters.com/article/us-singapore-economy-budget-autos-idUSKBN20C15D [Accessed 11 January 2021]; Jain, R. (2020). Top Upcoming Electric Cars in India. [online] Business Insider. Available at: https://www.businessinsider.in/business/auto/article/top-upcoming-electric-cars-india/articleshow/73983943.cms [Accessed 11 Jan. 2021]; PTI Agency (2019). India's 2030 electric vehicles target hard to achieve: TERI chief – ET EnergyWorld. [online] ETEnergyworld.com. Available at: https://energy.economictimes.indiatimes.com/news/power/in dias-2030-electric-vehicles-target-hard-to-achieve-teri-chief/70269882 [Accessed 30 Nov. 2020].

Apart from Japan the other developed economies in the region have really done a dismally appalling job at setting hard targets for the phasing out of ICE vehicles.

Australia had no official target as of mid-2020. The current political opposition, the Labour Party, had a pledge in the last election for the country for EVs to make up for 50 per cent new car sales by 2030. Subsequently, a study published by the government's Bureau of Infrastructure, Transport and Regional Economies projected that the number would likely be 27 per cent by 2030 and 50 per cent by 2035,[7] based on current trends. Australia's neighbour, New Zealand, had formulated in 2016 a short-term target, namely 64,000 EVs on its roads by the end of 2021. A few years later, however, it admitted that it would fail to achieve the short-term target. The nation only had 19,000 EVs as of February 2020, making the 64,000 EVs target hard to reach.[8]

The remaining two developed economies in the region are Hong Kong and Singapore, with Singapore being quite more advanced, aggressive, and progressive in its planning compared to Hong Kong, which embarrassingly had no target at all as of the middle of 2020. Embarrassingly because it is a special administrative region of China, itself a global leader in the growth of EVs. Hong Kong formulates its own environmental policies and, so far, has done little serious long-term planning. Hong Kong's Secretary for the Environment did say in early 2020 that it will engage with various stakeholders in arriving at the territory's first electric vehicle roadmap.[9] As such, one can be hopeful that sooner or later the special administrative region will follow in the path of China when it comes to EVs. On the other hand, during a budget speech in February 2020, Singapore's deputy prime minister declared that Singapore would completely phase out ICE vehicles by 2040. He also set some short-term targets including offering registration rebates for BEVs, furthering the expansion of the city's BEV charging infrastructure to 28,000 by 2030 from 1,600 points, as well as extending the existing Vehicular Emissions Scheme that comprises of tax rebates and surcharges depending on a vehicle's emissions.[10]

There are some factors or hurdles that could limit the possible upside of significantly more electric power consumption due to the sharp growth in the sale of BEVs globally. These potential impediments include the lack of charging infrastructure, the expense of switching out of existing of fossil fuel vehicle fleets, and the environmental impact of used batteries. There are some other hurdles, but these will be very short lived in the view of this author, being connected to the fact that as the economies of many Asian nations felt the sharp rise in the cases of COVID-19, during the second and third quarters of 2020, some very short-term negatives appeared. Given that many consumers were affected by unemployment or other forms of hardships, potential buyers were more sensitive to the high upfront costs of EVs. This was coupled with lower running costs for ICE vehicles given the lower gasoline prices after the fall in the price of crude oil. Also, during this period lockdowns of various forms as well as different priorities resulted in slower utilities and government funding for the charging stations build out.

The first hurdle is having insufficient infrastructure – or charging points – to charge the vehicles. This is an absolute priority. It sounds simple enough but the building of EV charging stations has been slow even in those countries where the government is aggressively promoting the switch from ICE vehicles. In Europe, based on data by the Union of the Electricity Industry – Eurelectric – a sector association with more than 34 full members, representing the industry in 32 European countries, there were about 20 nations with public financial support for public charging stations and about 22 had at least one public charging point per ten EVs in 2018. And, needless to say, the number is growing.[11] However, this author believes that this is simply a timing mismatch and definitely nothing close to a deal killer.

Another challenge is vehicle drivers and organisations running motor vehicle fleets perceptions of EVs. The consumer needs to be educated and the sentiment towards these new types of vehicles needs to change. The mind shift will happen, just like it did for the move from using landlines to cell phones, but the change may take some time. It will be progressive and may take a few years. The education process to raise awareness will continue to have to be done at the grassroots level. Car companies, electric power suppliers, and others in various countries will actively promote EV usage through such activities as EV displays and ride-and-drive experiences at shopping centres and other locations, for example. A sharp increase in charging points will help in the mind set shift.

Regarding the cost side of the EV growth equation, lower priced EVs and the savings thereof are an important consideration. The US Department of Energy's National Renewable Energy Laboratory (NREL) and Idaho National Laboratory (INL) published a study in mid-2020 indicating that BEV owners could save as much as $10,500 in fuel costs. The study did a detailed evaluation of the total cost of EVs in the US in 2019. It included the upfront purchase and installation expenses as well as the running costs, namely the utility tariffs that actually differ quite a bit region by region in the US. It found a national average of $0.14 per kilowatt-hour for PHEVs and $0.15 per kilowatt-hour for BEVs. It calculated that this translates into fuel cost savings ranging between $3,000 and $10,500 over a 15-year time period.[12] Also, many predict further sharp falls in the price of BEVs. A more conservative projection from the Massachusetts Institute of Technology (MIT) concludes that the key cost component of a BEV, the battery, which today are mostly lithium-ion battery packs, will drop by about 50 per cent by 2030 versus 2018. It expects the cost of the battery to decline to $124 per kilowatt-hour from a range of $175 to $300.[13]

Another concern that may also take a few years to be fully resolved, most probably through technological innovation, is the recycling of the batteries used by the EVs. A multitude of governments, organisations, and corporations are working hard towards addressing this issue. For example, in the EU, the European Commission launched the European Battery Alliance (EBA) in 2017. This grouping is dedicated to finding strategies addressing various parts of the battery value chain, especially recycling. The motivations here are environmental, economic, and strategic. The environmental motivations are to reduce the amount of materials that have to be mined. An economic motivation is that the recycling of batteries generates jobs and creates economic value. The strategic motivation is that it will 'allow the recovery of mineral resources which the EU does not exploit on its own lands, and which can be re-injected directly into EU industries'[14] argues a researcher.

The foremost protagonist of the global EV growth momentum are the EV makers. All of the major global brands have EV models, but the industry has also created new makers, the highest profile one probably being Tesla. To better understand this new sector and better understand how EV manufacturers are approaching the market, China's WM Motor Technology Group, one of the fastest growing new makers, was approached by this author. WM Motor was founded in 2015 by a team of automotive industry veterans, including Freeman Shen who was a senior executive at Geely Automobile Holdings, one of China's major car manufacturers and the company that acquired Volvo and Lotus.

WM Motor is an emerging leader in the Chinese passenger EV market, focused, according to Chief Strategy Officer Rupert Mitchell, on the design and manufacturing of affordable, smart, pure EVs under the Weltmeister brand, Weima in Chinese. He added that WM Motor's products are equipped with proprietary battery management systems, robust driving ranges as well as the full suite of Level 2 Advanced Driver Assistance System (ADAS) functions. The group was China's first EV start-up to have vehicles in mass production out of its self-owned Industry 4.0 manufacturing facility in Wenzhou, Zhejiang Province. It also recently completed construction of a second facility in Huanggang, Hubei Province, giving it full production capacity for 500,000 vehicles per annum.

Rupert thinks that WM Motor's vehicles are attractive from a consumer standpoint in large part due to their unique connectivity features, which turn the driver's personal car into a fully customised, interactive space (see Figure 2.3 for a sample car model from WM Motor). The in-cabin infotainment system, 'Living Engine', integrates many of China's most popular mobile-based apps into a dash-mounted 12.8-inch full high-definition (HD) touchscreen display. This includes navigation software from Baidu (China's leading Internet services provider); instant messaging, video streaming, and podcasting through Tencent; and Vehicle-to-Home connectivity through Xiaomi, a leading smartphone and electronics manufacturer. An in-cabin AI assistant called 'Xiaowei' responds to natural voice commands to control complex hardware and software functions such as the air-conditioning system, windows, sunroof, navigation, and various driving modes and will even activate home appliances when the driver is returning home. All of WM Motor's Android-based software is designed in-house and updated in new and bought vehicles via Over-the-Air (OTA) software updates.

Figure 2.3: Weltmeister EX6 Plus.
Source: Courtesy of WM Motor Technology Group Company Limited.

Rupert added that the convenience of on-board connectivity is a major consideration for new car-buyers in China. He referred to a consumer survey[15] conducted by consultancy McKinsey in 2019 that found that 61 per cent of Chinese consumers would be willing to switch car brands for better connectivity features alone, compared to 18 per cent in Germany. The automakers whose products promise the most connectivity in China are the upstart EV manufacturers, since they are unburdened by old supply chains, legacy design principles, outdated technology, and bureaucratic barriers to innovation. WM Motor, alongside several other successful domestic automakers, continue to take market share away from the larger original equipment manufacturers who are struggling to keep up with the shift towards better, faster, more convenient connectivity features and the subsequent 'softwarisation' of the vehicle operating system and drive units. This comes as more and more consumers are turning away from new cars that cannot receive software updates.

Rupert believes that the Chinese government's target of 25 per cent of all new cars sold by 2025 to be electric is achievable. He thinks that considering the fact that the most difficult aspect of the market shift towards electrification (customer preference) is clearly in favour of EVs in China, with the right policy support, infrastructure spending, and commitment by automakers, this target is entirely feasible. Rupert also referred to another McKinsey survey.[16] This survey found that 60 per cent of respondents in China said that they are seriously considering purchasing an EV as their next car. Notwithstanding the attractiveness of cost savings, high performance, and other benefits that electric vehicles offer over their traditional competitors, the trendy and 'futuristic' label attached to EVs seems to be one of the major drivers of the industry's growth in China and globally. Having understood this early on, Rupert adds, WM Motor is continuing to carve out its place as the most advanced smart electric vehicle maker in the mass market segment.

2.1.3 Energy Efficiency Could be Growth Containment Factor

2.1.3.1 Saving Money: The Regulation and the Implementation

The one factor that could curtail the gigantic rise in energy consumption in Asian markets is an increase in effective energy efficiency or conservation measures. The actual realm of the impact of this factor will chiefly depend on regulation, and specifically, on the actual comprehensiveness of the policy framework. It will also depend on how effectively the appointed government agencies actually implement the policies. Technical efficiency leads energy-intensity related developments according to the IEA studies – energy intensity being the number of units of energy per dollar of GDP. It calculated that improvements in technical efficiency in 2015–2018 cut energy demand by 4 per cent in 2018, the same amount of primary energy used by France plus Italy.[17]

The IEA has consistently recommended to governments that the specific energy efficiency areas they should be addressing include practical options for the transport sector, for buildings, for industry as well as suggesting prerequisite financing approaches.

In the area of transport, suggestions to governments include raising coverage and strength of transport policies for cars and trucks and non-road modes and creating structures to support the uptake and sustainable use of efficient vehicles. In the area of residential, commercial, and other buildings, recommendations revolve around generating efficiency-related rules and regulations for both new and existing building stock as well as appliances. This could come in the form of tax or other incentives to motivate end users to purchase appliances with higher efficiency and undertake deep energy retrofits. For industry, the suggestions are around rules forcing the deployment of higher energy efficiency equipment such as electric heat pumps. Also, motivating manufacturing facilities and others to strengthen the use of energy management systems.

The IEA specified that all of these policies must absolutely be backed with commercially viable financial investments that will ensure their sustainability. The financial tools should be both public and private. At the state level, programmes can be instituted to 'build scale and momentum in financing using programmes and incentives to increase activity'.[18]

All economies in Asia have in their various statutes energy efficiency rules and regulations, especially when it comes to buildings. However, strong and comprehensive policies and effective implementation is sorely lacking. In terms of global best practice, the EU probably has some of the most aggressive rules and targets of any region in the world. In Asia, the ASEAN Centre for Energy, an independent body within the Association of Southeast Asian Nations, is one of the many different public and private organisations that studies the most systematic energy management approaches and best suitable energy efficiency systems.

The implementation by governments of energy efficiency and conservation rules also translates into the requirement of significant capital outlays – especially capital upfront. One simple example are windows. Most people are aware that having double-glazed windows in one's apartment, for example, has many benefits, including better insulation, lower mould build up, and noise reduction. The cost is usually higher of course. For tilt-and-turn window models in the UK, double-glazed windows can be about 25 per cent more expensive than standard ones. However, the extra upfront cost will be offset by energy savings with a payback period of probably one to five years depending on the local conditions. This can be further evaluated by looking at lighting, buildings, and other energy management solutions.

2.1.3.2 Energy Savings Case Studies in the Region

The effect of energy savings and conservation efforts is massive when it comes to the usage of energy. There are examples in many countries in recent years that have shown that strong coordinated efforts by governments can actually reduce energy consumption. This can be measured by a country's energy intensity, computed by quantifying the amount of energy units needed to produce one unit of GDP. In theory, an emerging economy would have a rising energy-intensity number while a developed economy would have a declining one. The effect can be quite dramatic when combined with other elements such as a shift in energy consumption or self-generation, such as installing a rooftop solar photovoltaic system.

In Australia, energy intensity has declined by about 40 per cent and energy productivity has risen about 68 per cent in the 20 years through June 2018, based on data put together by the Department of the Environment and Energy of the Australian Government. Figure 2.4 shows GDP, energy consumption, energy productivity, and energy intensity, all rebased to 100 as of the fiscal year through June 1978. In the past 10 years alone, energy intensity fell 2 per cent annually to A$3,401 million ($2,424 million) per gigajoule while energy productivity increased 2 per cent per year to A$294 million ($210 million) per petajoule. The energy-intensity and productivity improvements were a result of Australia's GDP growing faster than energy consumption, the analysis argues. The department also highlights that the economy transitioning away from energy-intensive industries, such as steel manufacturing, as well as higher usage of renewable energy compared to fossil fuels consumption, also positively affected energy productivity.[19]

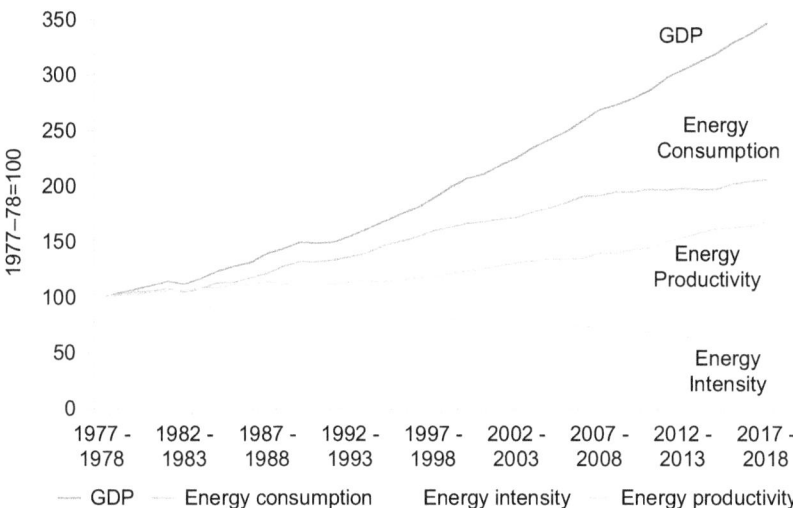

Figure 2.4: Australian Energy Intensity and Energy Productivity.
Source: Department of the Environment and Energy (2019). Australian Energy Statistics. Table B.

Interestingly, although the nation's GDP grew in the three fiscal years through June 2014, electricity output actually fell for the three years consecutively, by 1.1 per cent, 0.4 per cent, and 0.3 per cent (Figure 2.5). An influential think tank, the Australian Institute, attributed three reasons behind the fall in electricity

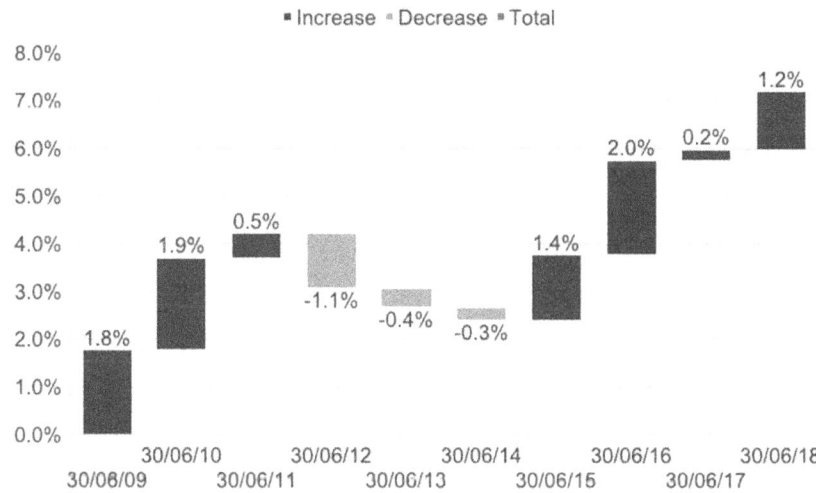

Figure 2.5: Ten Years to June 2018 Electric Power Output in Australia.
Source: Department of the Environment and Energy (2019). Australian Energy Statistics. Tables L and O.

demand to be the implementation of energy efficiency programs, a cut in usage by energy-intensive industries, and also, from 2010, the reaction by electricity users, especially residential households, from a jump in power prices.[20]

Japan wants to aggressively reduce energy intensity and become more efficient. The nation's Ministry of Economy, Trade and Industry (METI) set out a few years back a clear energy blueprint that included measures to reduce the overall consumption of energy by fiscal 2030, through March 2031. The country has long been concerned about its energy self-sufficient ratio, which is less than 10 per cent. It has to rely on imported fossil fuels for more than 87 per cent of its energy needs, especially after it had to shut down the bulk of its nuclear power fleet following the Tohoku earthquake and tsunami that caused the Fukushima Daiichi Nuclear Power Plant meltdown on 11 March 2011. It aims at reducing its total energy usage by about 6.1 per cent to 12.65 exajoules by the end of fiscal 2030 compared to the consumption of 13.47 exajoules in fiscal 2018.

METI's blueprint hopes to drive this through a variety of measures. One measure is to raise the rate of LED lighting usage to 100 per cent for all types of premises and consumers. Another measure targets all buildings to meet energy efficiency standards, including small premises. Yet another is to increase households use of high-efficiency hot water systems to 46.3 million, about 90 per cent of the country's total (Table 2.5).

Table 2.5: Progress of Measures to Improve Energy Efficiency.

Fiscal Year Though			31 March 2018	31 March 2031
Total Energy Usage			13.47 Exajoules	12.65 Exajoules
All Consumer Types	LED lighting	Rate	Industry: 56%; Services: 50%; Households: 55%	100%
Industry	Three-phase induction motors	Units in Use	2.07 million	31.2 million
Services	Buildings (Energy Efficiency Standards compliance)	Rate	Large: 100%; Medium: 91%; Small: 75%	100%
Households	High-efficiency hot water systems	Units in Use	14.57 million	46.3 million
Transport	EV, Plug-In Hybrid EV, etc.	% of New Sales	36%	50–70%

Source: Ministry of Economy, Trade and Industry, Agency for Natural Resources and Energy (2020). Japan's Energy 2019 Edition. March. [online] enecho.meti.go.jp. Available at: https://www.enecho.meti.go.jp/en/category/brochures/pdf/japan_energy_2019.pdf [Accessed 12 July 2020].

It is obviously colossally difficult to accurately predict the actual impact of energy efficiency and conservation on the future energy demand of various electricity markets but as explained with the example of the Australian and Japanese markets it can be a significant curtailment factor.

2.1.3.3 Energy Mega-Users: Buildings

The building sector is a massive consumer of energy. The consumption of residential and commercial buildings could be as much as 60 per cent of the world's electricity depending on how it is measured. Their consumption is from appliances, equipment, lighting, and space heating and cooling. The acceleration of urbanisation in Asia means that many more buildings are being constructed for commercial, office, and residential use. There are quite a few projections on the level of global urbanisation in the future. One of the many estimates is that it will be as much as 68 per cent of the total by 2050 from approximately 55 per cent in 2018. The situation in some of Asia's highly urbanised centres, such as Hong Kong, Singapore, and Shanghai, can be examined to attain a more concrete idea.

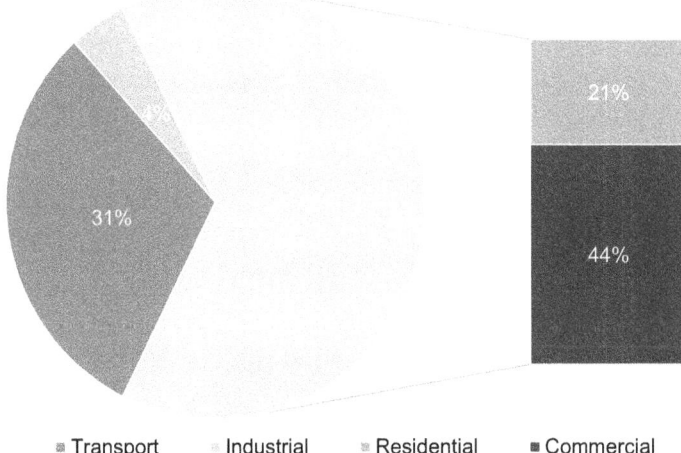

Figure 2.6: Hong Kong Energy Consumption by Sector (Adapted).
Source: Electrical and Mechanical Services Department, the Government of the Hong Kong Special Administrative Region (2019). Hong Kong Energy End-use Data 2019. September 2019. p.18 [online] emsd.gov.hk. Available at: https://www.emsd.gov.hk/filemanager/en/content_762/HKEEUD2019.pdf [Accessed 12 July 2020].

In Hong Kong, for example, the residential sector combined with the commercial sector accounted for 65 per cent, or 158,150 terajoules, of total energy consumption in 2017 while the electricity consumed by these two sectors is about 149,000 terajoules (Figure 2.6), in other words about 80 per cent of their total energy use.[21] Another developed economy in Asia, Singapore, presents a similarly high usage. In the city-state, commercial and residential users consumed about 51 per cent of total energy in 2019, showing that Hong Kong's energy consumption make up is not just an odd case. In another highly urbanised city, Shanghai, commercial and residential electricity use was 47 per cent of the total in the month of June 2020, 6.29 out of 13.37 terawatt-hours.[22] As such, any improvements that can be made in the energy efficiency of the premises will go a long way in terms of energy conservation and reducing the rate of future consumption.

To better understand what energy saving and conservation really mean on the ground, this author turned to one of the recognised leaders in the sector in Asia, one who is very passionate about sustainability. Hong Kong's Swire Properties Limited is one of the largest property developers and investors in Hong Kong with activities in other geographies as well. Its 23.2 million square feet in gross floor area (2.16 million square meters) in completed investment properties were valued at HK$276.8 billion ($35.7 billion) as of December 2019. Swire ranked eighth out of 23 of the world's top-tier real estate developers and ranked number one in Asia in the Dow Jones Sustainability World Index.[23] It is also a member of several other sustainability related indices, including the FTSE4Good, Hang Seng Corporate Sustainability Index, and the MSCI ESG Leaders Index. Also, Swire happens to be one of the pioneers and leaders in green and sustainable financing, a subject discussed in detail at the end of the book, in Chapter 4 'Financing the Growth'. The company raised $500 million through the issuance of a green bond with the proceeds raised earmarked for green buildings, energy efficiency, renewable energy, and sustainable water and wastewater management (Figure 2.7), reinforcing its industry-leadership status.

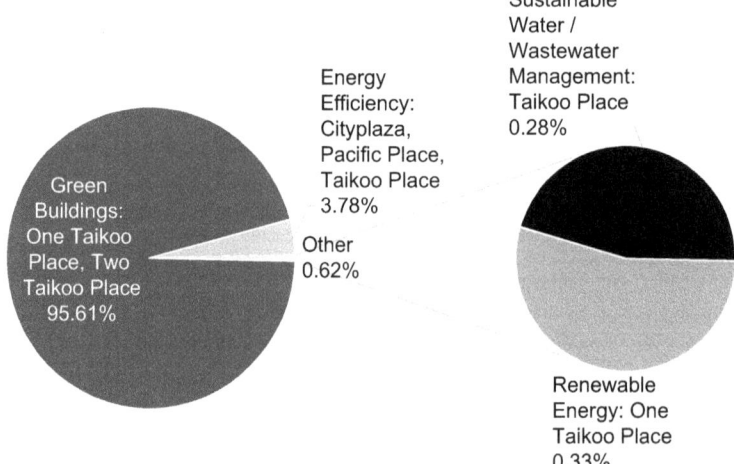

Figure 2.7: Allocation of Green Bond Proceeds (as at 30 September 2019).
Source: Swire Properties Limited. Green Bond Proceeds Allocation. [online] swireproperties.com. Available at: https://sd.swireproperties.com/2019/en/performance-economic/green-financing [Accessed 21 July 2020].

The property company has developed through several decades a cluster of commercial, office, and residential buildings on the eastern side of Hong Kong Island in a district called Taikoo; Taikoo is actually the company's name in Cantonese. The latest additions to the massive development in Taikoo are One Taikoo Place and Two Taikoo Place with 5.7 million square feet (about 543,000 square metres) of mostly office space (Figure 2.8).

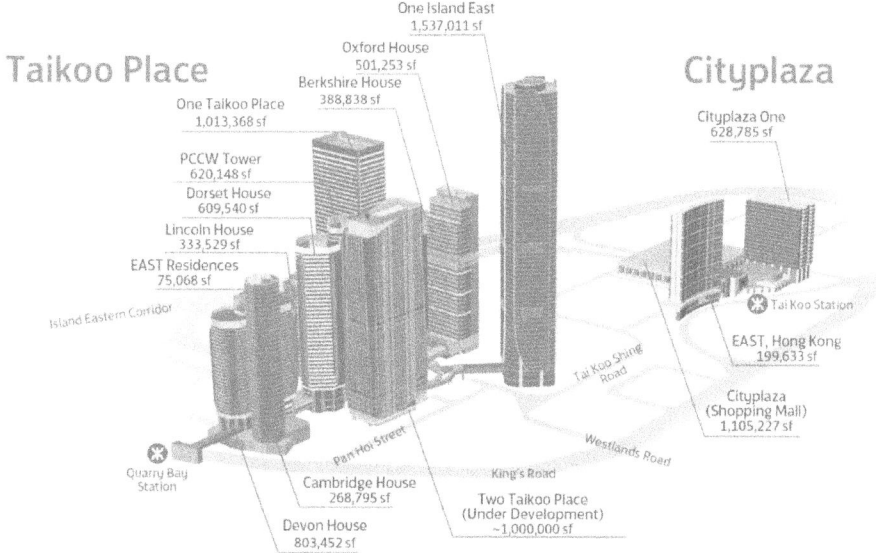

Figure 2.8: Swire Properties Taikoo Portfolio Map (Simplified maps are not to scale and for illustrative purpose only; Gross Flow Area figures are for reference only).
Source: Swire Properties Limited (2020). 2019 Final Results Analyst Briefing Presentation, 12 March 2020.

The first of the two is Swire's latest landmark example of its sustainability policies. After its completion in 2018, it became the first commercial building in Hong Kong to feature a bio-diesel tri-generation and adsorption system to supply heating, cooling, and electricity. This essentially is a closed loop waste-to-energy system to save energy and reduce greenhouse gas emissions.

On the green front, One Taikoo Place features a dual-level roof fitted with a combined green roof and solar photovoltaic power system. The vegetation on the green roof insulates the roof from the sun's heat, reducing building energy use associated with cooling loads. It will also serve to cool the solar PV system above, increasing the system's efficiency. The total renewable energy produced is estimated to contribute to approximately 4 per cent of the building's energy requirements.

One Taikoo Place adopted the latest technology of using Electronically Commutated Plug fans that are direct-current (DC) operated and can easily adjust speed.

The overall power consumption is about 40 per cent more energy efficient than a traditional centrifugal fan used in traditional air handling units. In addition, cutting-edge, high-performance, highly efficient centrifugal variable speed chillers with sophisticated optimisation controls achieve about 10 per cent energy saving compared to the standard chiller with traditional controls. The one-million-square-foot (929,000 square metres) office tower has already won multiple accolades and international green certifications.

Efforts such as these have helped Swire to reduce its annual energy consumption in its Hong Kong portfolio by 62.7 gigawatt-hours at the end of 2019 compared to 2008 levels (Figure 2.9). And all projects under development achieved the highest ratings of green building certification and 97 per cent of all existing buildings are certified green buildings, of which 84 per cent achieved the highest ratings, as of December 2019.

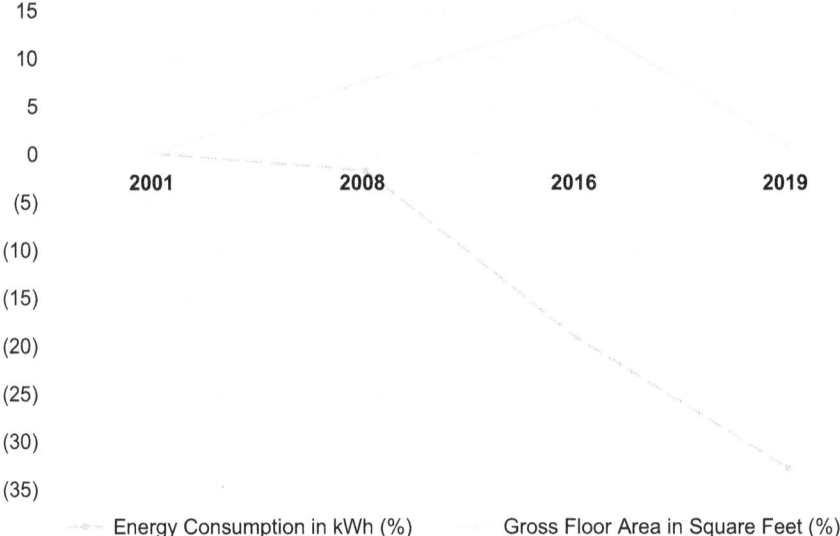

Figure 2.9: Swire Properties Change in Gross Floor Area and Energy Consumption.
Source: Replicated by author, July 2020. Swire Properties Limited (2020). 2019 Final Results Analyst Briefing Presentation, 12 March 2020.

2.1.3.4 Making a Business Out of Saving Energy

Entities involved in energy management, such as those providing consultancy services, provide companies with unique solutions to optimise the tail end of energy usage. The development of energy management systems, commonly known as EMS, has been stratospheric in Asia in recent years. This has probably been largely due to more acceptance on the part of the assets' operator or owner as they acquired a better understanding as to all of the benefits EMS can provide. In fact, globally speaking the number of EMS companies has boomed. These companies need to obtain an international certification known as ISO 50001. This is based on a standard published by the International Organization for Standardization (ISO) for the premises, be it a building or a factory, which uses EMS with the objective of consuming energy more efficiently. The number of certified companies has grown to more than 21,501 in 2017 (the last year where data is available), from less than 460 in 2011, based on data from ISO (Figure 2.10), or a compound annual growth rate of almost 90 per cent. Those companies incorporated in European countries accounted for more than 82 per cent of the total while those from the East Asia and Pacific region made up almost 12 per cent of the total in 2017, including 62 per cent from the Chinese Mainland and 15 per cent from Hong Kong, Macau, and Taiwan.

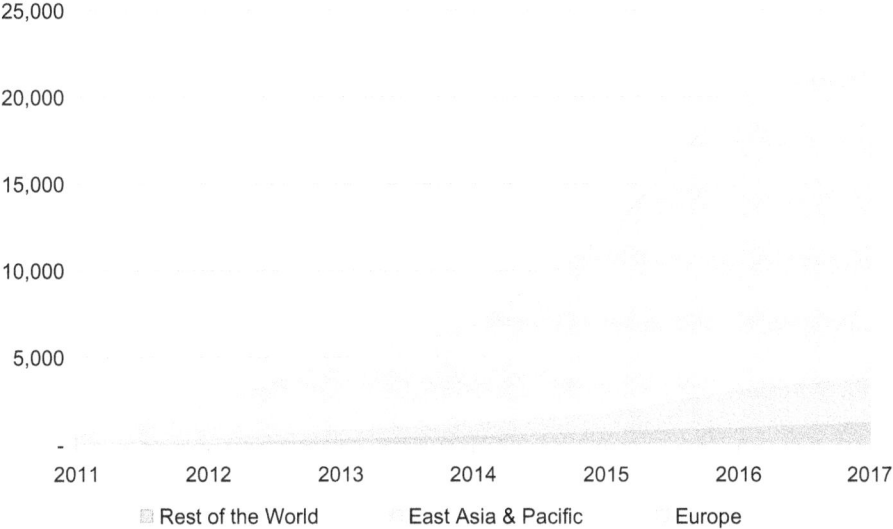

Figure 2.10: ISO 50001 Energy Management System Certifications by Region, 2011–2017.
Source: International Organization for Standardization. The ISO Survey. [online] iso.org. Available at: https://www.iso.org/the-iso-survey.html [Accessed 12 July 2020].

But, when EMS is discussed, what does it really entail in practice? How do companies save money by investing money? A great example to paint some colour on the subject comes from a pretty unique facilities management operator founded in Asia by two French entrepreneurs a couple of decades ago. Headquartered in Shanghai, the 20-year-old Aden has more than 26,000 employees and operates in 26 countries. It specialises in integrated facility management, which includes deploying its energy optimisation and automation platforms in partnership with the commercial, industrial, and private sectors. The general manager and co-founder of Adenergy, a subsidiary of Aden, Fulvio Bartolucci, was asked to provide some concrete examples on how EMS actually works in practice and its benefits.

Fulvio thinks that the way EMSs have been applied in most facilities generates a little confusion about the 'M' in EMS. Is it Energy Management or Energy Monitoring? He personally heard them used interchangeably by his clients and practitioners alike. He believes that this small confusion is a significant starting point to appreciate how EMSs are changing, merging with other automation tools and contributing to an unprecedented surge in energy efficiency. He explains that EMSs have been mainly used to collect and compile energy consumption data from multiple sources, and with diverse level of granularity – depending on the quality and quantity of the meters placed at the facility. Data was then analysed to highlight areas of improvement, yet the actual improvement required the operation team to do the job, hence the confusion between 'Management' and 'Monitoring'. EMS has essentially been used as a 'Monitoring' tool, for indirect 'Management' at best. So, he asks, what has exactly started to change in more recent times? Fulvio argues that on the data side, the evolution of IoT allows for sensors and meters to become much cheaper without sacrificing accuracy and security. This has in turn improved the quality and scope of the EMS data input. On the analysis side, the use of Machine Learning software engines enables the system to treat these much larger and more diverse data sets, and to discover relations and solutions not evident to the naked eyes. On the system side, the emergence of systems of systems[24] finally lets us close the loop from insight to action, where the system can automatically adjust based on the data coming from various sources internal or even external to the organisation, leading to EMS becoming true to its name, 'Managing' the energy system in real time.

Fulvio kindly offered to share some of the Adenergy work to better understand EMS. Adenergy applied the evolution of EMS in two of the main energy consuming areas in all kinds of facilities: heat and cooling systems and compressed air.

The first example is at one of Adenergy's clients, a multinational company in the tyre business. The plant consumes a massive amount of compressed air at a cost of more than RMB60 million ($8.56 million) per year. High-efficiency compressors were already installed, but there was still significant room for improvement. Adenergy designed an EMS platform to address both supply and demand side performance. It wanted to raise the efficiency of the production of the compressed air as well as enhance the monitoring of network distribution and leakage data. The combined implementation of supply side EMS Dynamic Controls and Upgraded leakage monitoring, with an advanced tool from Enersize, resulted in an increase of overall efficiency of 15.4 per cent, calculated as total energy per ton of product, bringing savings of more than 10 million yuan (about $1.5 million), just from better controls, without any equipment modification (Figure 2.11).

The second example is at the premises of another Adenergy client, a high tech and research and development (R&D) campus of about 650,000 square metres. There, Adenergy addressed efficiency issues with the dynamic boiler and chiller control system. This is a specific application of Controlling EMS, blending EMS and a Building Automation System, to automate controls of the utility supply at the site, based

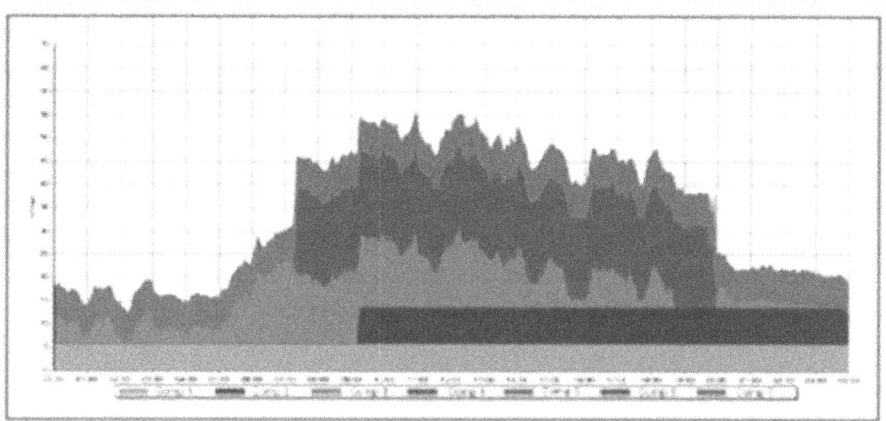

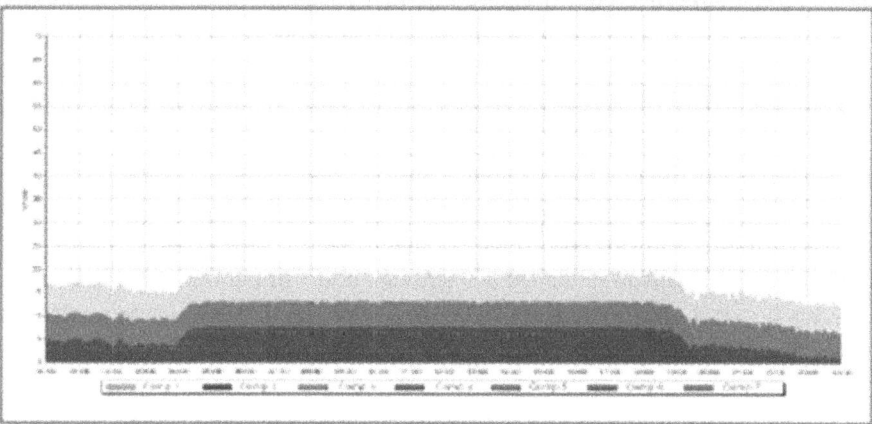

Figure 2.11: Compressors Using Optimisation: Compressor Room Before and After Adenergy's EMS and Leakage Work.
Source: Courtesy of Adenergy.

on environmental sensing and heating/cooling load consumption forecasts. The campus needs heating (steam) and cooling (chilled water) mainly for space comfort. Its annual energy bill only for heating and cooling was approximately 36 million yuan (about $5.38 million). The challenge for large sites like this one is the significant time lag between modification in heating, cooling, and ventilation needs in the various parts of the building on one side, and the time these changes are perceived by the control logic of boilers and chillers on the other. The solution Adenergy applied revolves around a Building and Energy Management System powered by a Machine Learning engine, accessing data from a Building Automation System, security network, weather data, and additional environmental sensors, providing temperature and air quality information. To simplify this with an image, let's imagine there are 500 people in a large auditorium with another 500 people attending a seminar during a hot summer's day. The lecture finishes and everyone leaves. The auditorium is then empty. However, the chiller will continue producing chilled water for a while, based on cooling needs during the lecture. This, unless a smart Building and Energy Management System can sense the change in the number of people in the auditorium and adjust the cooling load requirement in real time. Using such a Building and Energy Management System as well as supplementary sensors produced a reduction in heating and cooling consumption of about 20 per cent, equal to 7 million yuan ($1.05 million) a year.

2.1.3.5 Increasingly Brighter Lighting Prospects

The efficient use of energy and energy conservation are nothing new. This author remembers that in the common corridor of a building he used to live in when he was very young in France in the 1970s there was a light switch with a timer that turned off the light just a few minutes after being pushed. Surprisingly, more than 40 years later this is still not commonplace in many developed economies in Asia such as Hong Kong, Japan, or Singapore. Admittedly the move to having highly energy efficient LED lighting is reducing electricity consumption for lighting purposes.

Lighting technologies have made enormous advances in the past few years. The US Department of Energy, for example, estimated that LEDs for residential use will consume 75 per cent or less energy when compared to the traditional standard incandescent lights and can last about 25 times longer. It also expects that by 2027, LEDs could potentially save almost 345 terawatt-hours in the US. This is equivalent to 44 electricity generating units of 1,000 megawatts each, and much more than the consumption of many large economies, such as Italy, which consumes, for example, 320 terawatt-hours per year in total.[25]

LEDs are just one of many tools available to residential, commercial, and industrial users to save electricity. Many more technologies are currently being developed and in future energy conservation and saving could possibly have a massive impact on electricity demand.

There is of course a large amount of lighting-related technologies to save electricity, but this is only one tiny aspect of energy efficiency and conservation. There is an enormity of easy to implement solutions, many of which are easily affordable. What is imperative is strong government motivation to create adequate frameworks and to go out and actually implement this. In fact, very few governments in the region have done a decent job in the view of this author. China and South Korea are the rare success cases.

To get some more colour on energy savings through changing or upgrading lighting systems, the author approached a friend, Paul Snelgrove, who works for Energys

Group. It is one successful example of a company that has been able to benefit from the growing demand in energy efficiency solutions in Asia and the rest of the world. Originating in the UK, Energys built a strong reputation in energy savings solutions, with specialisation in low carbon lighting retrofit solutions. It built a track record of more than 15 years serving major UK government related customers including the National Health Service (NHS), the Ministry of Defence as well as British Telecom (BT). Initially a consultant to its customers, Energys found that the total solution model was what customers were willing to pay for and adjusted its business model accordingly. One of their slogans 'We don't over-complicate, we don't baffle with science, and we don't over-engineer', sums up their years of experience in the energy saving industry, as customers simply want results. Typically, this involves doing an energy audit by measuring the existing energy usage and designing a solution that achieves the lowest effective power consumption. Energys will then do the installations and offer after-sales service. Having been supported historically by a R&D, procurement, and assembly team based in Hong Kong and China, the company delivers customised solutions from factory floor to end-users' ceiling.

In terms of effective power-saving solutions, Energys retrofits traditional lighting into state-of-the-art LED products, and in combination with AI controls, projects typically achieve 70 per cent to 90 per cent energy savings. Their model is to deliver the right amount of light at the right time for the right purpose. It uses fully automated real-time technology for measuring and verification (M&V) to accurately gauge the energy savings and effective reduction on carbon footprint. This system is often retained by the customer for energy monitoring purposes. By knowing what one is spending, customers can easily achieve a further 20 per cent to 30 per cent savings by adjusting their energy usage behaviour.

The energy cost savings from lighting can be described as the low hanging fruit, but Energys also offers consulting and solutions for air-conditioning and other usage of energy. To make it an easy decision for the customer to create cost savings and have a positive impact to the environment, Energys offers a pay-on-savings option for its projects. In simple terms, this means that no upfront capital investment is required from the customer. Given that the repayment is a monthly fee that is just a fraction of the effective cost savings, it is cash-flow positive for the user from day one.

Since 2019, Energys has used its UK experience to successfully tackle opportunities presented by the growing demand for energy and sustainability in Hong Kong and the rest of Asia. So far, Energys has won a contract with the Hong Kong Government, providing an LED retrofit lighting solution for 12 indoor sporting venues. The effective 60 per cent lighting energy saving equates to an annual 900 megawatt-hours energy consumption reduction with a net carbon footprint cut of 730 tonnes. Another project was for the offices of leading Hong Kong architect, Ronald Lu & Partners. The work led to an 80 per cent reduction in lighting energy costs through a combination of T5 and T8 LED retrofit as well as implementation of their

IntelliDim controls to maximise cost savings (Figure 2.12). The overall impact was a 120 megawatt-hours reduction in annual electricity consumption and more than 96 tonnes reduction in carbon footprint.

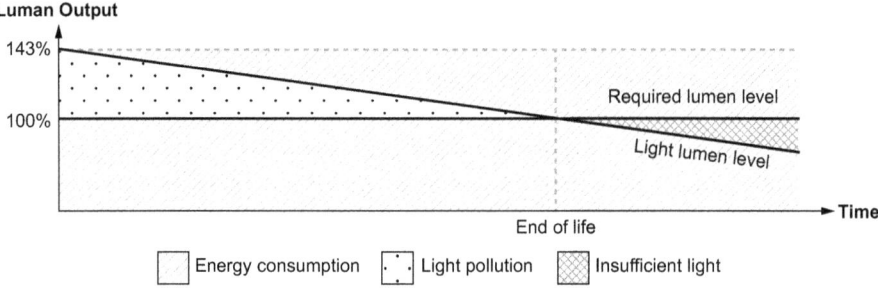

Figure 2.12: Typical Lighting System.
Source: Courtesy of Energys Group Limited, July 2020.

Standard lighting is constant; regardless of its surrounding environment, it will provide the same amount of light and consume the same amount of energy across its service life. Having said that, lights will also deteriorate over time (meaning a fall in lumen output), so the industry standard often refers to the end of service life when the lumen output reaches 70 per cent of its original output (L70). For this reason, lighting designers often design light usage with a 43 per cent margin to ensure sufficient light will be provided across its service life. Energys' IntelliDim is a three-in-one dynamic controller with Automatic Lumen Adjustment, Occupancy Sensor, and Daylight Harvesting functions. Instead of adding a 43 per cent margin upfront, lumen output can be commissioned to the precise requirement. As the light deteriorates, more energy will be consumed to drive the light to maintain at the precise level. This facilitates a huge amount of extra energy savings and prolongs the service life of the lights. Additional energy can be saved through dimming or switching off the light with the Occupancy Sensor when space is vacant, and Daylight Harvesting (Figure 2.13).

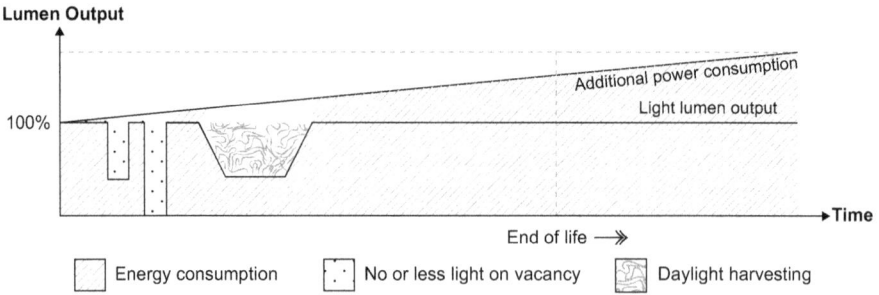

Figure 2.13: IntelliDim – Dynamic Lighting Control.
Source: Courtesy of Energys Group Limited, July 2020.

2.2 Huge Regulatory Shifts Transforming the Energy Industry

There has been a dramatic amount of regulatory disruption in the past few years in many Asian electricity markets. In the next five to ten years such changes will not only continue but they will most likely accelerate. The broad-based move towards the opening of the energy market to competition is accelerating and this transformation will progressively have a significant impact on the energy supply mix as well as affect energy prices in the region.

There are already many advanced models of the energy markets deregulating and opening up. Australia and New Zealand have deep experience in the liberation of their energy markets. Apart from market opening another significant regulatory disruptor is climate change. Climate-related policies principally aim at abating pollution when it comes to the energy supply sector. All of these trends, whatever the motivation, are highly positive at many levels. They force into the equation more transparent, and arguably fairer, energy-pricing mechanisms. They foster new business models. They allow the introduction of new market participants. They generate more jobs. And, also, they provide more value-added services improving the overall consumer experience.

Apart from all of these motivations, there are also many reasons why it simply makes rational sense to deregulate energy supply, in part because it should ensure that the company responsible for the supply of the energy is efficient at supplying it at the lowest possible cost. The oversight is not just a regulatory one. An open market means competition and with competition comes consumers' freedom to choose their energy providers.

The traditional model of energy supply is relatively simple and straightforward. The company sources the energy, and it then supplies it to the end user. It will typically have a monopoly, so the management does not have to worry about competition. In some jurisdictions, the company is monitored by one or more government agencies and is subject to regulatory scrutiny. When there is actual tight government involvement in the company's business, this traditional model is relatively all right. It typically delivers energy supply reliability. However, the fundamental problem with such a model is how to ensure that the company is operationally and financially effective and efficient. This is obviously easier said than done because the primary task of such a company is generating investment returns for its shareholders. And, when it is not a shareholder-owned energy company but a government-owned one, then there are even less efficiency checks.

There are a great number of in-depth academic studies as to whether the energy market should be fully opened and liberalised. One author well summarised the controversy.

One side of the argument is that the old utility model of regulated monopolies is now obsolete and that opening up the market to competitive forces will cut prices, raise efficiency, and have net benefits for the economy as a whole. The other side of the argument is that competitive electricity markets do not necessarily have a good track record as they have been subject to manipulation resulting in higher prices for users and also create dangerous volatility in the supply of the energy.[26]

The view of this author is that liberalised markets have a great amount of benefits to economies and to consumers alike. The old utility model has been around for six or seven decades if not longer. Liberalised markets are something much more recent. Because of the idiosyncrasies of individual markets, every time a market is opened up to competition it will have some teething problems. As mentioned, there are the believers and the non-believers and there is a great number of studies of the issue. For the purposes of this book, there is no need to delve into the pros and the cons of open power systems. Rather, the focus is on what is happening and evolving today with a rising number of the electricity markets in the Asia region. Some of these markets have already deregulated, many are well on their way to do the same, while a few are still pondering the option.

There are three types of changes driving sector transformation from energy markets reforms that need to be discussed and evaluated. New pricing parameters are being created. New business models are being established. New market entrants are participating in the reformed markets.

But first let us establish a quick concept. When talking about deregulation actually, we are not actually talking about re-regulation. The pricing parameters of the traditional utility model is relatively straightforward. The consumer pays the power supplier or at least the retail arm of the power supplier. The money received goes towards paying for the retail business (including, for example, the billing and consumer services), the transmission and distribution network (the wires), and the electricity production costs (the power generation plants). Under this kind of model, typically the pricing significantly lacks transparency. There are few jurisdictions in the Asia region where the billing by the integrated utility actually breaks out in detail the costs of the individual three segments.

The open or deregulated or re-regulated market is significantly more complex. In terms of participating entities, the wires must be independent from the retail or generation segments. The owners of the wires (or grid network) are paid something like a toll fee, just like when you have to pay a fee on a toll road. The fee is typically fixed and often highly transparent. Also, typically these wire companies operate under great scrutiny as regulators want to avoid one or more participants cheating the system. The returns on investment earned by the grid network company and other fees charged tend to be relatively fair. The example of such electricity transmission and distribution companies in Australia can be examined. Most companies in the country are governed by the National Electricity Rules created by the Australian

Energy Market Commission (AEMO) under the nation's National Electricity Laws. The permitted investment return earned by these companies is overseen by the Australian Energy Regulator and reset every five years. During these five years, the regulator monitors the efficiencies, costs, and other aspects of the operator. When it is time to reset the permitted return, it uses all of this information to reset the benchmark return.

When it comes to the generation and retail part of the equation there are a variety of business models. Some companies have both generation and retail business segments while some others have just one or the other. And for those companies that control both, they may not necessarily be 100 per cent self-sufficient in terms of the power they sell. They may be able to only generate a portion of the power they sell. The shortfall will be bought in the wholesale market. The Australian model is often used as a good blueprint for other markets in the region.

Put very simply the kilowatt-hour flows from the generator to the transmission networks, then to the distribution networks and finally to the consumer. Between the generator and the transmission networks there is a 'pool' called the National Electricity Market that sets the wholesale price. But it is the energy retailer who owns the consumer (Figure 2.14).

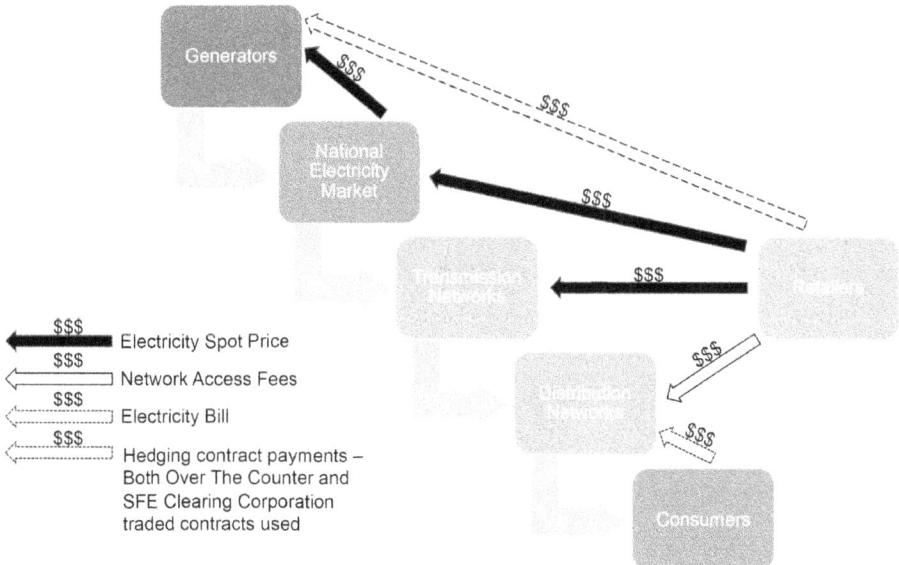

Figure 2.14: Australia Electricity Market Structure (Adapted).
Source: Nicholas Tan, Economic Group, Reserve Bank of Australia (2011). How Are Electricity Prices in Australia Set. 4 February. [online] rba.gov.au. Available at: https://www.rba.gov.au/information/foi/disclosure-log/pdf/101115.pdf [Accessed 13 July 2020].

Let us briefly put into context what exactly is happening in the key electricity markets in the Asia region, in general and then in Australia, Japan, Singapore, and China specifically.

The status of regulatory disruption can be broadly categorised into three individual buckets. First, there are energy markets that are already fully open or that will be fully opened in the short term, on or before 2025. Second, there are some markets where either the reform is in process or where reform is still on the drawing board – and in some cases has been on the drawing board for a great number of years, and, sometimes, many decades. Finally, there are a few markets where there is very little happening in terms of liberalisation discussions at any level.

The most competitive markets in the view of this author are Australia, New Zealand, and Singapore. They are followed by Japan and then by China and the Philippines as well as some states in India. Energy markets that are seriously pondering a reform process include Indonesia, Malaysia, Taiwan, Thailand, South Korea and Vietnam. The remainder of the Asia economies all seem to be a little bit stale in terms of restructuring their energy markets. These include Bangladesh, Cambodia, Laos, Mongolia, Myanmar, Pakistan, and Sri Lanka (Table 2.6).

Table 2.6: Status of Electric Power Market Reform in Key Asian Economies.

	Reform	Reform Plans	No Clear Plans
Australia	✓		
Bagladesh			✓
Cambodia			✓
China	✓		
Hong Kong S.A.R.			✓
India	✓		
Indonesia		✓	
Japan	✓		
Laos			✓
Malaysia		✓	
Mongolia			✓
Myanmar			✓
New Zealand	✓		
Papua New Guinea			✓
Singapore	✓		

Table 2.6 (continued)

	Reform	Reform Plans	No Clear Plans
South Korea		✓	
Taiwan (China)		✓	
Thailand		✓	
Vietnam		✓	

Source: Author, July 2020.

As mentioned, this author believes that one of the great advantages in having an open and transparent energy supply market is that the price will be competitive. In addition, there are other positives, which actually have proven to be a little less controversial. These include the fact that consumers get more value-added, more convenient, more reliable services from their chosen energy provider.

There are many examples of this. Here the examples will be limited to the Australian, Japanese, and Singaporean electric power markets. A bit of time will also be devoted to looking at the Chinese market. This is because not only is it the largest power market in the world but also it is one that slowly but surely has been reforming and gradually opening up direct energy supply to end users, or at least for some user segments. First, the reform history of each of these markets will be briefly summarised and then there will be a focus on lessons learned from each of these markets. Prior to tackling these four examples, it is worth mentioning another trend that is arising in part because of the liberalisation of the pending deregulation of some markets, and that is local utilities seeking growth opportunities abroad. This is important because these utilities can become formidable new market entrants (Table 2.7). The key takeaway is that the bulk of the companies are currently still at the staying-at-home stage, but they are clearly changing. Many utilities are likely to follow in the footsteps of those from Japan or Thailand, for example.

Table 2.7: Markets Where Domestic Utilities Invest Abroad.

	Yes	Yes But Limited	No
Australia			✓
Bangladesh			✓
Cambodia			✓
China		✓	
Hong Kong S.A.R.	✓		
India		✓	

Table 2.7 (continued)

	Yes	Yes But Limited	No
Indonesia			✓
Japan	✓		
Laos			✓
Malaysia	✓		
Mongolia			✓
Myanmar			✓
New Zealand			✓
Papua New Guinea			✓
Singapore		✓	
South Korea	✓		
Taiwan			✓
Thailand	✓		
Vietnam			✓

Source: Author, July 2020.

2.2.1 Australia – Deregulation Lessons Brutally Learned

The Australian reform experience was purposely chosen as the first region where other power markets in Asia can draw some lessons. This is because together with the experiences of Chile (which started reforms in the early 1980s) and of the UK (which started reforms in the late 1980s), the Australian electric power market reform is one of the most observed examples in the world. This even remains true today because the Australian market is heading towards full decarbonisation in 2050 despite the fact that fossil generation still occupies roughly three-quarters of the nation's total. So, it is a quoted example for its 20-year-plus experience and because experts are highly interested in the strategies the country adopts to tackle the energy transition to a low or zero carbon market over the coming few years.

Sector reforms were decided in the 1990s as part of the then government's drive towards microeconomic reform. This is widely known as the Hilmer reforms, named after Professor Fred Hilmer who was the chair of the National Competition Policy Review Committee created in 1992 and that produced a report that led to the Competition Policy Reform Act 1995, and led to the power sector reform. The Hilmer reforms 'sought to allocate the risks of investment in generation and retailing to market participants, and away from consumers and taxpayers, as a means to achieving allocative efficiency and greater productivity'.[27] Just like in most power markets, before the reform the electric power sector was government-owned and run, and was vertically integrated, i.e., the generation, transmission, distribution, and retail of power was all integrated. These were subsequently split into the competitive components (generation and retailing) and the natural monopoly components (transmission and distribution). In 1998, the National Electricity Market (NEM) was set up as the regional pool where the electricity is transacted. In 2009, the NEM was folded into the Australian Energy Market Operator Limited (AEMO), which became responsible for the planning and operations of electricity and gas. AEMO operates independently but is 60 per cent owned by the Federal Government as well as individual Australian states and territories; the remainder is controlled by industry members. Generators sell the electricity to retailers or directly to large-scale customers through AEMO's NEM, which works a little bit like a stock market. The NEM operates the Australian power system that today comprises all of the nation's states and territories except for Western Australia and the Northern Territory, chiefly due to the great distance between networks, which actually only account for about 11 per cent of the national population (Figure 2.15).

Most international and domestic experts view the NEM's first decade as having been relatively successful. Until the mid to late 2000s, it served the end users quite

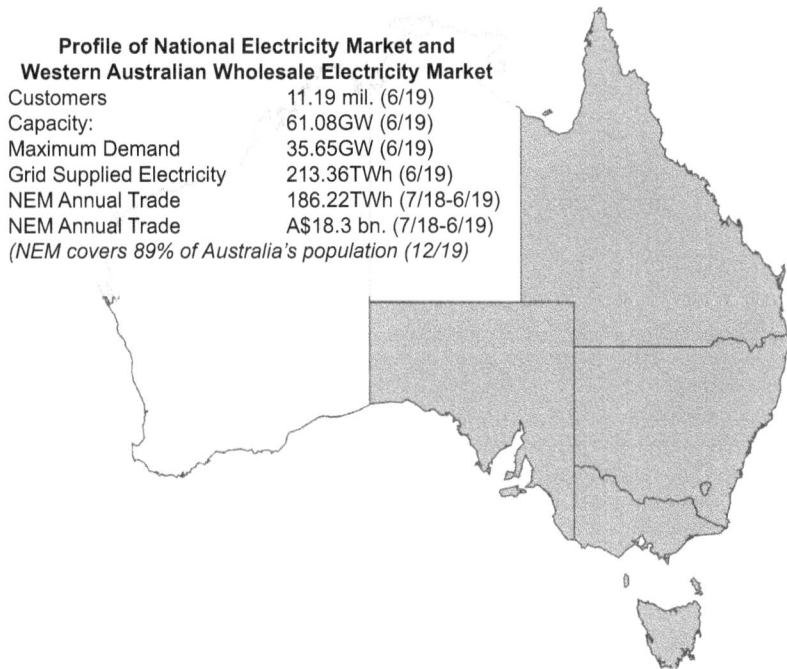

Figure 2.15: Profile of National Electricity Market and Western Australian Wholesale Electricity Market. Source: Author, July 2020. Data sourced from Australian Bureau of Statistics and the 2019 Annual Report of the Australian Energy Market Operator. Map courtesy of mapchart.net.

well. It gave them freedom of choice in terms of provider and lower prices and an improved system reliability. However, some problems began to appear that led to consumer prices actually rising from the late 2000s to the mid to late 2010s; the wholesale spot price trend in major states can be used as a proxy to show the effect (Figure 2.16). One author identified three sets of reasons for the rise. One was a sustained rise in power distribution costs. A second was the lack of a consistent and transparent government policy towards emissions reduction; this is something absolutely critical in any market as the investor needs to have long-term clarity on the market direction in order to assess whether it is worthwhile making investment as electric power assets have a lifetime of 20 to 30 years, or even longer in some cases. And third, higher expenditures for existing and new-entrant dispatchable plants[28] – generation plants that can be called up to meet demand spikes.

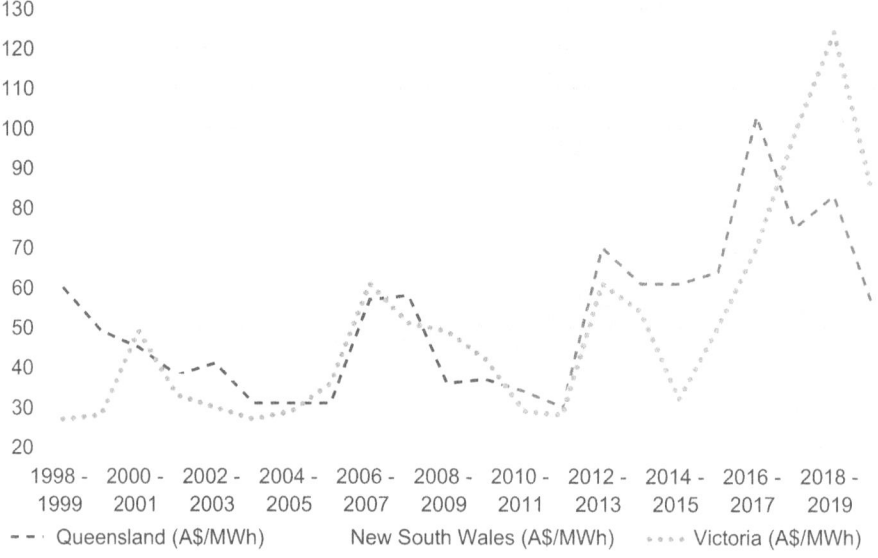

Figure 2.16: Annual Average Spot Prices in Australia's Three Largest States.
Source: Author, July 2020. Data sourced from the Australian Energy Regulator. Australian Energy Regulator (2020). Annual Volume Weighted Average Spot Prices – Regions. [online] Australian Energy Regulator. Available at: https://www.aer.gov.au/wholesale-markets/wholesale-statistics/annual-volume-weighted-average-spot-prices-regions [Accessed 2 December 2020].

Needless to say, during the period there were other challenges that the Australian power reform faced – challenges that occasionally even caused some in the industry to raise the possibility of scrapping the reforms altogether. But just like in other markets, higher end-user prices cause an extreme number of debates and frustrations, particularly when the power bill of residential users goes up. These challenges easily and readily grabbed big media headlines. Headlines such as 'Can Australia bring its sky-high energy prices down to earth?'[29], 'Australian households pay highest power prices in world',[30] and 'Low-income households hit hardest with thousands of Australians stuck on expensive electricity plans'.[31] The relatively good news is that most forecasters as of mid-2020 expect average Australian retail power prices to either remain flat or even fall in the coming years. Also, regulators and their advisors have closely analysed the reasons behind the price rises and have been instituting rules and guidelines to prevent some 'playing the market'.

Another key take-away from the Australian experience is that the retail power business is highly complex and is brutally competitive. The biggest energy retail companies (i.e., the highest profile ones) have not exactly posted stellar financial results in the past few years despite the rising wholesale prices. The three biggest domestic energy companies are AGL Energy, EnergyAustralia, and Origin Energy (Figure 2.17). To be fair, Origin's dismal financial results were in part due to booklooses (i.e., accounting losses not actual cash ones) related to its ownership of 37.5 per cent in the Australia Pacific LNG project, which has little to do with its domestic energy supply markets business; the project is one of the country's largest producers of natural gas and a major exporter of liquefied natural gas. Still all three have had a record of poor earnings due to all the reasons mentioned above as well as volatile electricity prices and market share gains and losses.

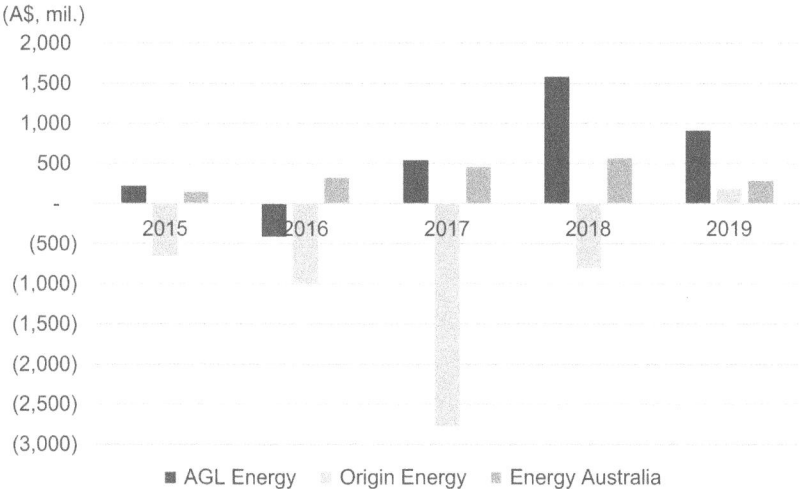

Figure 2.17: Earnings of Australia's Largest Electricity Retailers.
Note: AGL FY18 restated; AGL and Origin Statutory Profit after tax; EnergyAustralia is net income as reported by its parent company CLP Holdings AGL and Origin year through 30 June; EnergyAustarlia year to 31 December.
Source: AGL Energy Limited, CLP Holdings Limited, and Original Energy Limited company annual reports 2016–2019 which can be accessed at https://www.agl.com.au/about-agl/investors for AGL Energy Limited, at https://www.clpgroup.com/en/investors-information/financial-reports for CLP Holdings Limited, and at https://www.originenergy.com.au/about/investors-media.html for Original Energy Limited.

The relatively poor earnings across Australia's biggest three retailers, as well as others with large natural positions, also has to do with trading and portfolio optimisation. The trading of electricity is not too dissimilar from the trading of equities or commodities, albeit I would argue it is even more complex, with more variables. The energy companies have a specific department that trades and hedges electricity and gas natural positions. These trading desks must handle the volatility of wholesale prices. Also, they must manage risks such as carbon, regulatory policy, volatile weather, and fuel costs. How does this work? A simplified hypothetical example can be used. Let us assume that an energy company has entered into a contract to sell 100 units at $10 per unit. Let's also assume that it has a three-year power selling contract. The energy company will need to ensure that they will not be buying the units above the strike price of $10. If the price moves up to $13, then the energy company would have to decide whether it should wait for the price to come back to $10 or less or use hedging contracts to hedge part of the 100 units commitment. The hedging contracts effectively set the price in advance. However, hedging comes at a price. It is like purchasing an insurance policy. So, the trader must carefully manage risks and judge the amount to hedge and at what price. This must all be done within the company's preapproved commodity risk framework. If the judgement is wrong, then the company can incur losses (Figure 2.18). This has repeatedly occurred to many electricity retailers because in real life, special events such as heatwaves do occur and do disrupt the supply balance. For instance, an electric power plant suddenly failing

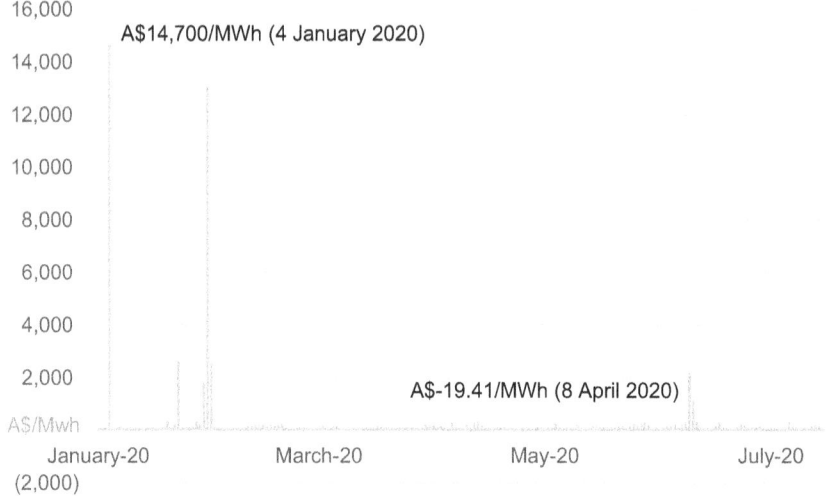

Figure 2.18: New South Wales Power Prices 1 January Through 14 July 2020.
Source: Author, July 2020. Data sourced from the Australian Energy Market Operator (AEMO). Australian Energy Market Operator (2020). Aggregated Price and Demand Data. [online] https://aemo.com.au. Available at: https://aemo.com.au/energy-systems/electricity/national-electricity-market-nem/data-nem/aggregated-data. [Accessed 2 Dec. 2020].

with no warning. This will reduce the total supply of electricity into the market, and assuming demand does not change, then the wholesale electricity spot price will rise. Annual once-in-a-hundred-years events due to climate change do not help.

Apart from that, the intensity of the competition is clearly shown by a very high churn rate, also referred to as retail transfer rate, that is, the number of customers who change provider. For example, in the four major energy consuming states in the NEM the one month annualised transfer rates were 16 per cent, 11 per cent, 11 per cent, and 20 per cent for New South Wales, Queensland, South Australia, and Victoria, respectively, in June 2020. In the twelve months period through June 2020, the highest churn rate was experienced by Victoria with 27 per cent in July 2019 and the lowest rate was also in Victoria with 10 per cent in June 2020 (Figure 2.19). This means that the retailers there were churning between one in three to one in ten customers each year. This makes it extremely challenging to run an energy business with long duration physical assets. It obviously calls for these energy companies to adopt new strategies and business models, which is discussed in detail later in the book (see 2.3 Changing Nature of Industry Players). One thing is sure, the intense competition over the past few years has significantly helped consumers in the form of better and more diversified service, again something that is discussed below.

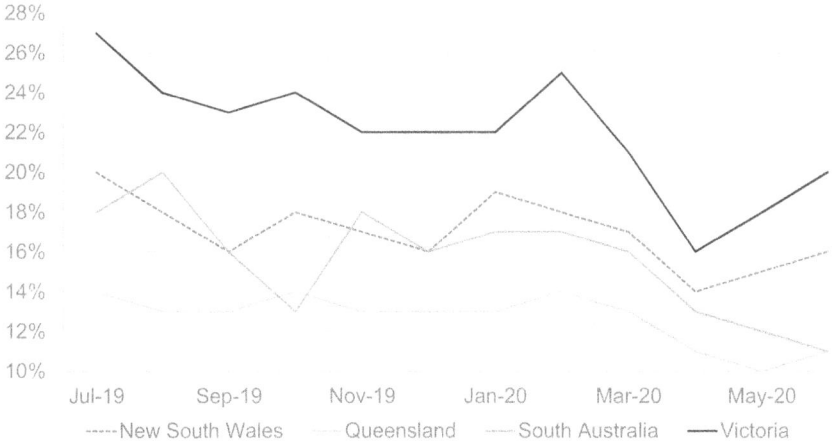

Figure 2.19: National Electricity Market One Month Annualised Retail Transfer Statistics (Churn Rate).
Source: Australian Energy Market Operator (2020b). Real Transfer Statistical Data. [online] https://www.aemo.com.au/. Available at: https://www.aemo.com.au/energy-systems/electricity/national-electricity-market-nem/data-nem/metering-data/real-transfer-statistical-data [Accessed 2 Dec. 2020].

Part of the ongoing national debate is over retail prices. This issue is a little cooler in 2020 as prices have fallen but will continue to be a highly sensitive one. Another area of fierce debate is with the energy resources and the management of these resources.

Those in the national debate favouring more green power, especially solar and wind, are vehemently against fossil fuel based generation. Those promoting using fossil fuels emphasise the unreliability of green power sources (no wind power with no wind and no solar power with on sun). But a major part of the debate must be on the energy resources impact on the entire energy value chain and the whole economy. Missing, is a reliable, transparent, conclusive long-term energy policy from the government, something that would probably be best handled by an independent entity with some teeth. This is because the existing elected administration may have its policies repealed from a subsequent one. This has already happened in the country in the past few administrations. As mentioned earlier, energy assets investments are long-term investments. Energy infrastructure planning cannot take U-turns after every election. Policy consistency has been behind the relative success of electric power reform and liberalisation in China and Singapore. Something Australia should emulate.

2.2.2 Japan – The Land of the Slowly Rising Competition

Japan's reform experiment took a quarter of a century. And the market structure as of 2020 is not as fluid as in other deregulated countries in the region such as Australia or Singapore. Still, it is a major achievement for the island nation. The incumbent electric power companies (EPCOs) in the past have been, and still are today, politically powerful and financially strong. As such it has been no easy feat to open up the electric power markets to competition. The general trajectory is that the 10 incumbent EPCOs may continue to lose market share in the short to medium term. Then the market will see some form of consolidation. This is a trajectory seen in the UK and in Australia, for example.

Japan's government has been trying to reform the electric power sector since the late 1990s. Officially it undertook three stages, starting with the passing of the Electricity Business Act Laws passed by the National Diet of Japan, the bicameral legislature, in 2003. However, the sector incumbents were quite reluctant to support the reform. As such, it was not until the 2010s that consumers started to benefit from the reform. Unlike Australia or other jurisdictions where the sector was reformed and liberalised, the Japanese electric power utilities were not state-owned entities. So, the implementation of the announced reforms met steep resistance. It was not until the extremely unfortunate, preventable Fukushima Daiichi Nuclear Power Plant meltdown following the Tohoku earthquake and tsunami that the government was really able to push through with sector reforms, in the view of this author. While the opening of energy markets in Japan was probably one of the slowest in the world, the slow and gradual approach has seemingly been breaking down barriers, barriers that many had expected to be unbreakable. The changes have encouraged, for example, the incumbent industry participants to restructure and transform themselves into total energy solutions providers.

It is worth pointing out that actually Japanese authorities slightly changed the narrative. It is a fact that the country launched reforms in 1995 that followed with promulgation of the Electricity Industry Reform Act 1998. The act called for the break-up of vertically integrated electric power companies that controlled every aspect of the sector (generation, transmission distribution, and retail). It is also a fact that this break up, with Japanese characteristics, had not been fully completed as of June 2020. The latest reform milestone officially started with the preparation of the issuance of the amended electricity business act from 2013. It was not until 2015, 20 years after authorities had decided on the market reform, that the process started gaining pace. The first significant step was the establishment of an independent electric power network system operator, the Organisation for Cross-regional Coordination of Transmission Operators, Japan (OCCTO). OCCTO's main job is safeguarding a stable supply of electricity. It is also there to strive to keep electricity tariffs as low as possible and encourage the expansion of consumers' choice as well as creating business opportunities. A year later, in April 2016, the market was fully opened to competition giving for the first time consumers the choice as to which company will sell them electricity. The last step, from April 2020, lies with the existing 10 vertically integrated electric power utilities. They have to unbundle their transmission and distribution networks from their generation and supply (i.e., retail) businesses (Figure 2.20).

Having explained the reform background, two areas can be addressed. One is evaluating the challenges the sector faces in its development as of 2020. Another is how the sector is likely to evolve in the mid to long term, the next five and the next 30 years.

Japan's electric power sector's first challenge is to raise the amount of electricity traded in its wholesale markets. The trend has been a positive one (Figure 2.21) but experts believe that more kilowatt-hours have to go through the market. Low liquidity in the wholesale electricity markets is actually a barrier of entry to independent power producers as well as power suppliers (i.e., retailers). The UK regulatory authority (Office of Gas and Electricity Markets or Ofgem) suggested that liquidity is important to consumers because it is needed for any new market participant, it raises price transparency providing opportunities for higher market competition and creates a level playing field for smaller market participants to compete with the larger companies. Better liquidity helps the electricity wholesale market to have longer term financial products to support hedging as well as generate robust reference prices. These are key to ensuring that the real market value of the electricity is based on the underlying economic conditions,[32] including the demand and supply balance. In Japan this has been slow but has been improving in alignment with the policy changes. Almost 800 gigawatt-hours passed through the Japan Electric Power Exchange (JPEX) in the year through March 2020, 40 per cent more than the 571 gigawatt-hours traded the previous year through March 2019, which itself was 257 per cent more than the year through March 2018. Also, a far cry from the 10

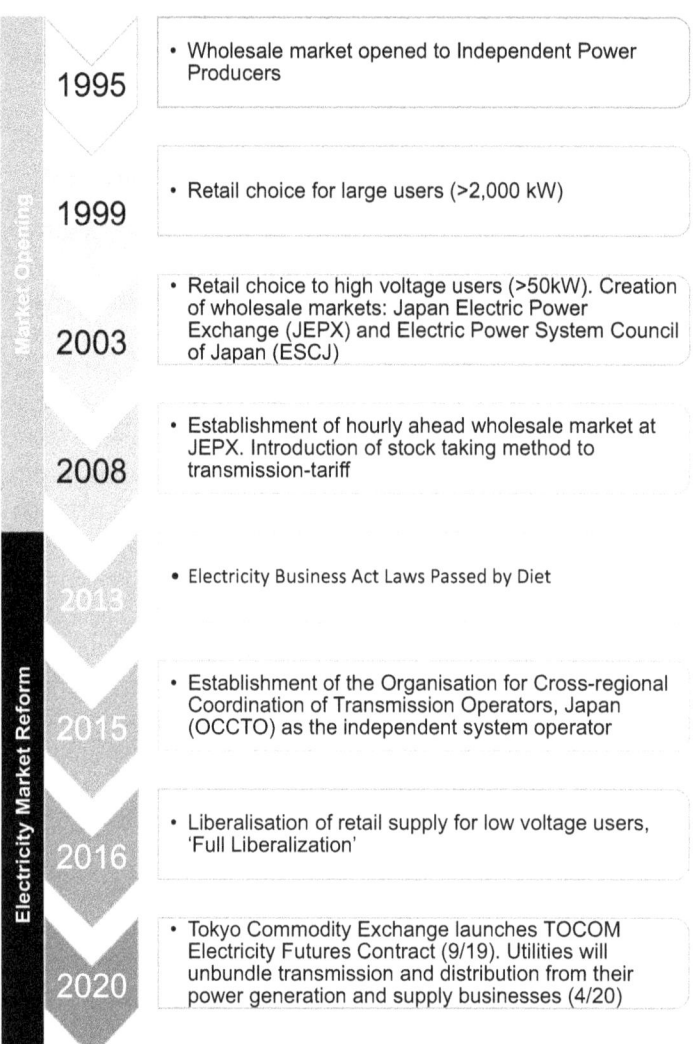

Figure 2.20: Japan's Electricity Market System Reform.
Source: Author, July 2020. Data sources include Tatsuya Shinkawa, Electricity and Gas Market Surveillance Commission, Ministry of Economy, Trade, and Industry (2018) Electricity System and Market in Japan. 22 January. [online] https://www.emsc.meti.go.jp. Available at: https://www.emsc.meti.go.jp/english/info/public/pdf/180122.pdf [Accessed 17 July 2020]; Overview of TOCOM Electricity Market, Tokyo Commodity Exchange, Inc. September 2019.

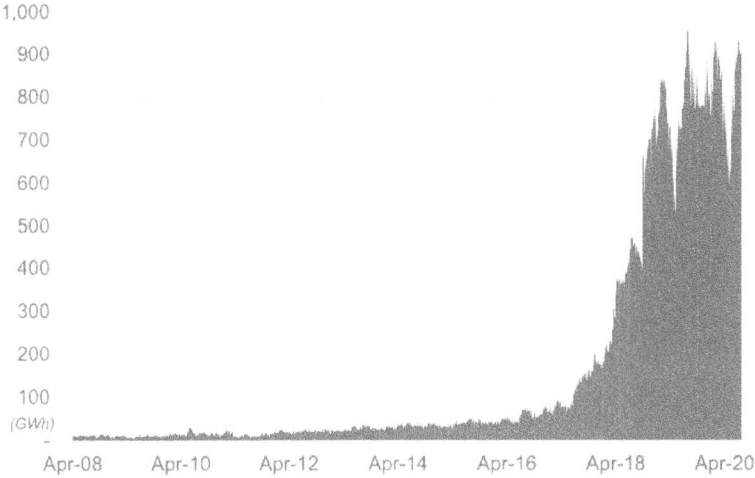

Figure 2.21: Japan Electric Power Exchange (JPEX) Total Transaction Volume (GWh).
Source: Japan Electric Power Exchange (2020) Trading Information: Spot Market/Intraday Market. [online] http://www.jepx.org. Available at: http://www.jepx.org/english/market/index.html [Accessed 17 July 2020].

gigawatt-hours traded in the year through March 2008, when the hourly ahead wholesale market was established at JEPX.

Early 2016 must have been an exciting time for those companies planning to enter the domestic power retail market. Authorities had confirmed that from 1 April consumers would be free to choose their electricity suppliers. So, many companies started to undertake very aggressive advertising. In the first few months, very few of Japan's 53 million or so households decided to change provider; actually only 511,000 had done so by the end of March, based on statistics compiled by OCCTO. This was most likely due to the lack of familiarity with the new scheme on the part of Japanese households. However, by the end of the year the number had grown to 2.57 million. Thereafter, the rate grew by an average of 2 million or so every six months. It reached a total of 16.85 million by the end of June 2020. It is almost certain that the rate of households switching their accounts will remain high, just like Australia's. Japan's big three electric companies bore the brunt of the change, about 80 per cent of the total in Japan. Tokyo-headquartered TEPCO Power Grid lost 8.14 million accounts. Kansai Electric Power Transmission and Distribution, based in Osaka, lost 3.45 million households. And Nagoya's Chubu Electric Power Grid saw the number reduced by 1.75 million (Figure 2.22). Electricity was not the only market that saw intense competition. Japan also opened its retail gas markets to competition one year after that of electricity. About 3.73 million households had switched gas supplier by June 2020, an increase of about 1.53 million over the number as of June 2019.

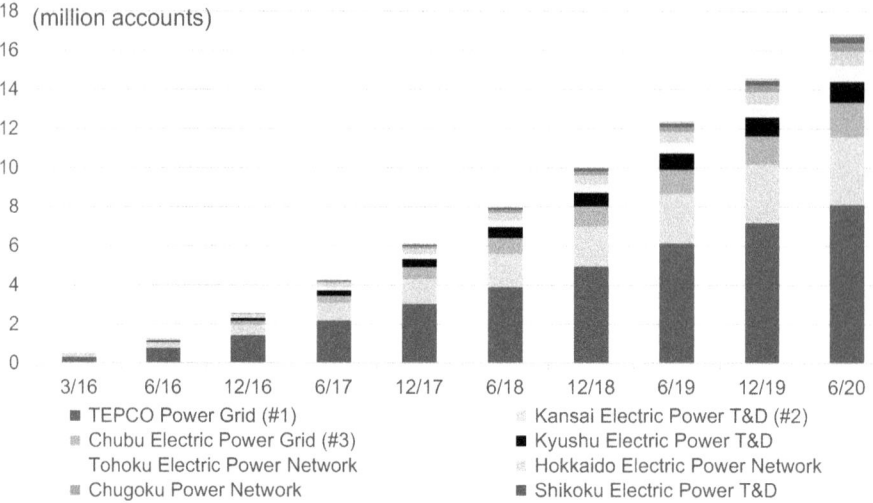

Figure 2.22: Number Electricity Accounts Switching Applications in Japan (March 2016–June 2020). Source: Author, July 2020. Data sourced from: The Organization for Cross-regional Coordination of Transmission Operators, JAPAN (OCCTO) (2020). Usage Status of Switching Support System. [online] occto.or.jp. Available at: https://www.occto.or.jp/system/riyoujoukyou/index.html [Accessed 17 July 2020].

To better understand the complexities of the Japanese energy market, this author called on highly experienced 'on the ground' experts to share their views.

The first set of thoughts are from Dr Duncan Barker, country manager for Mott MacDonald in Japan. Duncan is a chemical engineer with more than two decades of experience in engineering consultancy, helping companies to plan and deliver new electric power generation projects ranging from thermal assets, such as gas-fired and biomass projects, to renewable energy plants, including solar PV and onshore and offshore wind. He has worked in Asia for more than 10 years.

Duncan's list of challenges for the future of Japan's electric power sector include the decarbonisation process, future energy demand, the nuclear revival, energy security, slow market reform, and sourcing skilled labour, as well as entry barriers to foreign companies.

On decarbonisation, electricity is a big driver to comply with the nation's commitments under the Paris Agreement. There is also public pressure against coal and limited options on the financing side given that major Japanese financial institutions no longer want to lend money to coal-fired power plants. This is part of a global movement.

Another challenge, not seen so much in other Asian countries, is the ageing, declining population of Japan, which means that long-term energy demand may fall or remain similar to today. There may not be much growth in terms of grid capacity beyond the current installed capacity of approximately 300GW in the next decade or beyond.

The nuclear revival the Japanese government has been hoping for faces major hurdles. Nuclear could provide a low-carbon energy supply, but a restoration of the pre-Fukushima amount of nuclear generation is considered unlikely. If, and how, nuclear power will recover is unclear given that it is such a political 'hot potato'. Public perception domestically appears to be broadly against nuclear power.

Ensuring security of supply is another challenge for Japan. It is something Japanese authorities have been concerned about for decades. Natural resources, such as oil and gas, need to be imported. This is not an ideal situation if the nation wants to improve its energy security. How can an island-based grid be managed without its own key fuel resources and with no interconnection to a neighbouring country, yet still have energy security? Renewable energy is one answer and is one reason why Japan has embraced renewable sources of power.

Another feature testing a healthy future development of the market is the slow market reform, or what this author likes to call the 'Japanese cultural factor'. Japan has been trying to liberalise the market, but the big electric power companies – the market incumbents – continue to dominate. The country and the companies are known to be notoriously slow moving and bureaucratic, and, as a result, Japan may end up being one of the slowest countries to open its power sector to competition.

So far, the government and the major electric power companies have been reluctant to embrace disruptive technologies and business models, and one wonders if there will be any significant progress on this front in the next 10 years. As a practitioner, Duncan has witnessed first-hand what can at times be total paralysis in decision-making and moving things forward. Some of the big companies recognise that they should find ways or opportunities to move forward, but they seem to struggle to find answers and are unsure what their business plans should look like. Many are trying to embrace clean energy but are not sure how to go about it, for example. There is a mindset of trying to make their approach fit into the traditional, old-fashioned ways of doing things, rather than trying something new.

Most of the major vertically integrated electric power companies have large workforces, many with engineering backgrounds. Based on Duncan's experience, these are highly skilled and capable people; however, the model of doing everything yourself does not appear to be the right model for the future. Duncan senses that, although utility companies are struggling, they are reluctant to change and do something that is unconventional. This is the same whether one is talking about distributed energy or liberalised electricity supply. So, despite all of its wonderful technological skills, Japan could be one of the slowest markets to move forward. Another aspect is an unwillingness to seek best global practice in the electricity sector, including clean energy. Very few Japanese companies enter into joint ventures with companies from abroad, although doing so would bring the needed skillset and experience. This could change in the future, but the progress remains slow.

The lack of people with relevant skills is a major problem for the industry. This is partly to do with the ageing and declining population. Duncan has witnessed this, particularly with offshore wind opportunities in Japan. Offshore wind is experiencing a boom in project developments in the country, but, in Duncan's experience, it is difficult to recruit people domestically with the right skills.

A final challenge is something that has been going on for a long time. Foreign companies have great difficulty in accessing the Japanese market. The barriers to entry into the market are very high in Duncan's experience. This is partly because of the unique way that the Japanese conduct business. Even though there are critical issues, such as the lack of skills and market opportunities (for example, the current offshore wind projects boom), which international companies could help Japan solve, overcoming the cultural factor is a huge hurdle. There is no way that a foreign company can enter the market and operate in the same way as it does in most other parts of the world. It simply does not work in the current climate. They would need a local partner. Local companies do not have the capability, experience, and skills, so they need to leverage the international players. This could be resolved, and the sector could blossom if there were more agreements and joint ventures between Japanese and international companies.

If we could get to a stage where Japanese companies open up to new ways of thinking and work together with the best in the world in terms of capability, experience, and skills, then we could create a truly 'win–win' situation that leverages the best of both worlds. Duncan sees a real opportunity here – if there was mindset change, the sector would advance in the best possible way.

The second expert is one of Japan's energy sector legal eagles, Rupert Burrows. Rupert has lived and practiced in Japan for three decades, specialising in the development, financing, acquisition, and disposal of interests in energy and other projects, in particular helping Japanese energy and other companies with their overseas projects.

On the current challenges that the Japanese electric power sector, and companies in the sector, faces he identified several. First, there are hurdles revolving around the nation's drive towards clean energy, and thus, higher energy security. In agreement with Duncan, Rupert sees the acceleration towards building out the country's offshore wind, principally floating. Its objective is one gigawatt per year this decade for a total of 10 gigawatts, a massive and ambitious development. This is an area where Japan has lacked experience, and this could hinder the pace of the growth. Another hindrance is the lack of mega capacity batteries or energy storage impacting the management of the energy from offshore wind and from other forms of clean energy, including solar. On solar photovoltaic systems, he sees at least two issues. One is the impact of land scarcity for large-scale solar plants that curtails the growth from this energy source and forces the growth onto residential and commercial roof top solar. This will have a variety of impacts on the grids given that much will be for self-use and many of these systems will include battery storage, including EV charging. The other solar-specific issue is the reduction in solar feed-in tariffs. These were first set at 40 yen (about 37.34 cents US) per kilowatt-hour in 2011 after the Fukushima disaster. Now these have been basically reduced by half. The Ministry of Economy, Trade and

Industry in early 2020 announced the rate for the fiscal year through March 2021 at between 12 yen (about 11.2 cents US) per kilowatt-hour for large-scale systems to 21 yen (about 15.6 cents US) per kilowatt-hour for small-scale ones, of 10 kilowatts or less. Second, Rupert highlighted three other hurdles. The first, concurring with Duncan's view, is the fact that continued social antipathy towards nuclear power will hinder this power resource to make up a significant portion of the energy mix. He adds that the development of geothermal projects faces a similar antipathy from local society. Lastly, is the plan towards creating a hydrogen economy in the long term. The nation believes that hydrogen can be an ideal energy storage medium for renewable energy. Companies are trying to scale up this energy storage method to megawatt-scale hydrogen fuel cells. But it may take time before these are fully functional and also their cost is brought down to more competitive levels. Currently, there is an obvious lack of hydrogen-related infrastructure, although the idea is for this to be developed in parallel with hydrogen gas filling stations for cars. Japan will initially import the fuel from Australia and the first carrier has already been launched.

When asked as to his best guess as to how the electric power sector in Japan will evolve in the next 30 years, Rupert believes that all of the issues that he highlighted will linger on a 30-year timeframe. He believes that we can expect that the nation will be a heavily renewables focused economy supported by battery and fuel cell developments.

2.2.3 Singapore – Progressive but Paced, Highly Controlled Reform

The reform of the electricity market in Singapore has been a successful one I believe. It has been slow, more than 20 years in the making, but the market has opened up to competition successfully and is now a vibrant one with many new players, nonetheless. Other markets in the region that are going through the process can definitely learn many lessons from the city state.

The start to the opening up of the whole retail electricity market in late 2018 in Singapore marked the final step to a gradual opening that had begun about 20 years earlier. The approach by the city state has been quite unique compared to other markets around the world. In many ways, authorities tightly controlled the process. This measured approach may be a reason behind its relative success. Relative, because the initial market participants have not as yet really financially benefited from the reform, as shown by the annual weak earnings or financial losses they have been posting for many years. Consumers seemingly have benefited much more. They were suddenly offered more choice and more services and more of a say in the quality of the service they get. At the same time, new market participants have been able to enter and identify new business opportunities.

The reform of the Singapore electricity market first started in 1998 when the electricity pool begun to operate. The regulatory authority, the Energy Market Authority (EMA), was set up three years later and by 2003 the large-users portion of the market was open to competition. It was not really until 2008 that deregulation started to accelerate in earnest. This was when Temasek, an investment firm owned by the government of Singapore, sold to a variety of mostly foreign investors three power generation companies that account for the bulk of the city state's installed capacity. Importantly, an electricity futures market was introduced in in the middle of 2015. Also, the electricity consumer market was fully opened by May 2019 (Figure 2.23).

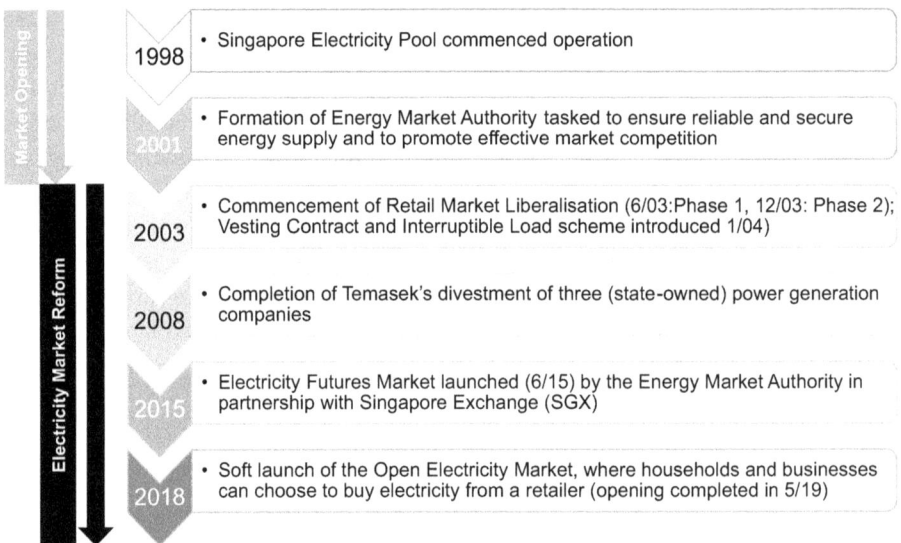

Figure 2.23: Singapore's Electricity Market System Reform.
Source: Author, October 2021. Data sourced from Energy Market Authority of Singapore (2010). Introduction to the National Electricity Market of Singapore. [online] ema.gov.sg. Singapore: Energy Market Authority of Singapore. Available at: https://www.ema.gov.sg/cmsmedia/Handbook/NEMS_111010.pdf [Accessed 15 Dec. 2020]. (Version 6).

In terms of short to mid-term challenges, they can be separated into some general ones and some more technical ones. The more general ones include overcapacity, intense retail power competition that affects the revenues of the sector participants, and legacy fuel contracts that affect the generators' costs. The more technical (or structural) ones chiefly relate to regulation and market structure and revolve around a stringent strategy by the nation on its energy security according to another Asia energy expert, Mark Hutchinson.

The first major challenge that is unlikely to be resolved in the next three to five years is over generation capacity. In any electric power system there needs to be a safety gap, or margin, between peak demand and the maximum generation capacity. One of the key priorities for any electric power sector is to ensure the safety and reliability of supply. If there is very little margin between the peak demand and the maximum capacity there is a danger that if there is a sudden surge in demand, there will not be enough capacity to meet this sudden surge – peak demand being the maximum demand a system may experience in a day over a 12 month period. Generally speaking, experts would recommend any system to have anywhere from 10 per cent to 20 per cent buffer. So, if the whole power system had a total capacity of 1,000 kilowatts then ideally the peak demand should not exceed 800 to 900 kilowatts. In Singapore, though, the authorities are even more conservative, as they regard the safety margin to be a minimum of 30 per cent. The small nation's total installed generation capacity grew about 1.25-fold or a compound time annual growth rate of 2.3 per cent, to 12,582 megawatts from 10,030 megawatts between 2009 and 2020. The nation's peak demand rose 1.23-fold or a compound annual growth rate of 2.1 per cent to 7,376 megawatts from 6,041 megawatts, during the same period. On the surface, this should be good. However, the problem lies in the fact that the gap between the peak demand of 7,376 megawatts and the total installed capacity of 12,582 megawatts (Figure 2.24) is obviously a lot more than 10 per cent to 20 per cent and is even more than the very high 30 per cent targeted by the nation's planners. This clearly produces an 'overcapacity' situation, or, put simply, more supply than demand. In such a situation, the likelihood is that the electric power tariffs will stay weak or become even weaker throughout most of the period because the buffer (or margin) between peak and capacity is so exceedingly healthy.

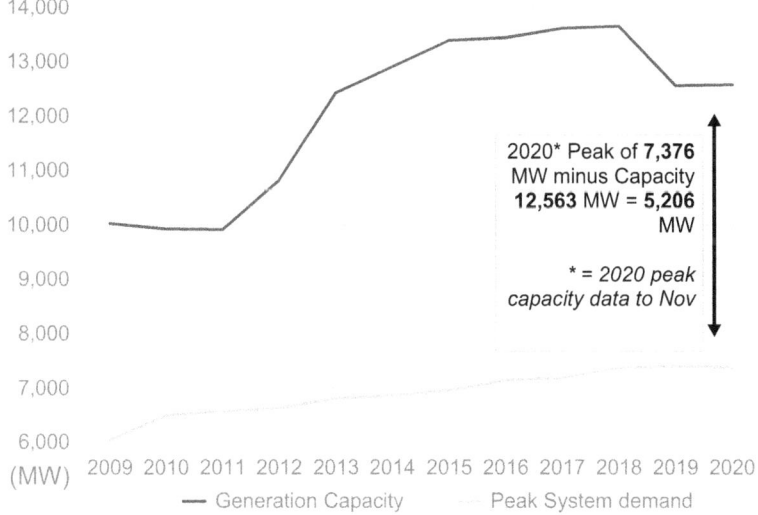

Figure 2.24: Singapore Peak Demand versus Generation Capacity.
Source: Author, July 2020. Data sourced from: Government of Singapore (2020). EMA: Singapore Energy Statistics – Energy Consumption. [online] www.ema.gov.sg. Available at: https://www.ema.gov.sg/singapore-energy-statistics/Ch03/index3 [Accessed 20 July 2020].

Other factors that affect the wholesale market electric power prices are always related to demand and supply. On the demand side of things, a sudden, sharp, and unexpected change in the weather or a sudden positive or negative economic shock will influence the amount of consumption and the amount of capacity required. On the supply side of things, a sudden shutdown of power generation plant will automatically reduce the supply available and prices should technically rise. When looking for a negative economic shock that would affect demand and hit power prices negatively, one actually occurred quite recently, in the first half of 2020. It was caused by COVID-19 (Figure 2.25). The economy fell into recession with GDP declining 41.2 per cent in the second quarter of 2020, the first recession in more than ten years, due to the impact on the economy from the government mandated lockdown to battle COVID-19. In the first quarter, GDP had fallen 3.3 per cent. COVID-19 recorded cases begun in mid-February but stayed at or below 11 cases per day until early February. During this first quarter, the market power prices progressively fell, especially in March. The calculated monthly average Uniform Singapore Energy Prices (USEP) were S$84.15 per megawatt-hour ($60.96) in January and S$88.64 in February, falling 21 per cent to S$69.72 in March. By April, COVID-19 recorded cases were a hundred or more a day causing more economic pain. The calculated monthly average prices based on USEP were S$55.23 in April, S$49.97 in May, and S$46.91 in June.

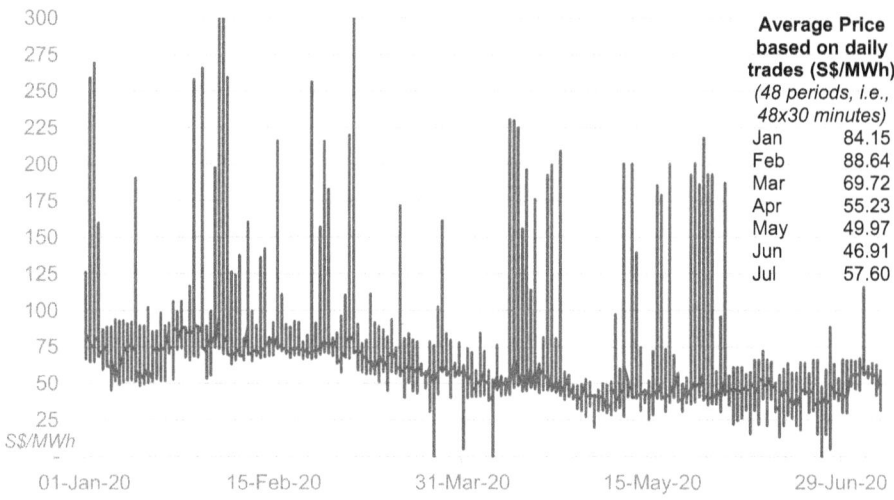

Figure 2.25: Uniform Singapore Energy Prices (USEP) 1 January Through 8 July 2020.
Source: Energy Market Company (2020). Price Information. Available at: https://www.emcsg.com/marketdata/priceinformation [Accessed 19 July 2020].

A cost challenge for the generators is the long-term natural gas supply contract that they have signed in the past. These contracts have take-or-pay clauses. This means that even if demand sharply falls for the generators' plants, maybe because of macroeconomic reasons such as the economic slowdown created by COVID-19 or for other reasons, they still have to pay for the set amount of volume as per the contract. These mostly take-or-pay gas contracts do have some 'banking' capabilities so they can defer and use later. Still, this hits cashflow negatively, although not necessarily causing a 100 per cent loss as the amount will depend on the value ascribed to the future gas to be consumed. In Singapore's oversupplied market it puts further financial strain on the generating companies. The contracts, known as vesting contracts, are for both electricity and gas. They have had an effect of somewhat distorting the markets and were one of the causes for the current oversupply. Arguably, this was partially an intended outcome. The Singapore government wanted to err on the side of security of supply, ensuring that the city-state would never be short of electricity. Within the framework established by the government, the generators were free to make investments, and many did, and this resulted in the over-supply.

Another challenge in the market is the relatively great number of electricity retailers. The electric power markets only became fully contestable in May 2019. There are three consumer groups: business users with an average monthly consumption of at least two megawatt-hours, those using less than two megawatt-hours, and those electricity retailers only serving residential consumers. The serving of a particular consumer type will depend on the type of licence they have secured from Singapore's Energy Market Authority. Customers can purchase the power from any retailer, choosing the provider's price plan that best meets their requirements, be it price or other needs. They can freely switch retailer anytime, unless they have a locked-in contract for a certain period of time. The number of retailers started to increase in the three-year lead to the soft opening of the residential consumers segment in 2018 with the total reaching 25 compared to just 12 by the end of 2015 (Figure 2.26). It is worth noting that not all retailers can serve all three consumer groups, as mentioned.

Competition among Singapore's 28 electricity retail licensees will continue to remain intense in the next few years (Figure 2.27). About 48 per cent of residential customers and 47 per cent of business users had switched to sourcing electricity from a retailer of their choice by October 2020. The share was relatively highly concentrated with the biggest four controlling over 70 per cent of residential accounts and 66 per cent of business accounts. The likely trend, if we use such markets as Australia and the UK as blueprints, is that the market will become broadly split into a two-tier market comprising same very large retailers and same very small ones. There should also appear some consolidation. Some companies will merge or be acquired. Others will fold. The one caveat is action by EMA. EMA closely follows the market and if competition is too brutal and if the number of licensees' dives, EMA or the government could, and is highly likely to in the view of this author, intervene.

To look at some of the more technical or structural challenges this author approached one of the region's energy markets thought leaders, Mark Hutchinson who I mentioned earlier. He has more than three decades of market experience, most of it acquired in Asia. He has worked in a utility firm, in finance and in consultancies,

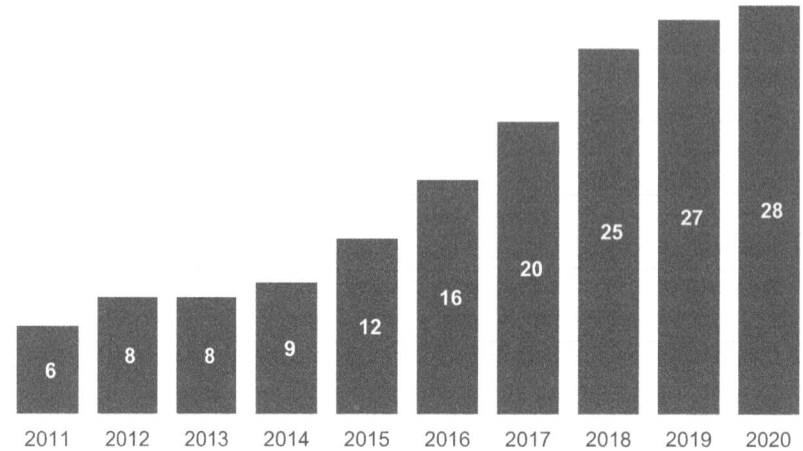

Figure 2.26: Singapore Electricity Retail Licensees.
Source: Author, July 2020. Data sourced from Energy Market Authority of Singapore. Available at: https://www.ema.gov.sg/Licensees_Electricity_Retailer.aspx [Accessed 20 July 2020].

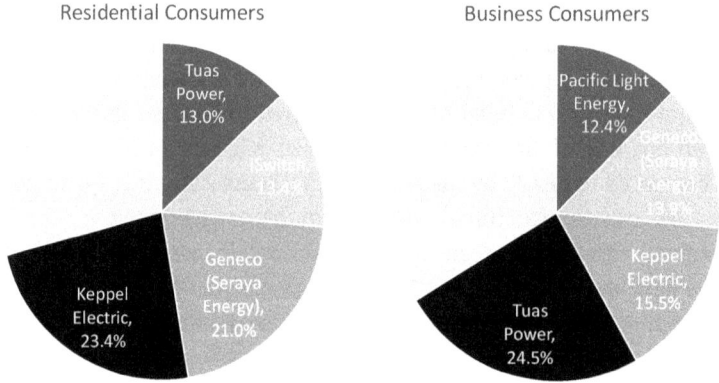

Figure 2.27: Singapore Electricity Retail Licensees' Market Share as of 31 March 2020.
Source: Author, December 2020. Data sourced from https://www.ema.gov.sg/Licensees_Electricity_Retailer.aspx [Accessed 15 January 2020].

giving him a broad and deep knowledge of the sector. Mark thinks that given Singapore's geographic and political circumstances, it is understandable why the government has a strong policy focus on energy security of supply. This policy has led indirectly to the over-supply currently seen through the implementation of vesting contracts and other mechanisms. The nation's geopolitical circumstances are in fact complex and challenging. For example, the city-state has had a multi-decade thorny relationship with its northern neighbour, Malaysia, regarding the price it pays.

Various Malaysian governments argued that the price was too low. Across its waters, Singapore has faced several challenges with Indonesia on problems ranging from air pollution caused by human-made fires to airspace issues.

2.2.4 China – a Giant's Long Build Up, Accelerated Adoption

The Chinese electric power sector reform and restructuring is incomplete as of 2020. But it is the most important in the world given its size and given its relative uniformity. Especially when compared to other large nations with many provinces or states, such as India or the US. The nation is already the single largest electricity market in the world in terms of power generation and installed capacity. Further, this electric power market will expand in the next 30 years. And, by 2025, the bulk of the market will be open to competition, in the view of this author. It is imperative to understand that the country can easily leap forward despite the size and the complexity of the market because it has learnt many reform related lessons domestically and also from abroad. All of this makes the next reform steps all the more fascinating and exciting.

The reform process of the electric power sector has been long and complex (see Figure 2.28). The full liberalisation will be reached during the fourteenth Five-Year Plan ending in 2025. The process can be split into seven phases. Phase one was between the establishment of the People's Republic in 1949 until 1988 or so. During this period the whole sector was state owned and controlled, and power tariffs were fixed. Phase two, between 1988 and 2001, started with a split up of the ownership from the central government to provincial and local government entities. It was a highly complex period with the central government changing the regulatory bodies several times – a theme I will address shortly.

Phase three, from 2002 through 2009, was when the first radical change occurred. This is when the State Council decided that it wanted to open the market to competition. Grid companies were carved out and generation companies were structured. Broadly speaking, China set up two major grid companies responsible for the transmission, the distribution and the sale of electricity. It also organised five nationwide power generation groups that at various times accounted for 40 per cent to 60 per cent of the nation's electric output; the remainder is mainly owned by provincial and local governments, some state-controlled enterprises, and a few private firms. Phase four, between 2009 and 2012, comprised evaluating the next steps to open competition starting with a pilot project where a power plant controlled by one of the big generation groups, China Huaneng, entered into a direct power sale contract with an aluminium smelter in Liaoning province.

Phase five, between 2013 and 2015, was when the second radical change occurred. A new regulator was established in 2013, the National Energy Administration or NEA. The following year more direct power sales pilot projects were introduced in

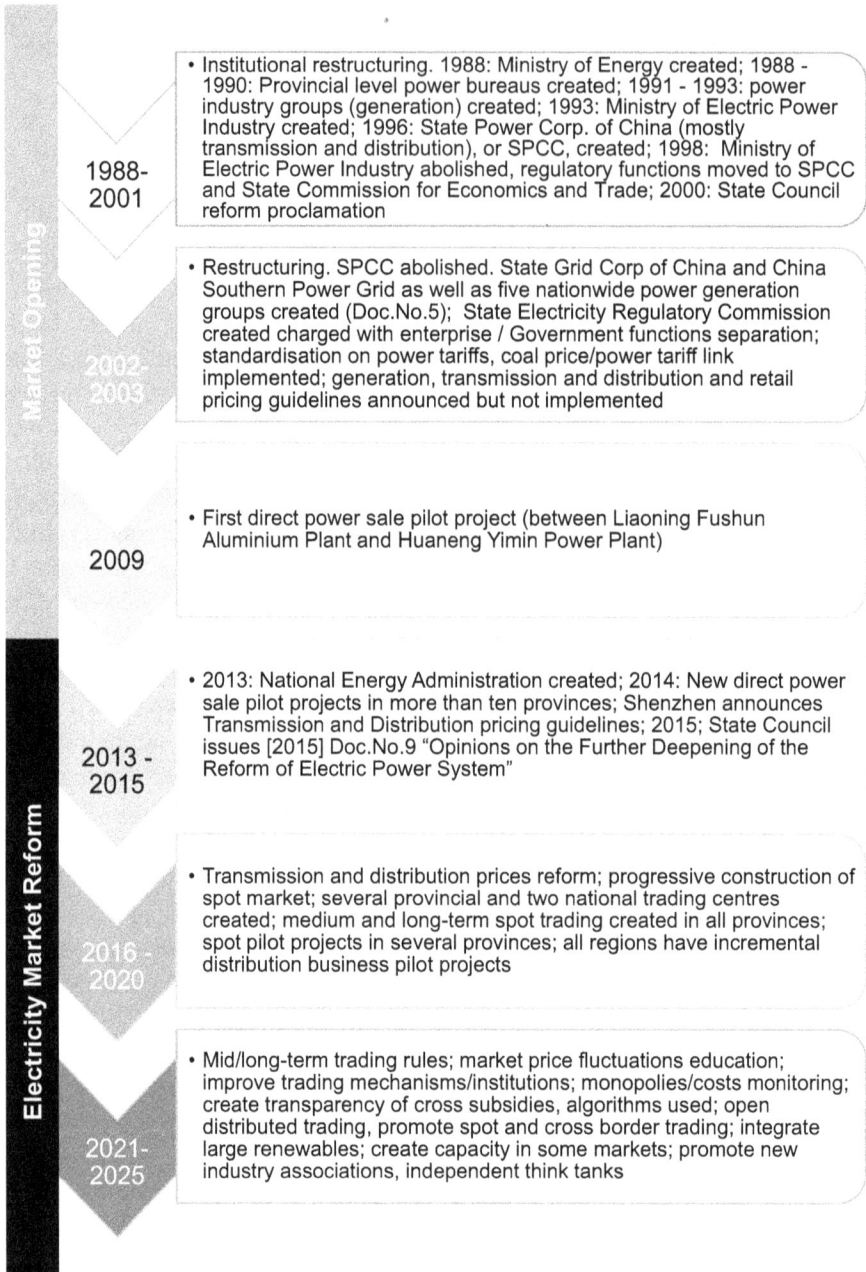

Figure 2.28: China's Electricity Market System Reform.
Source: Author, July 2020. Data sourced from wide variety of Chinese media and governments reports in 1994–2021.

about a third of the country's provinces. Also, more reform milestones and objectives were set by the State Council in 2015 to be implemented in the thirteenth Five-Year Plan, from 2016 to 2020. The thirteenth Five-Year Plan marks the sixth phase. During this period, many of the objectives were not quite met but the key features for the full market opening were successfully set. These included the price setting for transmission and distribution fees, the creation of power retail (or distribution) companies, the creation of two national electric power trading centres (or wholesale markets) as well as several provincial level ones, and significant pilot projects for short, mid- and long-term electricity spot price trading.

The fourteenth Five-Year Plan, from 2021 and 2025, will mark the seventh phase of development. Broadly speaking, it will be a fine-tuning phase. All of the foundation pieces were laid during the last Five-Year Plan. Now it is simply a matter of developing further the regulatory framework as well as the related institutions, as well as allowing the key actors, including the power retail companies, to grow and prosper. The necessary fine tuning includes setting up the trading rules for mid-term and long-term power trading, growing the key institutions such as the electric power trading centres, and raising price transparency. Interestingly, authorities will focus on fostering two areas. One is encouraging the establishment of industry associations as well as third party think tanks. The other is educating and training all of the market participants on the opportunities and risks with liberalised power markets, including financial products, such as futures.

In the past seven years or so, a gigantic leap forward has been made by the country in its vision to construct more competitive, effective, efficient, and transparent energy markets. China's hope to have the bulk of the electricity traded on power trading centres by 2020 did fall short of aspirations, but this can be largely forgiven given the size and complexity of its power markets.

In the view of this author, among the several challenges that China has is tackling a healthy long-term development of electric power reform. The first challenge is an old one, the monopoly powers of the two largest electricity networks. The second is relatively new, the integration of renewable energy sources. Another is the building up of rules, institutions, and financial products related to electricity trading. The final is power tariffs regime transparency.

A significant challenge to the development of an open and free electric power market is the transmission and distribution side of things. In open markets, such as the Australian or Singaporean markets, the setting and the transparency level of the transmission and distribution fee, just like the fee one would pay driving a car on a toll road, is high. China has several electric power grid networks. The biggest one accounts for the bulk of the market, namely, the State Grid Corp. of China (SGCC). The central government has been trying since about 2003 or so to set up a transparent grid network fee. Authorities had even issued some guidelines at the time. However, the SGCC is a gigantic company and it has not been easy at all to convince it to be more transparent. The Chinese media, and to a more limited extent the Western media, has brought up the subject of breaking up the SGCC in the past 18 years a few times. In 2012, an interesting English language article appeared from Reuters – again this is relatively rare – highlighting a story that mentioned that an electricity crisis causing blackouts in India at the time was used by the then head of the SGCC, Liu Zhenya, as tangible evidence why the SGCC should not be broken up. In India the transmission and distribution of electricity is not controlled by one single entity.[33] The SGCC was ranked third in the Fortune Global 500 companies in 2020. The service area comprises 26 of the nation's 32 national areas (provinces, autonomous regions, and municipalities under the central government) or about 88 per cent of the whole land area serving about 1.1 billion people. It was responsible for about four out of five kilowatt-hours transmitted in the country in 2019 and that year realised 77.03 billion yuan ($11.5 billion) in net profit, 4.14 trillion yuan ($619 billion) in total assets, and spent 447.3 billion yuan ($66.9 billion) in grid investments that year.[34] In short, it is a truly colossal enterprise. Trying to break up such giant monopoly is very difficult and is not unique to China. If one were to look at Europe, one could see that it was achieved in the United Kingdom but it failed in France where the state-owned utility Électricité de France (EDF) still pretty much enjoys the majority role in the French electricity system. However, one can be confident that the central government is very clear on this particular challenge and in the coming years it will find ways to achieve its goal of having transparent transmission and distribution fees.

The second impediment is the extremely sharp jump in renewable energy sources as a percentage of total electricity output in the nation. These sources are mostly from wind and solar facilities. Currently, such sources cannot provide a fully reliable and steady output in the same way that a coal-fired or nuclear power plant can. Also, the total cost per kilowatt-hour from renewable energy was higher than that for other major sources of electricity, especially coal-power and hydro-power. As such, it has been a challenge, just like in any other country in a similar situation, on the one hand to promote the development of renewable energy sources but at the same time to ensure that electricity supply is safe and reliable. This has been an impediment but concerns should dissipate in the coming five years as renewable energy production becomes increasingly competitive from a cost perspective and also as China introduces more energy storage solutions that renewable energy can piggyback on.

The third challenge to the long-term electric power sector development is with electricity trading and the related regulatory framework and institutions. These have made tremendous progress in the past five years. However, as alluded to earlier, the reform fell a bit short of the original objectives. Authorities suggested in the mid-2010s that the bulk of the electricity by 2020 would be going through electricity trading centres (or exchanges) and that all of the power would be exchanged under direct power sales contracts with the exception of electricity consumed by the agricultural sector and by households, which accounted for 1.1 per cent and 15.9 per cent respectively of total consumption in the first six months of 2020; during the period consumption from the manufacturing industry and that from the commercial and public sector consumed 67.1 per cent and 15.9 per cent, respectively.[35] So, there have been some successes, although there are still areas that need to be fine-tuned and developed.

In terms of successes, in the view this author, the first area of past success and future development was the rapid establishment of electricity trading centres; the more accurate description is 'exchanges' but here the Chinese has been translated directly. Starting from March 2016, 33 national areas had set up local level electricity trading centres and at the same time two national ones were created, the Beijing and the Guangzhou Electricity Trading Centres. The amount of market-traded electricity – i.e., not just directly sold to one single buyer – in the five years through December 2019 as a percentage of the total rose to 39 per cent from just 16 per cent in the first year of the opening of market traded power, or up 144 per cent. In terms of total amount, it rose to about 2,818 terawatt-hours from 888 terawatt-hours, or an increase of 217 per cent, this author calculates (Figure 2.29).

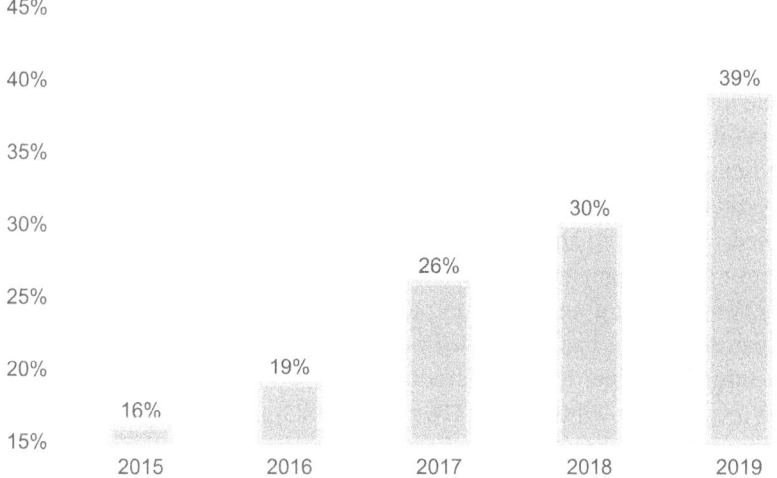

Figure 2.29: Percentage of Electricity Under Market-Oriented Trading.
Source: Wu, Z.H., et al., 'The Fifth Anniversary of Electricity Reform: Ten characteristics of electricity sales market', Energy Research Club, 18 June 2020. Available at: http://shoudian.bjx.com.cn/html/20200618/1082055.shtml [Accessed 18 August 2020].

This trend fully met some of the objectives of the reform, namely, to create relatively stand-alone electricity trading institutions and start growing fair and uniform trading platforms. The rules and regulations for market participants can be easily sourced by the market participants. However, they generally lack the other level of public transparency afforded by the Australian, Japanese, or Singaporean markets. This should change and evolve in the next few years. It is fairly certain that by the end of the new Five-Year Plan, namely 2025, the bulk of the power will be market-traded.

There is already evidence of this. There are some individual large companies that already have well surpassed the national average. China Resources Power, for example, announced in August 2020 that in the first six months of 2020, almost 70 per cent of the electricity volume it sold was based on market-pricing – amounting to about 40.2 terawatt-hours – although the management of the company thinks that the amount for the full year may be close to 65 per cent. One way that regulators want to improve the operations of the electricity trading centres is by mandating them to diversify their shareholding away from the State. As of the first quarter of 2020, nine of the centres already had put in place ways to facilitate diversified shareholding; they included those in eight provincial level ones as well as the Guangzhou one. Its northern counterpart, the Beijing Electricity Trading Centre, which was 100 per cent owned by the SGCC, had attracted ten shareholders in December 2019. These shareholders bought into 30 per cent of the equity of the centre. They include the subsidiary of one the country's oil and gas majors, Sinopec Asset Management, and most of the large nationwide electric power generation groups, such as China Huaneng, China Datang, China Huadian, State Power, China Three Gorges Group, State Energy, China Nuclear Power, SDIC Power, and China Resources Power.[36]

The second area of past success and future development was setting up structures to enable the creation of retail power companies. Nationwide, there were more than 4,000 retail power companies serving more than 100,000 registered users, mostly large and medium size enterprises as of March 2020. The concept of retail power simply did not exist and was not understood as recently as 2015 but has been readily adopted. The participants range from real start-ups to the subsidiaries of electric power or energy companies. Given that the electricity trading market is currently simply constructing a one-time contract, typically for six months or a year, these participants have not had to deal with all of the intricacies and complexities of electricity futures or other financial products. The setting up of a retail power firm has been relatively straightforward. The company simply acts like a middleperson. It first secures one customer or a set of users and gets the electricity requirement mandate, say 1,000 gigawatt-hours. It then negotiates a price with a power plant for the 1,000 gigawatt-hours. The firm will earn the price difference between the plant's price and the price the user or users will pay for the 1,000 gigawatt-hours. The idea is that the realised cost for the user will be lower than that it would directly pay a grid company. The registration is not too difficult. Some of the requirements include the need to state the firm's business scope, minimum electricity sales of 3 terawatt-hours, and registered capital of 20 million yuan ($3 million) There has to be a link between the assets base and the annual revenue, and there must be at least ten technical professionals with a minimum of five years' electric power sector experience. In the fourteenth Five-Year Plan, the objective is to enhance the service quality and scope, such as bundling other products such as gas. Basically, following the business models adopted in Australia or Japan.

The third area of development is to have fully open, market-driven, transparent electric power tariffs. This is by far the most sensitive and most complex of any area in the nation's drive for power reform, in the view of this author. It is sensitive because China is one of those markets where the government is highly sensitive to the social impact of increases in retail power tariffs. Allowing a completely freewheeling, market-led power tariff could cause some social discontent and pressures. China is not unique in the region. Examples of other electric power markets in the region where a similar sensitivity exists include the Malaysian, South Korean, and Taiwanese ones. This has to do with the historical development of end user power tariffs. In some other markets, users may be accustomed to fluctuations in power prices. These are typically due to the electricity production costs that have traditionally been driven by the price of energy commodities, namely thermal coal, gas, and oil. For example, consumers in the Philippines, one of Asia's least developed economies, and in Japan, Asia's most developed economy, are relatively accustomed to having monthly or quarterly changes in their power bills depending on the fluctuations in the price of energy commodities.

Historically, electricity volume and the price produced by power plants in China was tightly regulated by government authorities. The price of the power sold to a single buyer is referred to as the on-grid price in China. On-grid prices have been set by the nation's powerful economic regulator, the National Development and Reform Commission (NDRC), in combination with input from provincial governments. From the mid-2000s, the on-grid price setting mechanism broadly considered the average cost of similar units in a region and related charges including taxes. The result of this approach was that the on-grid price received by a coal-fired power plant in the province of Heilongjiang may be different from that of a similar plant built at the same time in Guangdong province. The nation begun experimenting the sale of a small portion of output through competitive bidding in 1999. Such pilot projects programmes came and went until the late 2000s when it was decided that the sale of electric power should be liberalised. The step taken was to allow a power plant to directly negotiate the power price with a large end-user. The direct transaction price would also comprise government set transmission costs. As discussed, during the thirteenth Five-Year Plan the proportion of electric power sold directly has accelerated with some of the larger companies such as Huaneng Power International and China Resources Power selling more than two-thirds of their output directly in the first six months of 2020.

Before the electric power sales are fully liberalised, the generation companies have to deal with different on-grid prices at different power plants in different regions and address direct power sales that themselves occur in a variety of formats. Some of the prices are determined by business-to-business negotiations, some are negotiated between the power plant and a retail power company, some of it is sold in competitive bidding, and some is directly sold in the emergent spot markets. It is quite a challenge for existing plant operators to navigate through all of these changes, in the view of this author. It is especially tough because the industry was built on the on-grid price mechanism. Rapidly transitioning away from this towards a market-driven system requires a sharp shift in mindset.

The next five years, through 2025, will prove even tougher as government and market participants tackle complex financial instruments: electricity trading, derivatives and carbon trading. The country will be developing the electricity spot market and thus will demand the introduction of related financial products to mitigate the risk of spot power prices fluctuations. Market participants, comprising the power generation companies, the retail power companies, and the consumers, are looking at the development of forward power price benchmarks and risk-management mechanisms so they will naturally promote the development of electricity futures and electricity over-the-counter derivatives trading. Looking at the Australian examples discussed in this section of the book, electricity futures will mean that the nation's traditional power generators will have to set up trading desks and navigate through the complexities of managing price risk. Authorities are also looking at introducing a carbon trading market. This, again, is not out of the blue. It has been planned and experimented for at least a decade, in the experience of this writer. By the end of the fourteenth Five-Year Plan one may only actually see the foundations being established due to the sheer size of the amount of preparatory work that is required. One expert,[37] one of the many, mentioned a very long list of some of the many challenges the country faces with the establishing of a working market. For example, China will have to set up complete, reliable, and transparent greenhouse gases accounting verification for the national carbon emission trading. Regulatory authorities are insufficiently staffed and financed. Another area is funding. Currently there is insufficient funding for subsidies to incentivise energy conservation and environmental protection. These are very critical and complex problems that will take a few years to sort through, on the view of this writer.

However, there is a high chance that some of the larger, and thus more sophisticated, players will be successful, in the view of this author, for three reasons. First, there was a long experimental period. The current reforms did not appear in a relatively short period of time like they did in Australia, Japan, and Singapore. Many of the generating companies had participated in pilot projects, on and off, for more than 20 years. Second, Chinese corporates have a high level of adaptation capacity and capability. The companies have been successful at adapting to changing markets and regulatory conditions, including the government-driven tectonic sector reforms. One of the many is the requirement to rapidly increase the amount of renewable energy as a percentage of total installed capacity despite the fact that the companies had the most experience with coal-fired generation. Third, both regulators and companies have established strong understanding of liberalised markets through broad and deep government-to-government and business-to-business relationships in the past two or three decades. The presence of French and Italian utilities in China is an example of this. The overseas investments in liberalised power markets by some of the generators is another example; these include Huaneng Power International's investments in Australia and Singapore, China Resources Power's investment in the UK, China Three Gorges' investment in Portugal, and Longyuan Power investment in Canada. A fourth factor is the relatively successful example of the price liberalisation seen in the natural gas distribution sector.

The majority of the prices of natural gas distributed in different regions in the country have been largely liberalised. The reform began during the twelfth Five-Year Plan, between 2011 and 2015. And it significantly accelerated during the thirteenth Five-Year Plan, through 2020. The 'city gate' gas price had a cap until November 2015 when this was changed to a more flexible price driven by demand and supply forces; the 'city gate' price refers to the sales price comprising the wholesale or wellhead price plus the pipeline transportation cost. While some caps remained, the pricing mechanism became largely more market driven. Authorities allowed for the first time companies specialising in natural gas distribution, such as ENN Energy, to use a cost–price linkage system, namely linking the cost of the natural gas and its retail price, at least for commercial and industrial users. A major breakthrough has been the establishment of a nationwide midstream pipeline operator at the end of 2019, the National Oil and Gas Pipeline Network Group (PipeChina). This will break up the monopolies held by the Chinese national oil companies, namely China National Petroleum Corp. (CNPC), China Petroleum and Chemical Corp. (Sinopec), and China National Offshore Oil Corp. (CNOOC). The pipeline assets of CNPC, Sinopec, and CNOOC will be folded into PipeChina; CNPC had controlled more than 75 per cent of the nation's gas pipeline with Sinopec and CNOOC controlling the balance. PipeChina's combined pipeline assets will enhance transmission network access as China continues to diversify domestic and international gas sources, develop a broader gas market, and promote private investment in the gas sector. Currently, many China gas sector experts and this author believe that natural gas prices for households will also be liberalised within the fourteenth Five-Year Plan, through 2025.

2.3 Changing Nature of Industry Players

The transition from vertically integrated, single-buyer markets to open, transparent, competitive ones allows for new market participants to enter the energy sector. These market participants are of all shapes, sizes, and backgrounds. They can range from spin-off of a previously vertically integrated utility to a brand-new start-up. The market changes have also highly motivated many incumbents to change their business models so as to meet the new challenges and the new competition. Many traditional utilities in Asia are reinventing themselves into total or comprehensive energy solutions providers. These companies no longer solely supply one single commodity or service, such as selling electricity or gas, to consumers. In some markets in the region, this model has become quite ubiquitous. Some are now supplying both key energy commodities, namely electricity and gas. Others have an even more comprehensive offering that can include, apart from power and gas supply, value-added services including providing other products like telecoms, water, and even home security systems and home insurance services.

This happens for at least two reasons. One is that the local regulators may no longer allow a single monopoly in a particular energy such as gas and power, opening the door to a power company to also supply gas or a gas distributor to also supply electricity. The key here is that all parties must be allowed to use the energy transmission or pipeline network; in other words all industry participants must have equal and unlimited open access to the network. Second, as the energy markets are opened, competition will intensify. This will usually lead to a decline in gross operating margins (the sum of revenues less costs divided by the revenues) and the energy players seek to maximise profits by providing a large offering of products to consumers. A better understanding of the general profile of the diversified services model is very important because it is a model that is slowly but surely becoming more widespread in Asia. Currently, the model does not exist in many Asian energy markets. It is limited to economies with liberalised or quasi-liberalised markets, including Australia, China, Japan, New Zealand, and Singapore. To better understand the trajectory, the earliest example of a multi-utility – a company providing a series of utility services – is evaluated. This was British Gas in the UK. Then specific examples are also discussed, including Australia's three largest energy retailers, Japan's Tokyo Gas and KDDI, New Zealand's Genesis Energy, Singapore's iSwitch, and China's ENN Energy and China Resources Power.

2.3.1 UK: Precursor of Change BG and its Complete Metamorphosis

The history of the multi-utility strategy is slightly nebulous as little has been written specifically on the subject. What is widely agreed on by experts and practitioners is that the UK was the first large economy in the world to liberalise its energy markets, a process it commenced in 1989, which transformed its nationalised industry into a completely competitive one at the generation end as well as the retail supply end. As the UK authorities were paving the way towards reform, some companies decided to take pre-emptive survival measures. They sought to lock in customers by offering, or bundling, more products so as to create a contained ecosystem that would offer customers more incentives to stick to one multi-utility provider. The first such attempt may have been made by Scottish Hydro Electric plc[38] in the early 1990s; the company became part of Southern Electric plc in 1998. Scottish Hydro in 1993 created a joint venture with the US's Marathon Oil Corp. called Vector Gas with the idea to supply commercial and industrial customers nationwide both electricity and gas.

One of the highest profile multi-utility strategy was established by Centrica, one of three companies created by the break up in 1997 of British Gas, which had controlled the midstream and retail ends of natural gas supply in Britain. The company introduced a variety of services from 1998. Centrica declared that its vision was to be a leader in the supply of essential services to customers in selected markets. In addition to its 70 per cent share of the British gas market, in the year 2000 it was supplying electricity to 4 million users and providing telecommunication services to 150,000 customers, vehicle assistance services to 11 million members through its acquisition of AA (a British motoring association), and financial services through its Goldfish-branded credit card (which boasted banking giant Lloyd as a shareholder). These, in addition to supplying 3.2 million service cover for home heating, drains, and kitchen appliances.[39]

Centrica decided to change strategy and focus on its core business from 2013 or so. Since then, it has progressively shifted its focus solely on providing its customers energy services and solutions. Its energy supply, branded British Gas, remains the nation's leading supplier with about 7 million customers, it is the second largest energy supplier to small and medium enterprises (SMEs) with 300,000 accounts, and it is also the number one home services provider with 4 million accounts; these comprise such services as home energy management, remote diagnostics, and home security. It did retain some of the services that it first established in the late 1990s, including equipment cover and home and building insurance. Importantly, Centrica has been able to retain its number one electricity supply market share, which is no easy feat given the intense competition in that market (Figure 2.30). Its share was 18.4 per cent

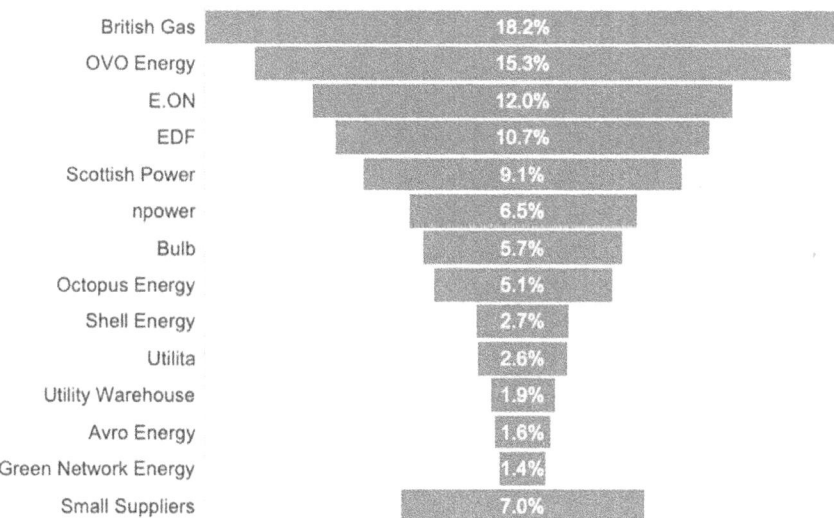

Figure 2.30: Electricity Supply Market Shares by Company in Great Britain (2Q 2020).
Source: 'Electricity supply market shares by company: Domestic (GB)', Ofgem analysis of electricity distribution network operator reports, information as of October 2020. Available at: https://www.ofgem.gov.uk/data-portal/electricity-supply-market-shares-company-domestic-gb [Accessed 16 January 2021].

as of March 2020. Its main rivals were OVO Energy (15.8 per cent share), E.ON (12.1 per cent share), EDF (10.6 per cent share), Scottish Power (9.2 per cent share), and Npower (6.7 per cent share).[40]

Looking at the history of Centrica/British Gas, it is important to understand that the company was not just a precursor but probably ahead of its times and Scottish Hydro with Vector Gas even more so. In an article in 2000, magazine *The Economist* pointed out that Centrica's multi-utility approach may have been obvious but was definitely not easy nor straightforward. Obvious because the chief asset of the company was its customers, almost 20 million of them. And the article praised the vision of the then management team, which was headed by CEO Roy Gardner, to have the sense to realise what its most important asset was. It remarked that 'the management team of the newly demerged business had the sense to realise this. Managers imbued with the culture of state industries might not have done so'.[41] This rings true today in Asia. Despite the different cultural background, a state industry culture is a state industry culture. It is what China, South Korea, or Malaysia are equally facing. Just like Centrica, the state-controlled utilities in those Asian countries that are yet to liberalise their power supply markets will go through two 'cultural revolutions' almost simultaneously, the demerger of the energy utility and its plunge into the unchartered territories of a liberalised market.

2.3.2 Australia's Major Energy Retailers: Similar Yet Different Models

Australia's market liberalisation history is roughly a decade shorter than that of the UK but it is by far the longest in Asia. The leading energy providers went through a lot of changes in the past two decades. Comparing and contrasting the existing business models and the recent changes of the leading three energy retailing companies, AGL Energy, EnergyAustralia, and Origin Energy, should allow one to form a better understanding of the future landscape in many electric power markets in Asia.

The three companies control more than 60 per cent of the electricity customers and more than 80 per cent of gas users in terms of market share in the NEM in the quarter ended March 2020 based on data from the Australian Energy Regulator. For the residential electricity customers segment, the three had a market share of 65 per cent. For SMEs it was a 62.2 per cent share. And, for large electricity users, it was 61.6 per cent. Origin Energy had the largest share in all three segments, AGL Energy the second largest, and EnergyAustralia was in third place. For two of three segments the gap between third and fourth place was wide, namely 16 per cent versus 9.5 per cent in residential and 12.2 per cent versus 9.9 per cent for large users. That was not the case for SMEs where the difference was 13.7 per cent versus 13.2 per cent, a relatively thin gap of just 54 basis points. Despite the high level of churn rates in Australia, discussed earlier (see 2.2.1 Australia – Deregulation Lessons Brutally Learned), the three energy retailers have been able to maintain their dominance, especially AGL Energy and Origin Energy. Based on the experience in the UK and Australia, the electricity-end of the energy retailing markets will most likely be composed of a few big players and a myriad of smaller ones. In Australia, there were about 50 for the residential and SME segments and about 40 for the large user segment. In terms of gas retailing, the three had a huge share, albeit largely dominated by Origin Energy, which has a stake in an upstream gas resource. On a combined basis (Figure 2.31) they controlled 83.5 per cent for the residential segment (27.3 per cent for Origin Energy), 92.5 per cent of the SME share (57.4 per cent for Origin Energy), and 94.6 per cent of the large user share (47.6 per cent for Origin Energy).

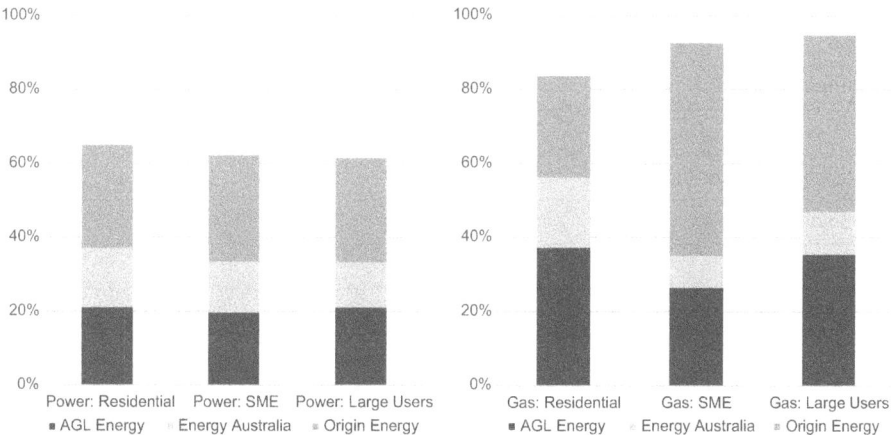

Figure 2.31: Retail Electricity and Gas Market Share for Quarter 3, 2019–2020.
Source: Author, August 2020. Data source: Retail energy market performance update for Quarter 3, 2019–20, Australian Energy Regulator, Release date 30 July 2020. Available at: https://www.aer.gov.au/retail-markets/performance-reporting/retail-energy-market-performance-update-for-quarter-3-2019-20 [Accessed 25 August 2020].

The three companies have very similar footprints in terms of regions where they offer electricity supply and offer gas supply, with Origin Energy having the slightly larger footprint (Table 2.8). They all offer green energy options to customers as well as solar photovoltaic systems installation – including panels, inverters, and energy storage solutions. Origin Energy and AGL Energy have slightly wider offering than EnergyAustralia. For example, they provide telecommunications services to their customers, which can be bundled with their energy services. Origin Energy also has a liquified petroleum gas (LPG) bottle supply business. Only Origin Energy has substantial exposure to upstream gas but all three have electric power generation assets. AGL Energy controlled 11,330 megawatts as of June 2020 and had produced 45.54 terawatt-hours in the 12 months through June 2020. Origin Energy's equivalent was 7,400 megawatts and 18.29 terawatt-hours. EnergyAustralia's was 6,520 megawatts and 9 terawatt-hours for the six-months through June 2020.

Table 2.8: Australia's Key Energy Retailers' Basic Services Profile.

	AGL Energy	EnergyAustralia	Origin Energy
Electricity	4	5	5
Australian Capital Territory		✓	✓
New South Wales	✓	✓	✓
Queensland	✓	✓	✓
South Australia	✓	✓	✓
Victoria	✓	✓	✓
Western Australia			
Gas	5	5	6
Australian Capital Territory		✓	✓
New South Wales	✓	✓	✓
Queensland	✓		✓
South Australia	✓	✓	✓
Victoria	✓	✓	✓
Western Australia	✓	✓	✓
Green Power Options	3	3	3
Green Rating	3.5	3	3.5
Solar	✓	✓	✓
Customer Rating	5	4	5

Source: Author, August 2020. Data source: Finder AU, as of April 2020. Available at: https://www.finder.com.au/ [Accessed 26 August 2020].

In terms of strategies all three energy retailers are working to add more green generation to their portfolios. All still have coal-fired generation plants as well as gas-fired generation ones. The coal ones will be retired at the end of their technical life or potentially a little earlier if they find it possible. The move from brown to green involves building wholly owned or partly owned plants, acquiring them or entering into long purchasing contracts with the green generation facility. At the same time, they are looking at raising their energy offering through providing energy storage and EV charging solutions. They are all spending a great amount of time and money on the digitalisation side of things. At the consumer-end, this involves shifting customers to electronic billing rather than using paper bills and offering smart home as well as energy efficiency and conservation solutions. The digitalisation efforts, especially on the part of AGL Energy and Origin Energy, also involve improving internal operational systems. While the two are seemingly more advanced on the smart consumer solutions front than EnergyAustralia, it is certain that the latter will be able to provide similar services in the near future.

In the view of this author, the large energy retailers in Australia will all gradually transition into becoming comprehensive energy solutions providers with progressively stronger data tech capabilities, with the ability to offer small and large consumers tailor-made products effectively and efficiently, as well as offering a significantly greener energy portfolio. These competencies will either be generated in-house, through acquisitions or joint ventures. In 2020, Origin Energy bought 20 per cent of Octopus Energy, a UK energy retailer for about A$507 million ($327 million).[42] Importantly, the purpose of the acquisition was not solely to gain a foot in the UK market. It was to secure an exclusive licence to use Octopus Energy's Kraken technology in Australia. Octopus Energy defines itself first and foremost as a software engineering and creative technology company. Its Kraken cloud-based platform is dedicated to interacting between the energy users through the web, mobile, or smart-meters and the energy supplier for data flows, consumption forecasting, and trading on the wholesale market.[43] Origin Energy said that it will be able to raise service quality and reduce costs by using the Kraken platform in Australia. The Octopus Energy tech investment is just one example of the type of efforts that Australia's leading energy retailers will undertake to become total energy solutions providers. This kind of strategy is actually absolutely not unique to Australia. Some energy retailers in other Asian power markets with liberalised energy sectors already are following a similar path, as will be evidenced in the discussion on Japan, New Zealand, Singapore, and China.

2.3.3 New Zealand's Retailers' Valuable Experience

The New Zealand power market is smaller than Australia's and its need to decarbonise is quite modest (Table 2.9). Comparing the twelfth months period through December 2019 for New Zealand and through June 2019 for Australia, New Zealand's electricity production was about 43.5 terawatt-hours and hardly experienced any increase over the previous ten years. Australia's was 264.4 terawatt-hours and grew at a compound annual growth rate of 0.7 per cent. Clean energy generation accounted for about 82 per cent of electricity production in New Zealand compared to just under one-fifth for Australia.

New Zealand's energy retailing market, though, is just as competitive as Australia's. For example, residential consumers have a choice of eight to 25 providers depending on their location. Just like in Australia prices varied greatly depending on the provider and the location. A residential consumer in Auckland, New Zealand's most populated

Table 2.9: New Zealand versus Australia Power Market.

	Production	Clean Energy
Unit	GWh	%
New Zealand	43,503	82.4
Australia	264,407	19.8

Source: Ministry of Business, Innovation & Employment, New Zealand (2020). Electricity statistics, Markets team, Evidence and Insights Branch. [online] mbie.govt.nz. Available at: https://www.mbie.govt.nz/building-and-energy/energy-and-natural-resources/energy-statistics-and-modelling/energy-statistics/electricity-statistics/ [Accessed 27 August 2020]; Department of Industry, Science, Energy and Resources, Australia (2020). Australian Energy Statistics, Table O Electricity generation by fuel type 2018–19 and 2019. [online] energy.gov.au. Available at: https://www.energy.gov.au/publications/australian-energy-statistics-table-o-electricity-generation-fuel-type-2018-19-and-2019 [Accessed 27 August 2020].

city, would pay between NZ$1,999 ($1,324) and NZ$2,412 ($1,597) for an annual consumption of 7,576 kilowatt-hours assuming no discounts as of August 2020.[44] About 73 per cent of the electricity retail market was controlled by just four energy retailers as of July 2020: Contact Energy, Genesis Energy, Mercury NZ, and Meridian Energy. The lion share of the four remained largely unchanged in at least the previous five years (Figure 2.32).

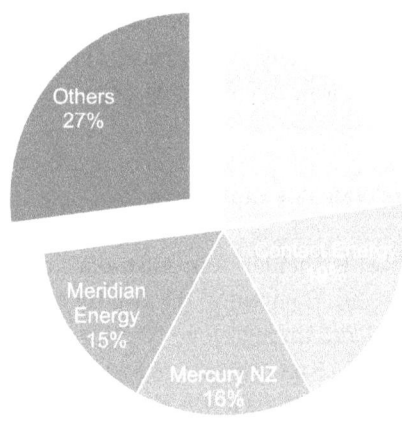

Figure 2.32: Electric Power Retail Market in New Zealand Market Share.
Source: Electricity Authority, New Zealand. [online] emi.ea.govt.nz. Available at: https://www.emi.ea.govt.nz/Retail/Reports/R_MSS_C?DateTo=20200731&MarketSegment=All&Percent=Y&seriesFilter=ALL&_si=v|3 [Accessed 28 August 2020].

From a top-down view, the energy retailers in New Zealand are quite similar to those in Australia. Especially the largest one. At the energy-end of their offering, most supply dual fuels (electricity and gas), and some also provide LPG as well. Given the level of clean energy available in the country, they are not as compelled to offer a full smorgasbord of services related to solar installation for their consumers, as the energy retailers in Australia have to do. At first glance, looking at the consumer-facing websites as well as the published financial reports of the leading four energy retailers, there is a sense, in the view of this author, that they may not be as aggressive as their Australian counterparts in terms of aggressively building a multi-utility capability in the way many UK energy retailers already have built and in the way many Australian ones are trying to build.

For the largest of New Zealand's energy retailers in terms of market share, Genesis Energy, the retail offering is seemingly simple. New customers are offered some billing discounts and reward points depending on the contract. The retailer also offers an energy management tool to help the customer to better monitor and control consumption. Contact Energy, the nation's second largest retailer, has a very similar offering with a couple of twists. The company is trying to aggressively become more digital and also is trying to offer a broader range of products and services tailor-made to meet different customers' needs. One of its customer service plans includes a package bundled with broadband, which had managed to accumulate about 25,000 users as of July 2020. Mercury NZ, the third largest company, offers its customers similar services to Genesis Energy and Contact Energy, while also offering electric mobility solutions including e-scooters as well as EVs rental, in partnership with other companies. This service comprehensively includes concierge delivery and pickup, insurance, roadside assistance, all maintenance, and registration fees. In other words, a worry-free EV experience.

2.3.4 Japan's Tokyo Gas, KDDI: Cautiously Progressive Approach

The Japanese energy retailing experience is one of the most interesting in Asia. In some ways it has been running for more than two decades for some of the companies but in other ways it is brand new. For some of the incumbent energy providers, specifically the integrated electric power companies, the experience of trying to retain and acquire customers is relatively new. They have only been forced to try hard since the opening of the electricity retail market to competition in April 2016. The nation's largest natural gas suppliers in terms of volume sold, Tokyo Gas and Osaka Gas, have more experience because they have been competing with the power companies for many decades given that both can provide energy to consumers for cooking, cooling, and heating, and even for some appliances. Other energy retailers include the subsidiaries of established companies, such as leading telecommunications services provider KDDI.

Founded in 1885, Tokyo Gas is Japan's largest gas supplier and is today also one of the leading electricity retailers. To put the company into perspective in terms of size, it realised revenues of 1.93 trillion yen ($18.2 billion) chiefly from the sale of 17.7 billion cubic meters of gas chiefly to 11.95 million customers and some wholesale customers, as well as 20.6 terawatt-hours of electricity to about 2.4 million users in the twelve months through March 2020. Tokyo Gas started to provide more energy services to its customers as far back as the 1990s. The provision of such services more aggressively accelerated in the 2000s. This is when a campaign was launched by the electric power companies to control the market for newly built homes. They wanted to ensure that such new premises would solely use electricity as the source of energy; this is known in Japan as All-Electric Houses. Tokyo Gas now provides not just a one stop for energy but pretty much for all of the basic needs of a residence as well as the needs for its commercial and industrial customers. It is worth looking into the variety of the energy and related household services that the company has been providing to better understand the multi-utility found in Japan, as a similar model is most likely going to be mimicked by energy retailers in other countries.

At the commercial and industrial end, the offering is relatively straightforward. Some of the solutions that it offers businesses include decentralised energy generation and storage as well as energy efficiency and conservation solutions. These can be from Tokyo Gas or from some of its business partners, whose number has been expanding. It offers, for example, dispersed (or decentralised) energy systems that can be gas-based cogeneration and solar photovoltaic systems and can turn these systems into Virtual Power Plants; this is an IoT based tool to run and control decentralised energy sources as if it were a single energy generating facility. It also supplies gas-based air conditioning equipment, which includes controls for office space. The company and its partners also can provide a myriad of digital technological solutions for energy management that use blockchain technologies and AI-based data analysis. The company's smart energy systems are enjoying robust growth. It is installing such a system in a variety of settings, which include hospitals, residential complexes, and even whole new districts. An example of the latter is Ekimachi Energy Create, which is a joint venture between Tokyo Gas and railroad giant East Japan Railway Co. (more commonly known as JR East), and a related company JR East Building Co. The joint venture will be responsible for the management of all of the energy that will be used by a whole district, which will be called Global Gateway Shinagawa and is currently been constructed in phases over the next decade near the Shinagawa rail station and Takanawa Gateway in Tokyo.

At the residential customers end, the offering is very abundant. The company offers a great variety of plans for residential consumers as well as several optional value-added add-on services. It has built over the past 10 to 20 years an extraordinary capability to provide households with a one stop in terms of services, in the view of this author. One can sign on to just get gas from the company or one can choose to bundle the gas supply with electricity and even with telecommunication services. The dual fuel plus Internet and smartphone plan, called Tokyo Gas Triple Discount, was recently launched. One of the packages, for example, uses So-Net, a leading Internet provider, part of the Sony group of companies, and the user will get a 4.5 per cent discount on their Internet use if they bundle it with gas and power. The company also offers a variety of 'living solutions', including smart services. For example, the company can place sensors in the residence and alert the occupant if the gas was left on, or if a window or the main door was left unlocked. The company sells a number of appliances which it promises to repair or maintain when needed. Through Tokyo Gas a consumer also can have easy access to household related services such as cleaning services, faucet replacement, and renovation, among many. Tokyo Gas has also been developing the fuel cell co-generation market for more than a decade. Back when it started in the late 2000s the large size of the fuel cell unit, called ENE-FARM, meant that it could only really be deployed at industrial and commercial premises, or detached houses. The company and its partners, Dainichi Kogyo and Kyocera, were able to develop in 2019 a fuel cell unit with a built-in storage that became the smallest size household fuel cell power generation system in the world. It is roughly the size of an outdoor air-conditioning unit. It can generate power even if there is a power failure.

Tokyo Gas two decades' experience in servicing customers other than selling them gas supply has enabled the company to rapidly gain new electricity customers after the full electric power market liberalisation in April 2016. Within the first 12 months, it managed to secure almost 700,000 new retail customers. It was subsequently able to add an average of about 138,000 new accounts per quarter, resulting in a total of almost 2.5 million new customers by June 2020. Retail electricity sales increased to 8.5 terawatt-hours in the 12 months to March 2020 from 2.2 terawatt-hours for the 12 months to March 2017 (Figure 2.33). So, its strategy has proven quite successful. Serious competition will mean that the company will continue to be quite imaginative in order to secure new users and retain existing customers.

Currently, the Japanese electricity retail market is still dominated by the incumbent vertically integrated electric power companies in terms of volume sold, such as TEPCO and Kansai Electric Power, and their subsidiaries. However, the newer market entrants are also gaining ground (Figure 2.34). Apart from Tokyo Gas that is ranked second as of April 2020, the other leading electricity retailers with no affiliation with incumbent vertically integrated electric power companies include ENNET (number 1 market share), JXTG Energy (number 3), KDDI (number 4), and F-Power (number 5). ENNET focuses on retail electricity, electrical services, and energy-saving consulting. It has several shareholders including telecommunications giant NTT as well as Tokyo Gas and Osaka Gas, Japan's second largest gas supplier in terms of revenues and volume sold. JXTG Energy is part of the ENEOS group of companies, the largest oil company in Japan with a market share of about 50 per cent in terms of fuel oil sold in the nation as well as a 43 per cent share of the country's service stations. F-Power, founded in 2009, purely focuses on electricity sales and

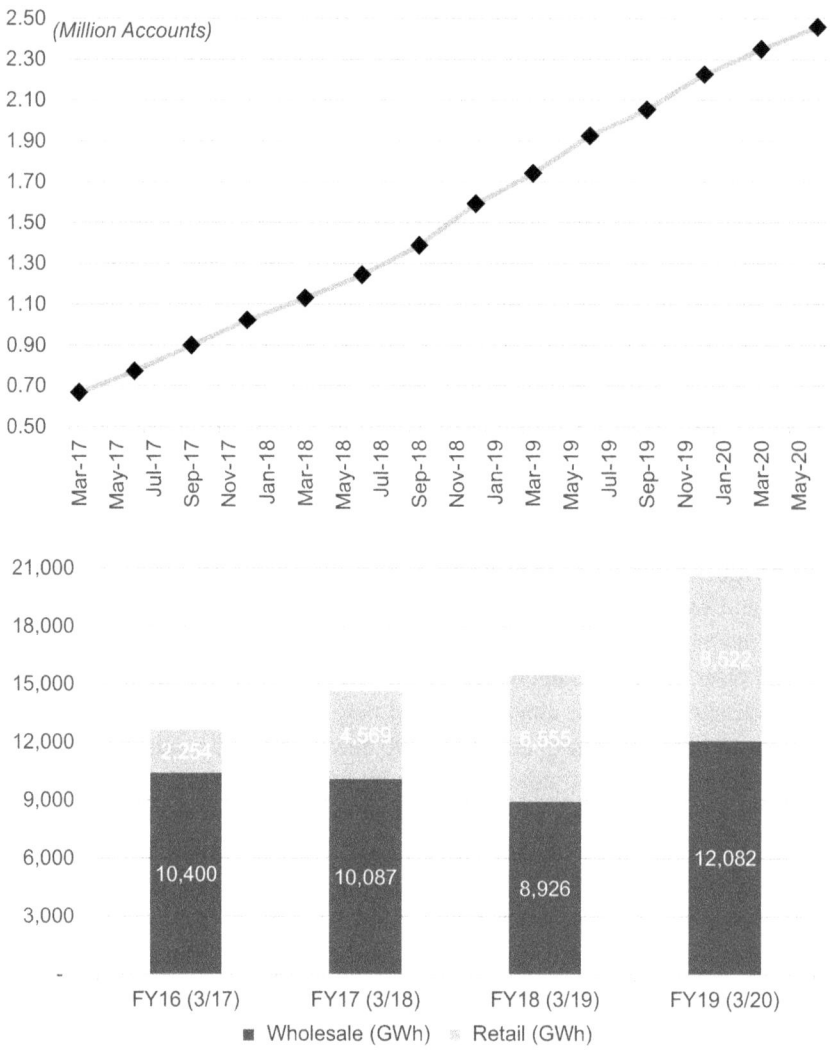

Figure 2.33: Tokyo Gas Electricity's New Contracts and Total Sales.
Source: Author, August 2020. Data source: Tokyo Gas Co. Ltd. (2020). Annual Reports. [online] tokyo-gas.co.jp. Available at: https://www.tokyo-gas.co.jp/IR/english/ [Accessed 30 August 2020].

related services such as supplying steam and hot water. KDDI, another telecommunications giant, is perhaps the most interesting case study.

KDDI provides telecom services in the country under the 'au' brand with a market share of more than 30 per cent and revenues of 5.2 trillion yen ($49.7 billion) in the 12 months through March 2020. In addition to smartphone and Internet services, KDDI has been gradually penetrating the electricity market. Its 'au Denki' service ('denki'

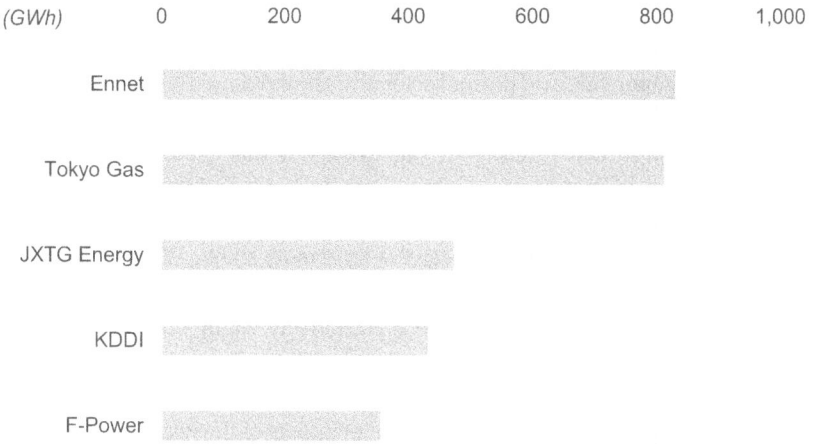

Figure 2.34: Major Electricity Retailers (ex-Major Electric Power Companies) in April 2020. Source: Author, August 2020. Data source: PPS-NET Available at: https://pps-net.org/ppscompany?ppskey=pps195 [Accessed 30 August 2020].

means electricity in Japanese) can deliver electricity to commercial, industrial, and residential customers using the power distribution network of existing electric power companies. au Denki had sold about 474.4 gigawatt-hours in April 2020 compared to 16,123.8 gigawatt-hours for TEPCO Energy Partner, the nation's biggest energy retailer and one of the former vertically integrated market incumbents.[45]

One commercial customer story on its website is about a property company. The company rents individual apartments, and the rental contracts differ substantially. Using au Denki's the company could better manage the billing of approximately 3,700 apartments as well as reduce its electricity bill. At the residential end, au's home IoT service boasts that its technology allows for the operation of household appliances through a smartphone app. Also, it cuts electric costs and allows the consumer to easily monitor electricity usage and get advice on how to conserve energy.

In its public filings, KDDI had not disclosed any profit target of the energy business, au Denki. However, the management did mention to this author that the gross margins of the energy business alone are not as high as for the telecommunications business, but it is definitely adding a rising stream of income to the company. au Denki had commenced offering services from April 2016. It had managed to secure more than 2 million contracts in the 12 months to March 2020 and had targeted 3.4 million contracts by March 2022. In the four fiscal years to 31 March 2020, au Denki had seen the number of contracts rise at about 20 per cent annually. KDDI has approximately 24 million 'au' service subscribers so it believes that au Denki may still have quite a lot of room for growth.

2.3.5 Singapore's Transition to Newborn Utilities

The reform of the Singapore electric power sector and the existing competition among energy retailers was evaluated in an earlier section (see 2.2.3 Singapore – Progressive but Paced, Highly Controlled Reform). The conclusion was that the market is indeed quite competitive but relative to the Australian, New Zealand, and Japanese markets, the competition may not be as brutal in Singapore after the initial two-year aggressive market share grab. The retail landscape became a little more stable, when retailers realised how funding intensive and complex such a business actually is. An interesting example of an energy retailer in the city state is a relatively young company, iSwitch Energy, which is now the number one green power retailer and number three retailer overall. The company does not have the same background as the other leading energy retailers in Singapore, which are basically the retail offshoots of the leading power generation companies. While they own generating physical assets and have to manage the costs and liabilities associated with that ownership, iSwitch deals in financial derivatives such as SGX listed Futures and Brent Oil Futures to manage their market risk.

Unlike its key competitors, iSwitch's background is with a company with a very long history in global commodities trading and investing, Jade Energy, which sits under The RCMA Group. iSwitch got its retail electricity licence in April 2016 and has grown to become one of the sector leaders, including being the largest green retailer. The company owes a great part of its success to its parent company's broad and deep experience in global commodity and energy trading, especially in Singapore and the rest of Asia. This experience has allowed its team to innovate and to offer new products to its users. But how can a trader with no generation assets become so successful? How does it distinguish itself from its peers? And, what are its long-term plans? To answer these questions, this author approached Andrew Koscharsky. Andrew is the chief commercial officer of Jade Energy, iSwitch's parent company, and is also the co-founder of iSwitch.

Andrew noted that strong risk management and wholesale market experience in other energy markets such as Australia, gave iSwitch the confidence to market-make, trade and retail in what was a 'newly liberalised business-to-business market' back in 2016. The company began predominantly as a trader in the SGX Futures market, which was listed in 2015, and this helped it get familiar with the technical and physical aspects of the spot market. He thinks it is a unique market, given its heavy correlation to oil prices, via LNG supply, and this means that they have oil and also electricity exposure to manage via energy derivatives trading. In 2016, the government provided the pathway for residential market liberalisation, slated for 2018. This would result in Singapore's 1.4 million households being able to leave the incumbent, SP Services, and chose a new retailer. Having worked for Origin Energy for 10 years in Australia, Andrew already had retail market experience and realised that this was an opportunity too good to miss. Given that Jade Energy now had trading and retail experience in similar markets, it was well placed to succeed.

All new retailers worked closely with the government to understand the information technology framework, marketing guidelines, and consumer protection laws for the 2018 residential launch. It was a daunting task to go up against billion-dollar global generating companies like Sembcorp, Engie, and Huaneng, but the iSwitch's team experience and ability to react quickly proved to be hard to match. iSwitch felt that using derivatives to hedge, rather than a physical asset, gave it a competitive cost advantage, given that the Singapore power generation market is over supplied with gas and also has excess capacity. This is highlighted by the recent financial demise of the Tuaspring generator, owned by Hyflux.

Andrew added that it is no coincidence that global oil majors have turned their attention to electricity retailing. Shell, Total, and BP have all made major investments in electricity retailing to help broaden their business model and lengthen their vertical integration. EVs, domestic and large-scale batteries, solar and smart home technology are all a threat to any upstream fuel provider, hence the desire to get closer to the consumer. Being 'in the home' gives them that broader conversation across multiple products, rather than just petrol. Given the LNG linkage to the Singapore market, electricity retailing in such a technologically advanced jurisdiction places iSwitch, with its significant market share, in the middle of the big energy transition.

2.3.6 China's Private and State-Owned Fast Changers

As per some of the other liberalised Asian electricity markets, we have discussed the history and the current status of the opening of China's energy markets. Many thousands of energy retailers have been created in the nation since 2016. Some are trying to follow the footsteps of energy retailers elsewhere in the world by targeting to transform themselves into comprehensive energy solutions providers that will supply their customers with a rich variety of services. Two examples were chosen. One is a natural gas supplier, ENN Energy, one of the largest ones in the country in terms of volume sold. The other is China Resources Power, one of the country's leading electric power-generating companies.

ENN Energy has a very modest background. It was born in a small city and was not a state-owned enterprise. It was founded by Mr Wang Yusuo in the early 1990s in Langfang, a small industrial city at a 60 kilometres drive south of Beijing and 70 kilometres drive north of Tianjin. Then called Xinao Gas Holdings, ENN Energy raised capital by issuing shares to investors by listing on the Growth Enterprise Market board of the Hong Kong Stock Exchange – this board is to accommodate companies whose track record or profitability do not meet the main board's requirements and has since been renamed the GEM board. Its business model was very straightforward. It bought natural gas from one of China's oil and gas juggernauts and sold it to industrial, commercial, and residential users through natural gas pipelines it had built. This author met with the management of the company many times since the early 2000s as an equity analyst, when it first became a publicly traded company. While investment analysts are supposed to ask a million questions and meetings are supposed to last at least an hour, this author actually struggled to conduct a meeting with ENN Energy longer than 30 minutes. The business model and the strategy were volume-driven growth. It was clear and straightforward. Moving the clock forward 20 years and the tiny gas operator (also known as a city-gas company) has become a complex, major energy markets participant and is set to grow even bigger in the next few years.

ENN Energy's revenue rose 291-fold to 70.2 billion yuan ($10.5 billion) in 2019 from 241 million yuan ($36 million) in 2001 or a compound annual growth rate of more than 37 per cent. Its growth profit jumped 79-fold to 11.3 billion yuan ($1.7 billion) in 2019 compared to 2001, a compound annual growth rate of more than 27 per cent. The number of commercial and industrial customers rose 419 times to 148,000, connected households increased 194 times to almost 21 million, pipeline length 419-fold to more than 54,000 kilometres, and gas volume sales 800-fold to about 27 billion cubic meters (Table 2.10). This is a staggering pace of growth for what was once a small company.

Table 2.10: ENN Energy Key Growth Indicators 2001–2019.

		2001	2019	Change (x)	CAGR
Total Revenue	(million CNY)	241	70,183	291.7	37.1%
Gross Profit	(million CNY)	143	11,265	78.7	27.4%
Number of connected households	(thousands)	108	20,920	193.7	34.0%
Number of Commercial and Industrial customers		355	148,761	419.0	39.9%
Length of existing pipelines	(kilometres)	464	54,344	117.1	30.3%
Total Natural Gas Sales Volume	(million m³)	35	26,963	779.6	44.8%

Source: ENN Energy Holdings Limited 2001 and 2019 Annual Reports.

But how has ENN Energy's business been shifting? In 2000, the year before its listing, it derived about 99.6 per cent of its revenue from the sale of gas to users. In 2001, it had begun to generate revenue from selling gas appliances its customers needed to use, such as cookers and water heaters. These were mostly manufactured by third parties. This author believes that the founder, Mr Wang, may have had an eye on what gas utilities did in developed regions such as Tokyo Gas in Japan and

Hong Kong & China Gas in Hong Kong. These companies first sold third party appliances to add to their gas sales revenues but then also sold some under their own name so as to raise their profits as the latter provided better operating margins than selling those from third parties. The gas income itself was mostly from what is called in China 'connection fees', a government-set one-time fee for the customer to pay the gas supplier to be hooked up to the gas network. It accounted for 82.8 per cent and 76.8 per cent of total turnover in 2000 and 2001 respectively and even a greater percentage of its net profit. These fees, now classified as the construction and installation business segment, accounted for 8.5 per cent of turnover and 25.9 per cent of operating profit in the first six months of 2020.

Two decades later, after growing more than 292 times, the revenue mix is much more diversified. It has moved away from making money just from the traditional selling of gas and the connection fees. Selling gas is of course still the main contributor but its operations have grown from just four cities, including Langfang, in 2001 to more than 217 cities throughout China covering a population of 104 million. This means that it had hugely grown its footprint, network, and number of bases. This is something that it can massively leverage. What does that exactly mean? Well, the more ventures it has, the more it can work its economies of scale. In other words, it can more effectively, efficiently, and cheaply add new ventures and new businesses to those ventures.

One of the businesses it added during the 20-year period is vehicle refuelling stations, which are basically gas stations that do not sell gasoline for internal ICE vehicles but sell compressed natural gas or liquified natural gas for vehicles that are converted to use natural gas instead of gasoline. Arguably, one could say that is simply an extension of the original piped gas retail business but if one thinks about it, it is a totally different business with a whole different set of risks and challenges.

Another business ENN Energy entered, in the second half of the 2010s, is the importation of liquified natural gas; in order to transport natural gas by ship it is first liquified, in places such as Australia or Indonesia, and then transported to a destination, such as China or Japan, and then turned back into natural gas before use. Also, it decided to sell natural gas in very large quantities to entities that would resale it so as to fully maximise the use of its dispatch system, logistics fleet, and midstream resources. In other words, it entered what is called wholesale energy or energy trading.

The company decided to enter this because in the past China's gas resources were tight. The difference between the peak of gas demand, typically in the winter, and the lowest point of demand, typically in the summer, was especially large. ENN Energy purchased a fleet of trucks designed to transport liquified natural gas, which were deployed to supplement the pipeline gas resource when the gas supply was tight in winter. However, in the summer, the truck fleet would be largely idle. So as to maximise their use, in order to revitalise these assets, the company started the wholesale business. The objectives were twofold. One was volume driven. It wanted to actively drive large-scale wholesale gas sales, increase its market share, improve the bargaining power with the upstream suppliers, and allow its city-gas projects to obtain lower cost gas. Another objective was to ensure supply reliability. Specifically, to manage to lock in some liquified natural gas resources in advance so as to effectively relieve the pressure and guarantee supply during the peak heating season. Interestingly, a company owned by ENN Energy's chairman, Mr Wang, actually became the first private company to invest in, build, own, and operate a liquified natural gas import terminal in the country, the Zhoushan LNG Terminal in Zhejiang province, which it completed in late 2018. ENN Energy signed three liquified natural gas long-term contracts with Chevron, Total, and Origin Energy in 2016, for the purchase of 1.44 million tons per year. The term of the contracts with Chevron and Total is for ten years. That with Origin Energy is for five to ten years. ENN Energy also signed a ten-year terminal usage fee contract with the Zhoushan LNG Terminal from 2018 to 2028. The contract of ENN Energy can be used, but it is not limited to the imports at Zhoushan wharf. It only needs to pay a terminal usage fee, which must be a fair market price; i.e., ENN energy does not purchase directly from Zhoushan LNG Terminal.

The newest business, and I would argue the most exciting one growth-wise, is integrated energy. This involves ENN providing a diversity of energy sources and services to its residential, commercial, and industrial customers. One of the primary target users are large industrial parks. These may have several or dozens of factories as well as industrial and related facilities that need gas or electricity for their manufacturing or other processes, and for heating and cooling solutions. In addition to these, ENN Energy will use digital technologies to ensure that it can provide more value-added solutions to the facilities in the form of raising energy efficiency and conservation. This, very much in the same way as those energy retailers in Australia, Japan, or Singapore. To accelerate the entry into integrated energy businesses, ENN Energy purchased from its parent a company called (strangely) Ubiquitous Energy Network Technology. The acquisition, in August 2018, means that ENN Energy can leapfrog into this area very rapidly. By June 2020, it had 108 projects in operation consuming about 4,806 gigawatts-hours and 2,055 megawatts in installed capacity. The company targeted revenues from the integrated energy business to reach 6 billion yuan ($897million) for 2020, as of August 2020, from 2,749 million yuan ($401 million) in 2019, and 1,005 million yuan ($150 million) in 2018.

So, energy markets deregulation has opened the doors to ENN Energy to get into other energy-related businesses. These will give the company new avenues to grow its income further and also manage risk. ENN Energy itself has underlined the importance of the deregulation for its growth. At the electricity end of things, it said that the policies of allowing companies like ENN Energy to directly distribute and sell power to end users is a brand-new area of growth of the company. On the heating and cooling side of things, these are energy solutions that can be very flexible and tailor made. Also, the company can earn higher profits from providing these additional services.

A second mini case study from China is China Resources Power. The company is the subsidiary of Chinese state-owned enterprise China Resources Holdings, which controls about 63 per cent of China Resources Power. China Resources Holdings was established in Hong Kong in 1938. It is a massive leading Chinese conglomerate covering a wide range of industries including commodities, manufacturing and distribution, real estate, infrastructure, public utilities, and pharmaceuticals in Hong Kong and mainland China. It had revenues of 654.6 billion yuan ($97.8 billion), profit of 72.6 billion yuan ($10.8 billion), total assets of about 1.62 trillion yuan ($242 billion) in 2019. China Resources Power, which listed on the Hong Kong Stock Exchange in late 2003, is also a sizeable company with revenues, profit, and total assets having amounted to HK$67.8 billion ($8.7 billion), HK$7.3 billion ($947 million), HK$215.7 billion ($27.8 billion) respectively. The equity-controlled operational generation capacity was 40.4 gigawatts, including 36 coal-fired plants, 96 wind farms, 21 photovoltaic plants, two hydro plants, and three gas-fired plants.

As soon as China commenced its progressive opening of the electricity retail market from 2015, China Resources Power jumped at the opportunity. This may seem surprising for such a large state-owned company that is the subsidiary of an even larger conglomerate. However, this author has followed the company as an equity analyst since its listing and can testify to the fact that the company has proven to be one of the most progressive companies among Chinese state-owned power companies. For example, it has been able to consistently earn better return on investments compared to its peers (Figure 2.35). In the past 15 years, the company managed to realise a return on equity of about 13 per cent, which is much higher than its direct peers, including China Power International (7 per cent), Datang International Power Generation (6.1 per cent), Huaneng Power International (8.7 per cent), and Huadian Power International (7.2 per cent); these companies are the nation's largest electric power companies listed on stock markets and coal-fired generation accounts for a high percentage of their respective installed capacity.

The company went about aggressively trying to become a market leader in the burgeoning Chinese electricity retailing. In March 2016, just a few months after the formal release of the market opening policy, the company set up a dedicated independent company in the province of Guangdong, solely responsible for the electric power sales and related services as well as ancillary products in Southern China, with a view to become the most competitive integrated energy service group in the region. By the first half of 2020, China Resources Power's direct power sales subsidiaries were market leaders in many of the provinces they operated in. In just four years, it had set up in 29 provinces and regions and was one of the top three electricity retailers in 11 of these provinces and regions. The bulk of the direct power sales contracts were for 12 months but the company has been successful at retaining these customers as it has been able to roll over the bulk of these contracts. As a percentage of the

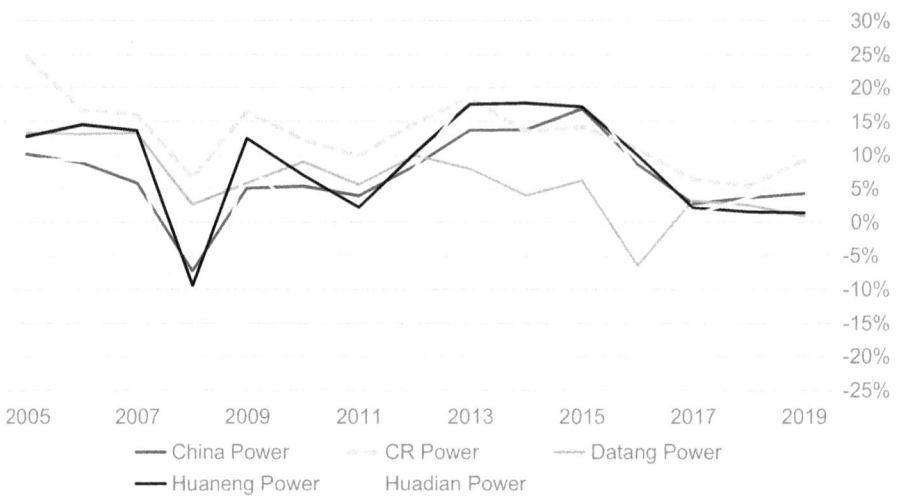

Figure 2.35: China Resources Power Return on Equity Versus Peers (%).
Source: author, August 2020. Data sourced from: China Power International, China Resources Power, Datang International Power Generation, Huaneng Power International, and Huadian Power International financial accounts.

company's total output 69.5 per cent went to the retail power market in the six months to June 2020, compared to 66 per cent for the whole of 2019, and 45 per cent in 2018. This is higher than the market average but where the market is heading over the next few years should definitely be crystal clear.

China Resources Power is also proving successful in providing customers with a rich variety of value-added energy-related services, which it calls its integrated energy business. In many ways similar to what ENN Energy is doing. For this business it had accumulated 6,856 customers by the end of June 2020, compared to 5,358 by December 2019, and 2,819 as at the end of 2018. It is worth briefly examining some of these business activities by explaining some of the projects. At smart energy project at the CR Circular Economy Industrial Park in Hezhou, Guangxi province, the company optimised the use of thermal, wind, and solar plants to reach net zero emissions, producing 60 gigawatt-hours a year and saving 250,000 metric tons of solid waste and 66,000 metric tons of standard coal. At a project at the Taigu Economic and Technology Development Zone in Jinxhong city, Shanxi province, the company is installing energy management systems that will be used by about 50 businesses in the area, in addition to supplying them with clean energy and saving 96,000 metric tons of standard coal per year. At the Liyujiang Power Plant in Hunan province, the company installed supply side management systems that includes energy storage thanks to a 12 megawatt lithium iron phosphate battery system. At a glass-making plant controlled by the Fuyao Glass Industry Group, which specialises in automobile safety glass and other industrial glass, the company set up an energy system capable of harvesting waste heat from the manufacturing process to generate electricity that is then used at the plant and office buildings; as well as serving for heating and cooling any remaining heat can be used to boil hot water for the living quarters and cool the air-conditioning system. Also, it will install rooftop solar systems to provide as much as 15 per cent of the plant's electricity needs saving 22,000 metric tons of standard coal and emissions reduction of 2,000 metric tons of sulphur dioxide and 54,000 metric tons of carbon dioxide.

Notes

1 Arto, I., Capellán-Pérez, I., Lago, R., Bueno, G., and Bermejo, R. (2016). The Energy Requirements of a Developed World. Energy for Sustainable Development. [online] 33, pp. 1–13. Available at: https://www.sciencedirect.com/science/article/pii/S0973082616301892 [Accessed 2 January 2020].
2 Market Report Series: Energy Efficiency 2018 – Analysis – IEA (2018). Market Report Series: Energy Efficiency 2018 – Analysis – IEA. [online] IEA. Available at: https://www.iea.org/reports/energy-efficiency-2018 [Accessed 31 May 2020].
3 International Energy Agency (2020). Global EV Outlook2020: Entering the Decade of Electric Drive? [online] International Energy Agency. Available at: https://webstore.iea.org/download/direct/3007 [Accessed 9 July 2020].
4 BloombergBNEF (2020). Electric Vehicle Outlook 2020. [online] BNEF. Available at: https://bnef.turtl.co/story/evo-2020/page/3/3?teaser=yes [Accessed 9 July 2020].
5 Deloitte (2020). Deloitte Insights: Electric Vehicle Trends. [online] https://www2.deloitte.com. Available at: https://www2.deloitte.com/uk/en/insights/focus/future-of-mobility/electric-vehicle-trends-2030.html [Accessed 5 November 2020].
6 Le Quéré, C., Jackson, R.B., Jones, M.W., Smith, A.J.P., Abernethy, S., Andrew, R.M., De-Gol, A.J., Willis, D.R., Shan, Y., Canadell, J.G., Friedlingstein, P., Creutzig, F., and Peters, G.P. (2020). Temporary Reduction in Daily Global CO2 Emissions during the COVID-19 Forced Confinement. Nature Climate Change. [online] pp. 1–7. Available at: https://www.nature.com/articles/s41558-020-0797-x [Accessed 5 November 2020].
7 'Half of all new cars sold in Australia by 2035 will be electric, forecast predicts', *The Guardian*, 13 August 2019.
8 *Newsroom* (2020). Govt to Miss Yet Another Electric Vehicle Target. 27 February. [online] newsoom.com. Available at: https://www.newsroom.co.nz/2020/02/27/1054554/govt-to-miss-yet-another-electric-vehicle-target [Accessed 11 July 2020].
9 Government of Hong Kong Information Services (2020). Electric Vehicle Roadmap Update. February 27. [online] news.gov.uk. Available at: https://www.news.gov.hk/eng/2020/02/20200227/20200227_180629_280.html [Accessed 11 July 2020].
10 'Singapore aims to phase out petrol and diesel vehicles by 2040', Reuters, 18 February 2020.
11 Eurelectric (2019). EV Public Charging Infrastructure and EV Rollout: State of Play. Eurelectric (https://cdn.eurelectric.org/). [online] Available at: https://cdn.eurelectric.org/media/3805/charging-infrastructure-factsheet-2019-h-1BC6E8C5.pdf [Accessed 11 Jan. 2021].
12 Borlaug, Brennan & Salisbury, Shawn & Gerdes, Mindy & Muratori, Matteo. (2020). Levelized Cost of Charging Electric Vehicles in the United States. Joule. 10.1016/j.joule.2020.05.013.
13 FutureCar (2019). MIT Study Finds EVs to Cost More Than Regular Cars Until 2030. 2 December. [online] futurecar.com. Available at: https://www.futurecar.com/3636/MIT-Study-Finds-EVs-to-Cost-More-Than-Regular-Cars-Until-2030 [Accessed 11 July 2020].

14 Energy Post (2020). Europe Needs its Own EV Battery Recycling Industry. 25 March. [online] energypost.eu. Available at: https://energypost.eu/europe-needs-its-own-ev-battery-recycling-industry/ [Accessed 11 July 2020].
15 Holland-Letz, D., Kässer, M., Müller, T., and Tschiesner, A. (2018). Profiling Tomorrow's Trendsetting Car Buyers, McKinsey Center for Future Mobility, McKinsey & Company, December.
16 Heineke, K., Holland-Letz, D., Kässer, M., Kloss, B., and Müller, T. (2020) ACES 2019 Survey: Can Established Auto Manufacturers Meet Customer Expectations for ACES? McKinsey Center for Future Mobility, McKinsey & Company, February.
17 International Energy Agency (2019). 'Energy Efficiency 2019'. November. p. 44.
18 International Energy Agency (2019). 'Energy Efficiency 2019'. November. p. 44.
19 Department of the Environment and Energy (2019). Australian Energy Statistics. Table B.
20 The Australia Institute (2013). Power Down: Why is Electricity Consumption Decreasing. Institute Paper No. 14, December.
21 Electrical and Mechanical Services Department, the Government of the Hong Kong Special Administrative Region (2019). Hong Kong Energy End-use Data 2019. September 2019. p.18 [online] emsd.gov.hk. Available at: https://www.emsd.gov.hk/filemanager/en/content_762/HKEEUD2019.pdf [Accessed 12 July 2020].
22 Shanghai Development and Reform Commission (2020). 'In June, Shanghai's Total Electricity Consumption Was 13.38 Billion kWh, up 9.5% Year on Year'. [online] www.cctd.com.cn. Available at: https://www.cctd.com.cn/show-19-204509-1.html [Accessed 2 December 2020]. (Original in Chinese).
23 The Dow Jones Sustainability World Index is a float-adjusted market capitalisation weighted index from S&P in partnership with sustainability investing specialist RobecoSAM, which measures the performance of companies selected with economic, environmental, and social criteria using a best-in-class approach.
24 Systems of systems are platforms that consent previously siloed systems to exchange data and issue commands based on a more complete picture.
25 IDC (2020). IDC Lowers Forecast for Worldwide IT Spending to a Decline of 5.1% in 2020, but Cloud Spending Remains Relatively Resilient. [online] IDC. Available at: https://www.idc.com/getdoc.jsp?containerId=prUS46268520 [Accessed 2 December 2020].
26 Owens, Sam (2017). Measuring the Effect of Electric Utility Deregulation on Residential Retail Prices in a Midwestern State. MPA/MPP Capstone Projects. 280. [online] uknowledge.uky.edu. Available at: https://uknowledge.uky.edu/mpampp_etds/280 [Accessed 2 December 2020].
27 Rai, Alan and Nelson, Tim (2019). 'Australia's national electricity market after twenty years', *The Australian Economic Review*', vol. 53, no. 2 (June), pp. 165–182.
28 Rai, Alan and Nelson, Tim (2019). 'Australia's national electricity market after twenty years', *The Australian Economic Review*, vol. 53, no. 2 (June), pp. 165–182.
29 PowerTechnology (2018). *Can Australia Bring Its Sky-high Energy Prices down to Earth?* [online] www.power-technology.com. Available at: https://www.power-technology.com/features/australia-energy-prices/ [Accessed 2 Dec. 2020].
30 Potter, B. and Tillett, A. (2017). *Australian Households Pay Highest Power Prices in World*. [online] Australian Financial Review. Available at: https://www.afr.com/politics/australian-households-pay-highest-power-prices-in-world-20170804-gxp58a [Accessed 2 Dec. 2020].
31 Krishnan, S. (2019). *Lauren Is in Debt from Her Power Bills, but like Thousands of Australians, Is Reluctant to Shop around*. [online] www.abc.net.au. Available at: https://www.abc.net.au/news/2019-12-04/some-energy-consumers-not-seeking-out-a-better-deal/11753818 [Accessed 2 Dec. 2020].
32 Ofgem (2013). Liquidity. [online] Ofgem. Available at: https://www.ofgem.gov.uk/electricity/wholesale-market/liquidity [Accessed 12 Jan. 2021].
33 Zhu, C. and Lague D. 'China's other power struggle – breaking state monopolies', Reuters, 17 October 2010.

34 'The total profit of State Grid in 2019 is 77 billion yuan, and the investment plan is soaring this year', Surging News, 27 March 2020. Available at: http://www.cspplaza.com/article-17633-1.html [Accessed 17 August 2020].

35 'From January to June, China's electricity consumption was 3,354.7 billion kwh, down 1.3 per cent year on year', Chia Electric Power Enterprises Council, 17 July 2020. Available at: http://www.escn.com.cn/news/show-1070848.html [Accessed 17 August 2020].

36 Sohu (2019). 'Beijing Electric Power Trading Centre's share reform breakthrough: Introducing 10 state-owned enterprises, energy investors, new shareholders holding 30 per cent'. 31 December. [online] sohu.com. Available at: https://www.sohu.com/a/363999810_764234 [Accessed 18 August 2020].

37 Huang, C.Y. (2020). 'Current situation and countermeasures of carbon verification in China'. 20 August. Basic Unit Construction. [online] huanbao.bjx.com.cn. Available at: http://huanbao.bjx.com.cn/news/20200820/1098642.shtml [Accessed 21 August 2021].

38 Reference for Business (2020). Scottish Hydro-Electric PLC – Company Profile, Information, Business Description, History, Background Information on Scottish Hydro-Electric PLC. [online] referenceforbusiness.com. Available at: https://www.referenceforbusiness.com/history2/44/Scottish-Hydro-Electric-PLC.html [Accessed 23 August 2020].

39 Centrica plc (2001). Annual Report and Accounts 2000. [online] centrica.com. London: Centrica plc. Available at: https://www.centrica.com/media/1367/centrica-annual-report-and-accounts-2000.pdf [Accessed 15 Dec. 2020].

40 'Electricity supply market shares by company: Domestic (GB)', Ofgem analysis of electricity distribution network operator reports, information as of July 2020. Available at: https://www.ofgem.gov.uk/data-portal/electricity-supply-market-shares-company-domestic-gb [Accessed 23 August].

41 'Business Britain: The fixit business', The Economist Newspaper Limited, 9 November 2000.

42 Paul, S., 'Australia's Origin Energy to buy 20 per cent stake in UK's Octopus Energy', Reuters, 1 May 2020.

43 Octopus Energy (2020). Backend Developer. [online] octopus.energy. Available at: https://octopus.energy/careers/back-end-developer/ [Accessed 25 August 2020].

44 Powerswitch (2020). Consumer NZ. [online] powerswitch.org.nz. Available at: https://www.powerswitch.org.nz/trends/auckland [Accessed 28 August 2020].

45 PPS-NET. Available at: https://pps-net.org/ppscompany?ppskey=pps195 [Accessed 30 August 2020].

Chapter 3
Twin Transformations: New Fuel Mix and New Tech

There are two major transformations that will drive the energy markets in the Asia region, which will also impact energy markets all around the world in the next 30 years. The first trend is the dramatic transition to clean, green, zero-carbon energy and away from polluting fossil fuels. Albeit different economies will make the shift at different speeds. The second trend is the gradual acceleration of the introduction of digital technologies and solutions in the energy sector. This will sharply raise the efficiency and effectiveness in the way energy is produced, distributed, consumed, and exchanged. These two trends will dramatically alter the future of energy in Asia. They will provide an infinite amount of new business and investment opportunities and in this chapter I will present a variety of facts and examples with two thoughts in mind. One is to allow the reader to be able to judge for herself or himself the viability as well as the potential of these opportunities. Another is to provide food for thought in terms of new opportunities.

In Chapter 2, I already established that in the next 30 years, the Asia region will exponentially consume more energy than the rest of the world combined. That in itself is already a major driver to change in the future of the industry in the region. The manifestation of this will be through the two aforementioned fuel mix and tech transformations, both of which will develop and evolve pretty much at the same time. But first let us touch first on the basic concept of time in the energy industry. Do not worry we are not diverging into something highly philosophical or something in the realm of quantum physics. When thinking about energy we need to look at the long term. Think of time in terms of several decades. The whole value chain necessary for growth needs a lot of complex long-term planning. First, energy consumption requires the actual energy, much of which comes from energy commodities, including oil, coal, and natural gas. Getting extra resources takes some time. Imagine that a country or a company forecasts it will require 10 per cent more energy five years out. For this it will need 10 per cent extra energy resources. It will have to decide what type of energy resource it will need and will have to carefully plan for the fuel availability and supply channels for the additional fuel energy resources. It will need time to do the planning, which currently takes months or years, and also will need time to construct the new energy supply facilities, such as a solar power plant or a wind farm. Such facilities can take between a year to seven or more years in planning and construction depending on the energy source. Even more difficult is the fuel choice. Clean or zero carbon energy, including nuclear, is more manageable on that front. The process of using fossil fuels for energy generation is more complex. When the planning is centralised, such as in China or South Korea, it is also a little easier. There are occasions when planning is left to market forces and things go badly wrong like in the UK in the 1990s with the 'Dash for Gas' period. This is crucial to understand as it will impact the price of the energy in the mid to long term. Let us use a simple analogy. Imagine you enter a five-year telephone service plan. Then just after a few months there is a major price war among telephone service providers and the very same telephone service plan could be obtained for half of the price. Unfortunately, the service contract is for five years and cannot be changed. Very much the same can happen in energy markets.

The Dash for Gas started in the early 1990s. In just a decade or so gas-fired electric power generation's share of the energy mix rose from just 1.6 per cent in 1990 to 39.3 per cent in 2000, to 46 per cent in 2010. Throughout this period electricity generation only rose from 320 terawatt-hours in 1990, to 377 terawatt-hours in 2000, to 382 terawatt-hours in 2010. This is equivalent to a compound annual growth rate of just 0.9 per cent, which I calculated by using data from the *Statistical Review of World Energy* published annually by oil giant BP.

Many factors drove this energy mix change. A study by the European Centre for Energy and Resource Security summarised that factors included

> the privatization of the energy sector promoted by successive British conservative governments, regulatory changes at the EU and national levels that facilitated the use of gas in power generation, technological changes that increased the efficiency of combined cycle gas turbines (CCGT), the availability of cheap gas on the British market as a result of the increase in production of British North Sea gas, as well as the reduced amount of carbon dioxide and sulphur dioxide emissions produced in gas-fired power stations by contrast to coal-fired stations.[1]

But this created quite a few negative side effects. Especially, a shortage in gas and higher electricity prices. Since then, this has been repeated elsewhere in the world and so it is particularly pertinent to understand what happened and learn some lessons from it. To get a better understanding I turned to another friend. He is a highly regarded Asia power and financial professional who actually witnessed the crisis in the UK first-hand. After a long career in Asian investment banking and private equity, Robert McGregor is now executive director and chief investment officer at AboitizPower, a leading electric power company in the Asia region, headquartered in the Philippines. In the 1990s, Robert was initially head of sales and marketing for Scottish Hydro-Electric and latterly the head of corporate strategy for the newly privatised nuclear business British Energy. Robert recalls the changing electricity market in the UK and offers these reflections in an interview with me.

> Firstly, Joseph, I am going back almost 30 years with these observations – so I will do my best to recall the thought processes – certain things do remain deeply embedded in my mind from my time in the UK electricity industry. There were so many factors all colliding at one time – so, before you can address the questions around the Dash for Gas in that era, I think you need to understand the various pieces of the picture that came together. I was going to call them building blocks – but that would imply something way more systematic and considered than I can remember happening. It really was much closer to 'unintended consequences'. A number of things crashed together and before anyone really realised, the UK was fundamentally changing its fuel mix. And not through any form of energy strategy, fuel mix policy or centralised planning.

The first thing that springs to my mind is the profound effect of privatisation. Thirty years later, there are so many privatised electricity markets globally that it is easy to forget that this was an industrial and social revolution of global significance driven by a strong political ideology. The UK Conservative Party, on the back of many years of poor opposition, had a platform to pursue privatisation and, with the strength and will of their leader, Prime Minister Margaret Thatcher, this policy diminished unions, reduced the dependence on UK coal mining, and saw the privatisation of oil, coal, steel and the energy industries of electricity and gas. I was actually recruited from the oil and gas industry because I had experience of competitive markets. I had been with British Petroleum for almost eleven years and competitive markets were a norm to me. I was used to being in trouble with the competition authorities if my market share was above 25 per cent and here I was going into an industry that was terrified of loss of any significant market share from 100 per cent. But privatisation brought with it a profit imperative. A need to grow share prices and a need to deliver on dividends. I saw at first hand, the need for industry executives to reduce costs, to increase revenues, to build more, to build cheaper, to build better. Some of the bigger UK generating companies quickly looked to the international markets and embarked on some ambitious growth plans (that funnily enough remind me of some of the Asian corporates who have found freedom in the last decade or so and have shown an interesting level of ambition to take positions in overseas markets). But, the UK Dash for Gas sprung out of more domestic ambitions.

The company I joined was Scottish Hydro-Electric, a vertically integrated power utility in Scotland with generation interests in hydro, coal, and nuclear and ownership interests in both transmission and distribution. The company was the smaller of the two in Scotland and it had shown some commercial savvy in recruiting a young CEO, Roger Young, from outside of the industry and an internationally experienced commercial director in Alan Young (no relation) to lead their new privatised business forward. Under them was an ambitious and hungry commercial team led by David Sigsworth, an industry veteran who had grasped the privatisation opportunity with both hands. This smaller company came out of the blocks sprinting – but in a very disciplined and determined way that reflected the steely character of the young Gordonstoun educated CEO. We had a profit imperative, we had financial accountability drummed into us, we had growth targets demanded of us – and, to the south, lay an England and Wales market of much more massive demand and growth potential than Scotland. A place where we could expand our electricity sales in a way that was meaningful for us but which would not really impact too much on our competitors in that market. We had a team that focused on that market and, after exhausting the available capacity that we could export to England, we found an opportunity to build a greenfield power plant in Keadby, Yorkshire. A 749 megawatt plant. We needed to grow, we needed to build, we needed to sell into a competitive power market that used competitively priced Contracts for Difference between generators and regional electricity suppliers and the balance had to be able to sell into the Pool market. We needed to secure a source of fuel. We needed to build as fast as we could. Thanks to legislation that had passed some years ago, gas was available for power generation. The technology of the plants created very flexible units that could ramp up and down more readily than coal. And so, it just made sense. It had to be gas. And it continued to make sense – for others too – and I was actually surprised to see some recent information stating that the UK had run for more than a day with no coal power being produced at all. That is just amazing because, when I started in the industry, gas accounted for about five per cent of the capacity.

Now I know that some more academic studies go beyond the politics, the effects of privatisation, the technology of CCGT, and the availability of gas and these studies take us into the impact of high interest rates and the environmental benefits of gas versus coal power plants. I think these were factors – but I really cannot remember those issues being at the forefront of our thinking – this is looking backwards and making rational sense of something that was much more basic and primal and less systematic and organised. For sure, we would have compared the financial impact of interest rates and compared what can be done with each of the technologies but I honestly do not think it was a driver in the same way as the many other factors that collided. Interest rates would have been considered but would not have been the decision driver. In addition, although I would not have been involved in many of the more detailed technical discussions regarding the plant, the choice of the technology was not driven by the environmental impact of one technology versus another. The fact that there were such benefits to be claimed was an aside. These were different times. We did not have banks that were reluctant to finance coal, we had no incentives in place for renewable energy, nor did we have penalties for carbon. Our decision was very pragmatic, we had to grow, we had to compete, and this plant, with this fuel, delivered an internal rate of return that we liked using technology that the engineers liked. And we could get on the system faster than with coal. Privatisation, competition, and the profit imperative were the drivers that I can remember. And the Dash for Gas was therefore completely logical because no commercial company, operating in a competitive market, for a commodity product, could ignore something that was faster, better and cheaper. They had to keep up.

The thing I find interesting now, Joseph, is that I continue to see governments securing the benefits of energy sector privatisation and then realising that they no longer can dictate the fuel mix; fuel diversity as a national security concern is now lost; renewable energy has to be encouraged and embraced with subsidies but cannot yet completely replace the thermal plants; foreign investors come into the country; local companies invest overseas; and the introduction of carbon policies can adversely affect not only the investors of thermal power plants but also the population at large that benefits from the low price economics of such plants at this time. Environmental concerns are not so easy to address in a competitive market that delivers power at the lowest cost. You said to me at the start, Joseph, 'sometimes when planning is left to market forces, things can go badly wrong'. But I think, on balance, market forces have delivered better standards and lower prices than we saw in nationalised hands. And I do not think any of these markets can go back to central planning. So, it looks like another Dash for Gas may be on the cards for the UK! And I am going to have to puzzle more on the entry of LNG into the Philippines!

3.1 Impact of Changing Energy Fuel Sources and Price Trends

The first of the two megatrends in Asia is the shift to renewable and clean energy resources and move away from polluting fossil fuels. My thesis goes a little beyond the currently hotly debated topic of the energy transition towards a decarbonised world. This is of course addressed, but the focus will be looking at the different evolutions in three regions. One is Australia, partly because of its status as the region's largest fossil fuel exporter and partly because of its clean energy resources huge exploitation potential. Another region is East Asia given the importance of two of its economies, Japan and South Korea, and sharp clean energy capacity addition upside. Last but absolutely not least, we shall look at the evolution in China, the world's largest energy consumer and the undisputed leader in clean energy in Asia and the world. In the following section I will turn to a discussion pointing out why the decline over the next 30 years in the amount of coal, gas, and oil consumption is likely to be faster than what is currently expected by most forecasters. The massive shift in energy resources will be particularly poignant in Asia given that the largest energy-consuming region in the world will become even bigger. The energy resources mix will have very profound effects on the volume of energy fuels that is supplied, especially oil, natural gas, as well as thermal coal. It will of course also affect the supply channels and the price.

3.1.1 Fuel Mix Metamorphosis to Green and Clean from Brown and Dirty

I think that there is already a lot of literature around that has highlighted the trend that most economies around the world are turning to cleaner resources to meet their energy needs. Some have been more aggressive and better coordinated, such as the European Community nations. Others are more unwilling at the centralised level but are still shifting to green from brown, such as the US. The same applies to economies in Asia. Many key markets in the region have clear plans in changing the sources of where the energy they use will come from. The change is chiefly in favour of solar, wind, and other zero carbon energies. Reliance on coal and oil electric power generation will sharply decline in the next five to ten years in most countries. And, gas-fired power should also be reduced in the next 20 to 30 years, especially in China and the East Asia economies.

A sharp transformation in the energy mix of a country or of a region is not common of course but it is in no way extraordinary. I already brought up the example of the UK and the various repercussions well explained by Robert McGregor, someone who actually was there and lived it. Another country that experienced such a shift is Japan. It has actually experienced it not once but twice in the past 50 years. Once was during the oil crisis in the 1970s and another was in the 2010s after the horrendous Fukushima Daiichi Nuclear Power Plant meltdown disaster caused by the Tohoku earthquake and tsunami. So, energy shifts or transitions can happen despite all of the costs involved; we must remember that electric power in infrastructure is not cheap. We can turn to some highly regarded sources to get an idea of the massive shifts that are about to happen. I note that the methodology and what is included in the numbers is different with different forecasters so we should not be comparing these on an apples to apples basis.

The International Renewable Energy Agency (IRENA) constructed two sets of forecasts for key renewable energy resources (Table 3.1), including hydro, pumped storage, solar, and wind. The base case scenario is one that IRENA refers to as the Planned Energy Scenario (PES), and uses existing plans and targets by governments, including the Nationally Determined Contributions (NDCs) or other climate and energy objectives; NDCs are the planned climate-related actions to cut emissions and address climate change of a country to achieve the goals in the Paris Agreement.[2] The higher case scenario, which it calls the Transforming Energy Scenario (TES), adopts more aggressive assumptions targeted to allow the country to allow its energy system to keep the global temperatures rise below 2 degrees Celsius. The PES points to the four renewable sources' installed capacity increasing about two times by 2030 to more than 5 terawatts and 3.5 times by 2050 to 8.8 terawatts from the 2018–2019 levels of 2.5 terawatts. The TES estimates the gain will be about 2.9 times by 2030 to 7.4 terawatts and 6.8 times by 2050 to more than 17 terawatts. IRENA expects renewable energy to make up 19 per cent and 36 per cent of total electric power generated by 2030 and 2050 respectively from just 10 per cent in 2018–2019 based on the PES, and for 35 per cent and 61 per cent respectively based on the TES.

Table 3.1: IRENA's Renewable Energy Forecasts.

	2015	2018–2019	Planned Energy Scenario		Transforming Energy Scenario	
			2030	2050	2030	2050
VRE Share in Generation	4.5%	10.0%	19.0%	36.0%	35.0%	61.0%
Hydro	1,099	1,189	1,356	1,626	1,444	1,822
Pumped Storage	112	121	200	300	225	325
Solar	222	582	2,037	4,474	3,227	8,828
Wind	416	624	1,455	2,434	2,526	6,044
Total (Gigawatts)	1,849	2,516	5,048	8,834	7,422	17,019

Source: IRENA (2020). Global Renewables Outlook: Energy transformation 2050. pp. 27, 29. [online] /publications/2020/Apr/Global-Renewables-Outlook-2020. Available at: https://www.irena.org/publications/2020/Apr/Global-Renewables-Outlook-2020 [Accessed 25 September 2020].

Another forecast also expects tremendous growth in renewable energy sources, though a little less bullish than the estimates from IRENA. In its reference or base case scenario (Table 3.2), the US Energy Information Administration (EIA) expects electric power output from geothermal, hydro, solar, wind, and other energy resources will rise 1.5 times by 2030 and 2.7 times by 2050 compared to the estimated 2020 level of 8,131 terawatt-hours. This would represent 40 per cent and 49 per cent of total electricity generated globally in 2030 and 2050, respectively.

Yet another estimate is even more bullish than IRENA's when it comes to renewables' share of global electricity production. In its *McKinsey Energy Insights' Global Energy Perspective* published in January 2019, the global consultancy forecasted (Table 3.3) that hydro, solar, and wind as well as biomass, geothermal, and marine energy would contribute as much as 51 per cent of the total by 2035 and 73 per cent by 2050, compared to 18 per cent in 2005 and 27 per cent estimated for 2020.

Oil giant BP presented three different 2050 scenarios for renewable energy sources in its 2020 Energy Outlook, a well-regarded annual analysis; BP includes biomass,

Table 3.2: World's Net Electricity Generation from Renewables.

	2010	2020	2030	2040	2050
Renewables Share in Generation	21%	31%	40%	46%	49%
Geothermal	65	81	182	219	236
Hydro	3,402	4,702	5,406	5,770	5,989
Solar	33	1,298	3,474	5,409	8,331
Wind	339	1,726	3,171	5,221	6,708
Other	347	323	283	301	392
Total (Terawatt-hours)	4,186	8,131	12,515	16,920	21,656

Source: U.S. Energy Information Administration (2019). International Energy Outlook 2018. [online] Eia.gov. Available at: https://www.eia.gov/outlooks/ieo/ [Accessed 26 September 2020].

Table 3.3: Global Power Generation and Renewables.

	2005	2020	2035	2050
Renewables Share in Generation	18%	27%	51%	73%
Total Generation ('000 Terawatt-hours)	18	27	36	49

Source: McKinsey (2019). Energy Insights by McKinsey: Global Energy Perspective 2019 – Reference Case. [online] mckinsey.com. McKinsey Solutions Sprl. Available at: https://www.mckinsey.com/~/media/McKinsey/Industries/Oil%20and%20Gas/Our%20Insights/Global%20Energy%20Perspective%202019/McKinsey-Energy-Insights-Global-Energy-Perspective-2019_Reference-Case-Summary.ashx [Accessed 17 Jan. 2021].

geothermal, solar, and wind in its renewables' forecasts. The company concluded that in the next 30 years, whatever the scenario used, the fastest growing type of energy generation resources will be solar and wind. The company estimated that renewables' share of total primary energy (including all types of energy not just electric power) will rise to 20 per cent, 45 per cent or 60 per cent by 2050 from 5 per cent in 2018 based on three scenarios: business-as-usual, rapid transition, and net zero. It underlined that in the first half of the outlook period – i.e., roughly through 2035 – the world would add 350 gigawatts of wind and 500 gigawatts of solar annually under the rapid transition and net zero scenarios. This compares to just 60 gigawatts per year since 2000.[3]

So, in essence both bearish and bullish scenarios all concur in terms of the growth of renewable energy sources, especially solar and wind, globally in the next 30 years. I now will try to capture briefly but comprehensively the individual situation in three key geographies in Asia, Australia, East Asia, and China. Their circumstances are slightly different albeit the differing evolutions are intimately intertwined.

3.1.1.1 Australia – Vast Energy Resources, a Curse in Disguise?

Australia is a vastly energy-rich nation, one of the richest in Asia. The energy mix shift in Australia actually has a number of dimensions and complexities – something that various Australian governments have struggled with in their policy setting. The dimensions include the massive amount of fossil fuel resources, the role of these commodities in domestic energy production, and their role in the economy, especially when it comes to the country's exports. In past decades that was a blessing but in today's decarbonisation-driven world, the blessing may turn out to be a curse. At the same time, the land down under is quickly learning the value of a little tapped form of energy, clean energy. It has vast amounts of geothermal, ocean (wave and tidal), solar, and wind energy resources.

Fossil fuels account for about 98 per cent of Australia's energy production, as of the 12 months through June 2018 (Figure 3.1). Crucially, black coal (also known as thermal coal or sub-bituminous, bituminous, and anthracite coal) accounted for almost 67 per cent and natural gas for more than 25 per cent. Production of the two fossil fuels has risen at a compound annual growth rate of 3.7 per cent and 6.8 per cent respectively chiefly on the back of exports to China and other countries.[4] Exports of the two commodities earn the country tens of billions of dollars annually. The exact amount will vary based on the fluctuating market prices; for example, in the 12 months for June 2021 the Australian government forecasted that gas exports will amount to A$38 billion ($26.7 billion), down about 25 per cent over the previous 12 months.[5]

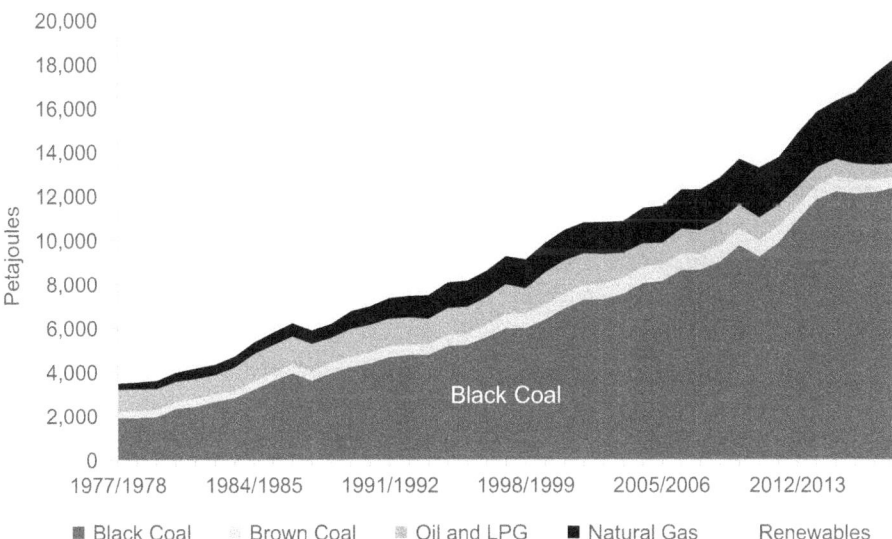

Figure 3.1: Australian Energy Production, by Fuel Type.
Source: Department of the Environment and Energy, Australian Government (2019). Australian Energy Update 2019 | energy.gov.au. [online] Energy.gov.au. Available at: https://www.energy.gov.au/publications/australian-energy-update-2019 [Accessed 26 September 2020].

Polluting fossil fuels also took up a large portion of electricity production. Data for the 12 months to June 2018 underline this. Coal and gas accounted for four out of five units of electricity generated in the country with coal responsible for 60.4 per cent of generation and natural gas about 20.6 per cent. Power from renewables and other energy sources were responsible for 17.1 per cent and 1.9 per cent, respectively.[6] As I will explain later, the percentage contribution from coal and gas is coming down but it is unrealistic to expect a big sudden drop. The decline will be gradual. And, in the meantime, emission will still be very significant. But first let us discuss briefly the existing controversy (as of September 2020) over Australia's version of Dash for Gas.

Today's heated debate within Australia over gas emanates from the current government's conviction that the production and export of natural gas will save the nation's recent erosion in commodities and resources exports, in other words a gas-led economic recovery. The discussion was particularly fired up when Prime Minister Scott Robinson gave an ultimatum to industry to build a one gigawatt gas facility to replace AGL Energy's Liddell coal-fired power plant station that the company will shut in 2023. If industry does not address this, government will intervene and facilitate the construction of a one gigawatt gas-fired power plant. In essence energy industry experts and think tanks all believe that intervention is a bad idea. They think that the nation should not be over relying on gas, which is set to become a progressively declining commodity. They believe gas prices cannot be pushed down to a competitive level. They argue that the potential one gigawatt gas station to replace Liddell would be under-utilised.[7] Finally, they worry over energy policy certainty, something crucial when making an infrastructure investment meant to operate for 20 or 30 years. In other words, government may be pro-gas now, but what happens if a future government changes its mind. Would the new one gigawatt gas then become out of favour?

I personally think that in the next 30 years the consumption of coal and gas will decline not just in Australia but in the whole region as well. The lower demand will severely cut Australia's coal and gas exports, or, at a minimum, drive down the price of the two commodities to levels where it would not make any economic sense to export them. This path is already evident today with thermal coal. While gas is less polluting, it is still harmful to the environment. Independent energy research and business intelligence firm Rystad Energy is one of the entities that concurs with my belief, which of course is by no means universal. A Rystad Energy analysis concluded that the combined output from wind and solar power facilities will overtake that of coal and gas by 2026 and that actually production as a percentage of the total already peaked in 2020.[8] I also agree with the arguments of Bruce Robertson. The Institute for Energy Economics and Financial Analysis LNG/Gas analyst stated that the use of natural gas for electric power generation within the National Electricity Market dropped 58 per cent between 2014 and 2020. He argues that the decline was chiefly driven by the fact that gas has not been a price-competitive fuel source. He emphasised that Australians pay a price for the fuel way above international prices. He believes that the country does not need to look for a new supply, such as exploiting new gas fields, as the use of the fuel has declined not just for power production but for industry as a whole. And now that the unit production price of energy from renewable energy sources has sunk, investments by energy market participants is flocking away from gas in favour of renewables.[9] It is worth noting that the price per megawatt-hour produced with standalone solar and wind (i.e., not including storage) is already cheaper than gas-fired generation as of 2020 and this price is set to fall even further in the next 30 years, based on a study by the Australian think tank Commonwealth Scientific and Industrial Research Organisation (CSIRO) together with AEMO.[10]

Now let us turn to the final area: Australia's shift to renewable energy and some potential new outlets for this clean energy. Almost all government agencies, think tanks, and consultancies expect that Australia will be adding massive amounts of renewable energy in the coming years. The big question is the actual amount and the pace. Unlike markets like China, South Korea, or Vietnam, the capacity additions will be made by private investors, and these investors will make the investment decision based on the potential economic returns. As mentioned a few times throughout, policy clarity and consistency are absolutely critical to energy infrastructure planning. And not just the investment but the financing as well. The nation has made some commitments to the world as a whole to cut carbon emission so we know that 'something' will be encouraged. But, how certain can investors be that a current or future government will not change its mind? The discussion today is for potentially more gas. Lots more. This, in my opinion and that of others is unlikely to happen. Still, where is the certainty. In my opinion, Australian authorities have one of the worse track record on this front in the region. Still, the change will happen because fossil fuels are polluting and will increasingly become more expensive relative to the average price of renewable energies including solar and wind. And, with the price and reliability of energy storage also expected to sharply fall in coming years, the renewable energy supply will actually be dispatchable and reliable.

In its latest suggested proposal for the future Australian electric power network, the 2020 Integrated System Plan, AEMO offered five scenarios (Figure 3.2). The Slow Change scenario is the most bearish scenario renewables' cost, consumer and politics wise. Then there is the Central scenario and three higher case ones: Fast Change, Step Change, and High Distributed Energy Resources scenarios.[11] I think we can safely assume that the Central scenario-based forecasts are more of the low base given renewables' increasing cost advantage.

In its Central Case scenario, AEMO sees black coal, brown coal, CCGT, and peaking gas plus liquids falling to 15 gigawatts by June 2042 from an estimated 34.3 gigawatts by June 2022 (Figure 3.3). At the same time, distributed solar, utility scale solar, wind, as well as dispatchable storage and behind-the-meter storage will rise to 85.9 gigawatts from 27 gigawatts; as a reminder fossil fuel plants per kilowatt dispatch more output so the kilowatt-hour increase will be lower. Hydro will only see a small increase to 7.2 gigawatts from 6.9 gigawatts given that the majority of hydro resources have already been mostly fully exploited.[12] The green (including hydro) versus brown (all coal plus gas) ratio would thus go to 86:14 from 50:50, based on my calculations from the data.

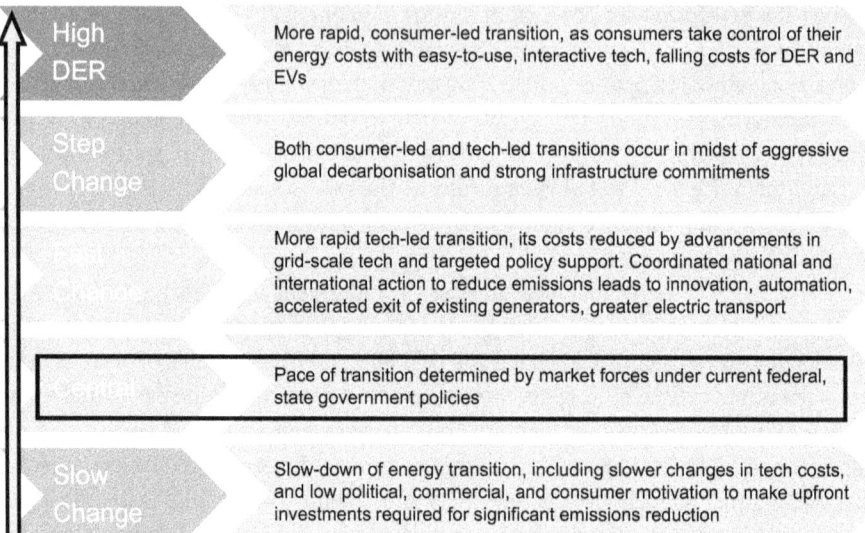

Figure 3.2: AEMO's Five Capacity Forecast Scenarios Definitions.
Source: Australian Energy Market Operator (AEMO) (2020). 2020 Integrated System Plan (ISP): For the National Electricity Market. [online] aemo.com.au. Available at: https://aemo.com.au/energy-systems/major-publications/integrated-system-plan-isp/2020-integrated-system-plan-isp [Accessed 27 September 2020].

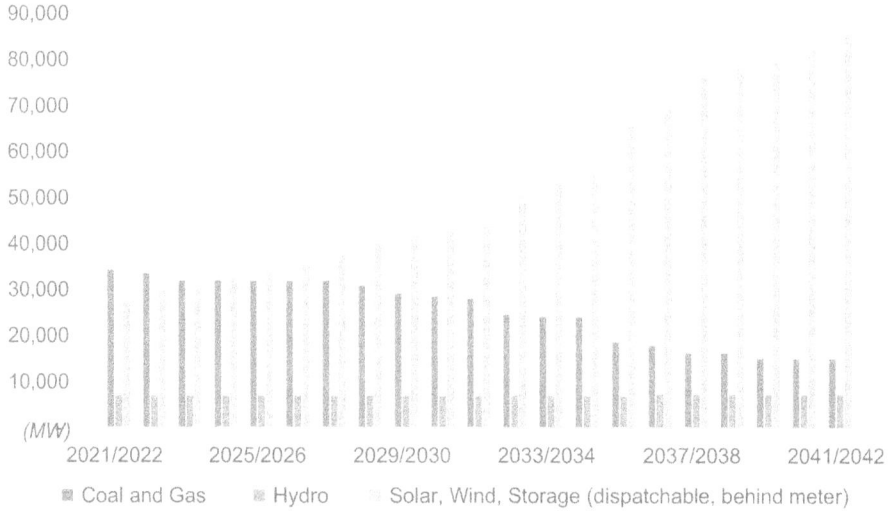

Figure 3.3: AEMO's Australia Central Case Scenario Capacity Forecast.
Source: Australian Energy Market Operator (AEMO) (2020). 2020 Integrated System Plan (ISP): For the National Electricity Market. [online] aemo.com.au. Available at: https://aemo.com.au/energy-systems/major-publications/integrated-system-plan-isp/2020-integrated-system-plan-isp [Accessed 27 September 2020].

The upside to the amount of clean energy, including storage, which the nation can add is formidable based on AEMO's forecasts (Figure 3.4). The amount could rise from 24.8–34.3 gigawatts by June 2022 to 46.4–123.5 gigawatts by June 2042 depending on the scenario.[13] This translated into an average annual addition of 1,082–4,616 megawatts or a compound annual growth rate of between 3.2 and 7.1 per cent. There are an enormous number of variables, especially the political variables, but if I were to best guess I would say that the higher end of the forecasts is more likely, chiefly for cost reasons.

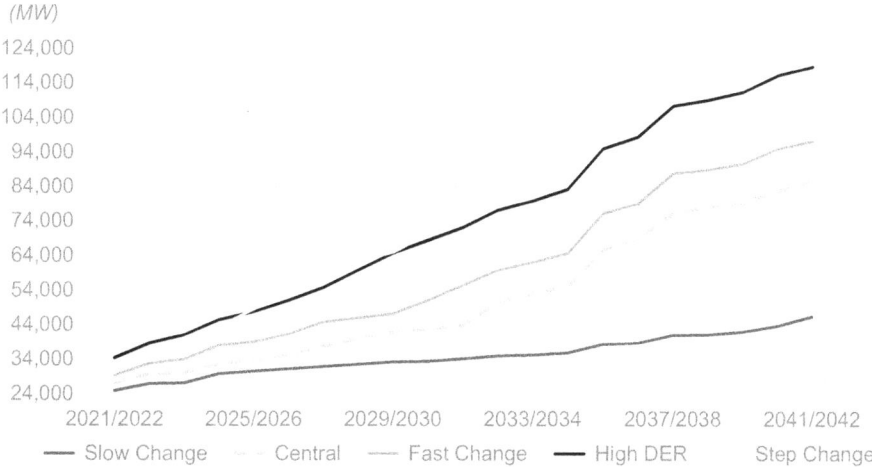

Figure 3.4: AEMO's Australia Renewable Energy Capacity Forecast.
Source: Australian Energy Market Operator (AEMO) (2020). 2020 Integrated System Plan (ISP): For the National Electricity Market. [online] aemo.com.au. Available at: https://aemo.com.au/energy-systems/major-publications/integrated-system-plan-isp/2020-integrated-system-plan-isp [Accessed 27 September 2020].

Another dimension of the country's vast clean energy potential is the value the country can create by overbuilding (yes, overbuilding) clean energy capacity and then directly or indirectly exporting the output or by-products. This may sound far-fetched but there actually are two real projects currently being planned that clearly demonstrate the immense potential. One is the Asian Renewable Energy Hub project and the other the Sun Cable project. The projects combined capital expenditures is currently expected to be about $41 billion and they could only be ready to go to market by the second half of the 2020s. Both are trailblazer projects, are at a relatively early stage, and face many hurdles. Still, they do offer two potential future energy-export-related money earners for Australia.

The Asian Renewable Energy Hub will take ten years and approximately A$36 billion ($25.2 billion) to develop. It will involve building massive wind and solar facilities of about 26 gigawatts in the East Pilbara region of Western Australia. It has already received permission for the first 15 gigawatts. About 3,000 megawatts would go towards consumption by industry in the region, potentially including mining and mineral processing. The rest could go towards various uses including the production of green hydrogen products, which in turn could be exported to Japan and other countries. The Asian Renewable Energy Hub had received Major Project Status from both the Australian federal government as well as that of the Western Australia state government, as of October 2020. If it goes ahead, construction may start in 2026 with the first exports of green hydrogen ready by 2027 or 2028.[14]

The Sun Cable project, called the Australian–ASEAN Power Link, will take about seven years and will also cost about A$22 billion ($15.4 billion) to develop. It aims at building a 10 gigawatt solar farm with energy storage near Tennant Creek in the New Territories. The output would be used domestically, and sold to Singapore via a 4,500 kilometres subsea cable, and possibly also to Indonesia. As of September 2020, the Australian–ASEAN Power Link had received Major Project Status by the Australian government.[15] Should either or both come to fruition they would generate hundreds of jobs in the country as well as revenue for the government. Importantly, it could trigger a few more similar ones and make Australia into a major direct and indirect clean energy exporter.

3.1.1.2 East Asia – Road to Less Energy Dependence

The three east Asia electricity markets of Japan, South Korea, and Taiwan have at least three key features in common. All three have limited indigenous conventional energy resources and heavily rely on imported energy commodities to power their economies, exposing them to supply disruptions. All three have to address the reduction in nuclear power generation consumption, which has been an important source of domestic energy. All three have to, and want to, add massive amounts of renewable energy assets in the coming years, although they face, to different extents, some challenges in terms of limited land space or suitable areas. Still, the lower and stable production cost and indigenous nature of renewables should mean that ways will be found to add tens of gigawatts of clean energy in the next three decades.

The three markets' primary energy consumption is heavily reliant on imported energy commodities (Table 3.4). The share of oil accounted for between 40 and 43 per cent of the primary energy consumed, natural gas (mostly liquified natural gas) was 16 to 21 per cent, and coal was 26 to 34 per cent in 2019. So, the three polluting fossil fuels accounted for about 87 to 92 per cent of the total. This exposes the three to supply disruptions or price volatility caused by geopolitical or weather-related factors and obviously makes the decarbonisation path all the harder. Looking just at oil, their reliance is almost shocking. I added together the amount the three imported in 2019, which was about 7,570 thousand barrels per day, and on this combined basis, they would rank as the third largest importer of oil in the world after the US (19,400 thousand barrels per day), China (14,056), with India then ranking fourth with 5,271 thousand barrels per day.[16] In terms of liquified natural gas the three were responsible for 38 per cent of the world's total, well ahead of the second largest importer, China, at 21.7 per cent. The more energy they can source and produce domestically the better for Japan, South Korea, and Taiwan's import costs and decarbonisation.

Table 3.4: Japan, South Korea, Taiwan Energy Mix.

(2019)	Exajoules			Percentage		
	Japan	South Korea	Taiwan	Japan	South Korea	Taiwan
Oil	7.53	5.30	1.93	40.3%	42.9%	40.2%
Natural Gas	3.89	2.01	0.84	20.8%	16.3%	17.4%
Coal	4.91	3.44	1.63	26.3%	27.8%	33.9%
Nuclear	0.59	1.30	0.29	3.1%	10.5%	6.0%
Hydro	0.66	0.02	0.05	3.5%	0.2%	1.0%
Renewables	1.10	0.29	0.07	5.9%	2.3%	1.5%
Total	18.67	12.37	4.81	100.0%	100.0%	100.0%
Fossil-Fuels	16.33	10.76	4.40	87.4%	86.9%	91.5%
Nuclear	0.59	1.30	0.29	3.1%	10.5%	6.0%
Clean Energy	1.76	0.31	0.12	9.4%	2.5%	2.5%
Total	18.67	12.37	4.81	100.0%	100.0%	100.0%

Source: bp p.l.c. (2020). Statistical Review of World Energy. [online] BP Global. Available at: https://www.bp.com/en/global/corporate/energy-economics/statistical-review-of-world-energy.html [Accessed 29 September 2020].

One obvious strategy to lower the reliance on imported fossil fuels is to use nuclear power generation. The Japan, South Korea, and Taiwan energy markets made many efforts in the past three or more decades to build up their nuclear power capacity (Figure 3.5). At the peak nuclear energy contribution to the total energy mix was in the high double digits. For different reasons this percentage has now declined and is mostly set to fall further. In Japan, the decline was due to a reaction to the Fukushima Daiichi Nuclear Power Plant meltdown following the Tohoku earthquake and tsunami. All nuclear reactors were stopped and their restarts – impacted by new, severe safety standards – has been much slower than the government would have hoped, and some reactors are unlikely to restart at all. Also, new construction approvals are near impossible given social-political pressures. In Taiwan, the antinuclear current ruling party lobbied and then legislated for the retirement of the existing three facilities and for preventing the start up a completed fourth one. It launched a campaign titled '2025 Nuclear Free Homeland project', and even included the closure of existing nuclear reactors in its amendments to Taiwan's Electricity Act, which regulates the sector. In South Korea, ambitions for nuclear additions were high but the current ruling party has reversed the clock on nuclear ambitions in favour of promoting clean and renewable energy. Based on current plans, new reactors additions will peak in 2024 but the total number of reactors will fall by more than a third a decade later. So, a sharp nuclear uptake to offset fossil fuel imports is highly unlikely in all three energy markets. Actually, the opposite is true. For example, Taiwan will progressively head for no nuclear contribution to the primary energy mix.

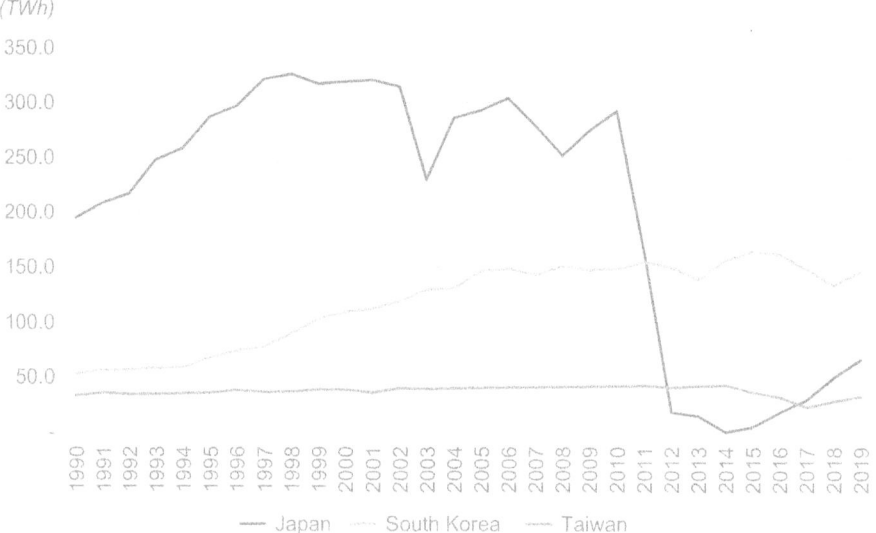

Figure 3.5: Japan, South Korea, Taiwan Nuclear Power Production.
Source: bp p.l.c. (2020). Statistical Review of World Energy. [online] BP Global. Available at: https://www.bp.com/en/global/corporate/energy-economics/statistical-review-of-world-energy.html [Accessed 29 September 2020].

The planning authorities in Japan, South Korea, and Taiwan are all looking at sharply raising electric power capacity from renewable energy assets in the coming years. Japan could add as much as 114 gigawatts by March 2031, South Korea at least 68 gigawatts by December 2030, and Taiwan probably more than 36 gigawatts by the end of 2035.

Japan should double or triple its solar and wind installed capacity by 2030, according to the Renewable Energy Institute, a Japanese independent think tank. It looked at the 2030 potential based on two scenarios: one using current policies and the other assuming that some policies will be changed in favour of a faster decarbonisation (Table 3.5). The base case scenario sees solar and wind rising to more than 125 gigawatts by the fiscal year ending March 2031 from 60 gigawatts as of March 2020. In the high case scenario, the total reaches almost 174 gigawatts.[17] It is highly likely that the amount for March 2031 would be higher than the base case and would

Table 3.5: Japan's Solar and Wind Capacity Forecast by the Renewable Energy Institute.

Year to 31March	2020 Actual (GW)	2031 Base Case1 GW	Change vs. 2020	2031 High Case2(GW)	Change vs. 2020
Wind Total	4.4	23.3	430%	29.3	566%
Onshore	4.3	16.6	286%	19.2	347%
Bottom-mounted	–	6.7	n/a	10.0	n/a
Floating	–	0.1	n/a	0.1	n/a
Solar Total	55.6	102.1	84%	144.6	160%
Residential (roof-top)	11.3	20.3	80%	25.8	128%
Industrial (roof-top)	n/a	15.7	n/a	36.1	n/a
Commercial (ground)	44.3	66.1	49%	82.8	n/a
Solar and Wind	60.0	125.4	109%	173.9	190%

(1) Current policy scenario; (2) Transition promotion scenario

Source: Renewable Energy Institute (2020). Proposal for 2030 Energy Mix in Japan. [online] Renewable Energy Institute, Tokyo, Japan: Renewable Energy Institute, pp. 7–8,11. Available at: https://www.renewable-ei.org/pdfdownload/activities/REI_Summary_2030Proposal_EN.pdf [Accessed 30 September 2020].

potentially even surpass that of the higher case if developers manage to reduce offshore wind construction costs more than expected – something not impossible given the European experience.

South Korea is set to raise its renewable energy capacity several folds. The actual target was not out as of September 2020 as planners were in the midst of fine tuning the nations' Ninth Basic Plan for Long-term Electricity Supply and Demand; every few years the government produces and updates demand and supply for the country, taking into consideration macroeconomic and other changes. Based on the published Eight Basic Plan 173 gigawatts in capacity was targeted for 2030 from about 125 gigawatts as at the end of 2019 (Table 3.6). The plan details additions and closures of coal, gas, and nuclear facilities. Crunching the numbers, I concluded that South Korea would have almost 68 gigawatts in renewables by 2030 from about 22 gigawatts in 2019. Numbers have also been published in 2020 regarding the renewables contribution to the overall electricity mix for 2034, namely 40 per cent from a current base of approximately 15.1 per cent, a target taking over the one for 2030 of 33.1 per cent. Similarly, to the Japanese and Taiwanese power markets, South Korea is highly likely to sharply expand offshore wind power.

Taiwan targets to raise solar and wind capacity to almost 27 gigawatts by 2025 from about 8 gigawatts it estimates it will have by the end of 2020 (Table 3.7). The sharpest addition is 13.5 gigawatts in solar generation, an amazing feat for the small island. It

Table 3.6: South Korea Capacity Addition Plan.

(Gigawatts)		2019	2020	2021	2022	2023	2024	2020–2030	Total	2030
KEPCO Group	Nuclear	23.25	1.4	1.4	n/a	0.8	0.5	(6.9)	(2.9)	20.4
	Coal	34.3	(2.1)	0.5	(1.0)	n/a	(1.1)	(1.0)	(4.7)	29.6
	LNG	16.5	0.2	n/a	n/a	(1.8)	n/a	1.1	(0.5)	16.0
	Oil	2.9	n/a	n/a	(1.2)	n/a	n/a	n/a	(1.2)	1.7
	Renewable	6.7	0.1	0.4	0.5	n/a	n/a	n/a	1.0	63.2
	Sub-total	83.7	(0.4)	2.3	(1.7)	(1.1)	(0.7)	(6.8)	(8.3)	67.7
IPP	Thermal	26.7	2.1	3.1	4.1	–	0.9	5.1	15.4	42.0
	Renewable	15.0	–	–	–	–	–	–	–	–
	Total	125.3	1.7	5.4	2.4	(1.1)	0.3	(1.7)	7.1	173.0

Note: Renewable energy capacity target under planning. Current capacity target is 173 GW so 67.72 GW in renewables assumed. Source: Korea Electric Power Corporation, '2019 fiscal year annual results presentation', March 2020.

Table 3.7: Taiwan'S Government Solar and Wind Capacity Forecast.

Year to 31 December	2020 Target (MW)	2025	Change vs. 2020	2035	Change vs. 2020
Wind Total	1,790	6,938	288%	n/a	n/a
Onshore	814	1,200	47%	n/a	n/a
Floating	976	5,738	488%	15,000	1,437%
Solar Total	6,500	20,000	208%	n/a	n/a
Solar and Wind	8,290	26,938	225%	n/a	n/a

Source: Taiwan Power Company (2020). Taiwan Power Company 2019 Sustainability Report. [online] https://csr.taipower.com.tw/en/index.aspx, Taiwan Power Company, p. 59. Available at: https://csr.taipower.com.tw/upload/132/2019110109130980581.pdf [Accessed 30 September 2020]; Global Wind Energy Council (2020). From 0 to 15GW by 2030: Four Reasons Why Taiwan is the Offshore Wind Market in Asia. [online] Global Wind Energy Council. Available at: https://gwec.net/from-0-to-15gw-by-2030-four-reasons-why-taiwan-is-the-offshore-wind-market-in-asia/ [Accessed 29 September 2020].

will also exploit its offshore wind resources by adding almost 5 gigawatts. There is a plan also to build more offshore wind between 2026 and 2035, more than 9 gigawatts. Assuming no new onshore wind and solar addition through 2035, its solar power together with wind power installed capacity will rise 4.4 times to 36.2 GW.[18] Significantly, overshooting these targets may prove difficult given topological limitations.

Larger renewables capacity numbers are possible but not as easily surpassed as in some other countries such as Australia or China. These three electric power markets face a unique challenge; they do not have easily accessible and abundant wind and solar resources like Australia, which I discussed in the section above. They face challenges in terms of limited land space or suitable areas, which also impacts development costs; building a solar array on unoccupied barren land will cost much less than trying to build it up a difficult-to-reach mountain peak. If some clean energy technologies, such as marine energy and energy storage, are able to reduce their cost per kilowatt in the next few years, they may prove particularly beneficial for these markets.

3.1.1.3 China – Dash to Clean Energy, Pressures from Pollution and Emissions

China has been the undisputed global leader in clean energy additions in recent years. The nation's recent new decarbonisation commitment is one of the many signs that it will retain the leadership. A key driver, in my opinion, has been its pollution problem. This has proven not just a serious issue to address but a gigantic one. The environmental near-disaster has damaged the health of the population as well as the economy. Thankfully, the central government has recognised this and has been implementing a variety of strategies for a while. It has been trying to aggressively tackle the very roots of the problem, especially pollution from electricity production, chiefly from thermal-coal power generation plants. While the successful execution of policies is the most challenging element of resolving the equation, the country has already built a strong track record. I think that there are several elements that will allow the reader to appreciate the Chinese clean energy circumstances and its prospects. The country renewed – or updated – its commitments in September 2020 with an announcement that surprised almost all observers, a commitment to net zero carbon emissions on or before 2060. Thereafter, I will look at two elements. After discussing the emissions issues and the nation's commitments I will focus on China's formidable feat in shifting the energy mix in favour of low to zero carbon generation from coal power. The growth has been impressive and is set to continue, if not accelerate, in the next three decades.

China renewed its commitment to cut emissions on 23 September 2020. These were made during an online speech at the 75th session of the United Nations General Assembly by the nation's president, Xi Jinping. President Xi said that the COVID-19 pandemic was a good reminder for all humankind to adopt a greener way of life, including investing in conservation.[19] He then went on to highlight China's own new sustainable commitments and said:

> The Paris Agreement on climate change charts the course for the world to transition to green and low-carbon development. It outlines the minimum steps to be taken to protect the Earth, our shared homeland, and all countries must take decisive steps to honour this Agreement. China will scale up its Intended Nationally Determined Contributions by adopting more vigorous policies and measures. We aim to have CO_2 emissions peak before 2030 and achieve carbon neutrality before 2060.[20]

As mentioned earlier, the Intended NDCs are the nation's declared climate actions to slash emissions and address climate change of a country to achieve the Paris Agreement's goals.

There have been many debates as to whether authorities had done enough, and these debates continue to this day. Many of the debates unfortunately generated politically fuelled and highly irrational statements. One of the many opinions argued that China should simply just shut down all of its coal-fired generation. The bulk of these arguments are unrealistic and never made much sense to me. China cannot just stop producing the bulk of its electricity given that this would be impossible economically and socially. The irreversible damage caused to the economy would most likely put all of the people who were lifted out of poverty in the last three decades back below the poverty line. In my opinion, China could have done better in the past decade, but it did try its best, and it did build a unique and significant track record in emission reductions in its energy sector.

Let us now look at this major motivator, environmental pollution, to try to better understand that the policy of driving the shift in the energy mix away from polluting thermal coal was not a short-term knee-jerk reaction but rather a long-term, convincingly sustainable policy. As an aside note, albeit an important one, researchers published an estimate in a well-respected journal that concluded that in 2017 alone air pollution in China killed about 1.24 million people (range of 1.08 to 1.40 million), deaths that included about 0.85 million people (range of 0.71 to 0.99 million) from ambient $PM_{2.5}$ pollution.[21]

So, exactly how bad were emissions in the Middle Kingdom and how bad are they now? The brief answer is that they were horrendous. In the 1990s and the 2000s, I was travelling throughout China to visit companies and facilities for research. But, during the periods of high pollution, these trips turned out to be extremely tough. If one was unfortunate to have a cold while travelling around the country, one was guaranteed to get a cough that would last many weeks. But apart from my personal experiences let us look at some actual data compiled by highly regarded expert institutions and organisations. Basing some calculations on the one of the more reliable databases run by the European Commission (Figure 3.6),[22] I derived that between 1990 and 2018 the world's total carbon dioxide (CO_2) emissions rose 1.7 times while China's increased 4.7-fold. The sharp difference is even more highlighted when looking at the compound annual growth rate. For the world and for China between 1990 and 2018 it was 1.9 per cent and 5.7 per cent, respectively. For the 1990 to 2000 period both the world's and China's rate were 4.4 per cent but for the 2001–2018 period the world's total rate was 607 basis points lower than China's, specifically 2.9 per cent versus 9 per cent. The same data set indicates that China's share of the world's total CO_2 emissions reached almost 30 per cent in 2018, compared to 27 per cent in 2010, 14 per cent in 2000, and just under 11 per cent in 1990.

A similarly reliable website, that of the World Bank, using data from the US's Carbon Dioxide Information Analysis Centre of the Environmental Sciences Division

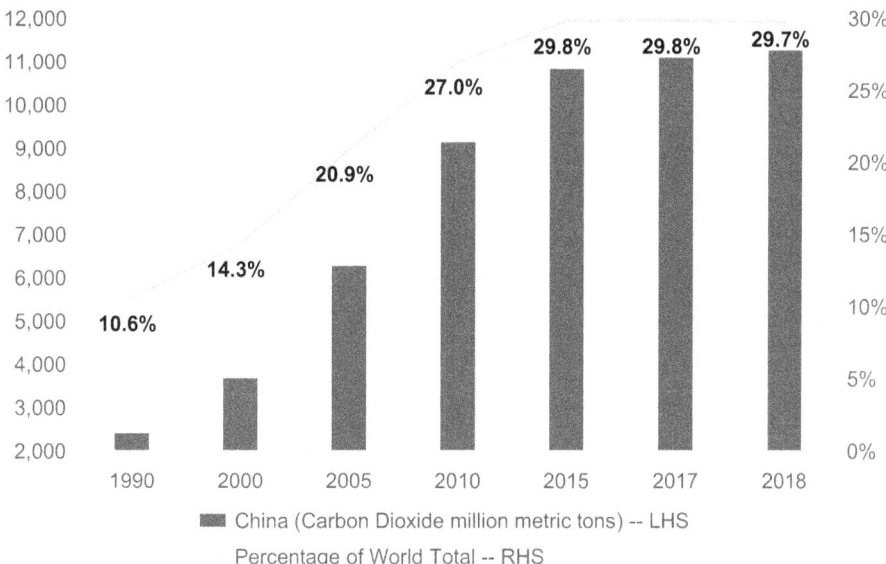

Figure 3.6: China's CO_2 Total Emissions.
Source: Crippa, M., Oreggioni, G., Guizzardi, D., Muntean, M., Schaaf, E., Lo Vullo, E., Solazzo, E., Monforti-Ferrario, F., Olivier, J.G.J., and Vignati, E. (2019). Fossil CO_2 and GHG Emissions of all World Countries – 2019 Report, EUR 29849 EN, Publications Office of the European Union, Luxembourg, ISBN 978-92-76-11100-9, doi:10.2760/687800, JRC117610. Available at: https://edgar.jrc.ec.europa.eu/overview.php?v=booklet2019 [Accessed 6 October 2020].

of the Oak Ridge National Laboratory (Figure 3.7),[23] provides the same conclusions, although the time series of the data is longer than that of the European Commission. It highlights not only the sharp surge from the 2000s but also confirms that the increase has slowed down.

Another well regarded source, the Climate Action Tracker,[24] which is operated by groups of scientists, produced data showing that CO_2 emissions in China rose 47 per cent in the ten years through 1999 and 96 per cent in the ten years through 2009, but slowed in the 2010s with an increase of 26 per cent in the eight years through 2018 to 13,442 metric tons of CO_2 equivalent. Evaluating reported data by various international media, journals, and think tanks I concluded that emissions may have risen about 2 per cent in 2019 compared to a year earlier and that the change may be plus or minus 1 per cent in 2020. In short, they have sharply slowed. The Climate Action Tracker as well as other organisations and experts believe that not only will China meet its CO_2 emissions commitments for 2020 but also that they will peak sometime in the 2020s, perhaps as early as 2025. China 2020 commitments were made when it signed up to the 2009 Copenhagen Accord at COP 15 and signed up to the 2016 Paris Agreement at COP 21 (Figure 3.8); COP refers to the

China carbon dioxide emissions in kilotons

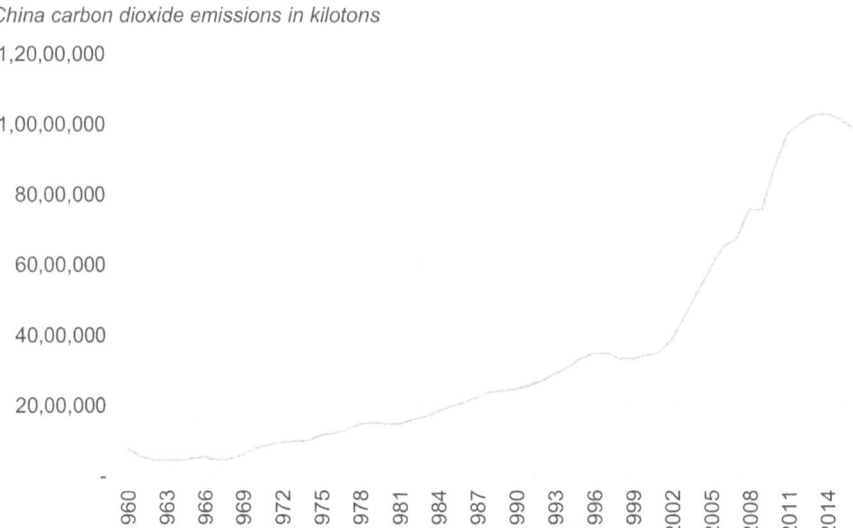

Figure 3.7: China's CO$_2$ Emissions (Kilotons).
Source: Carbon Dioxide Information Analysis Center, Environmental Sciences Division, Oak Ridge National Laboratory, Tennessee, United States (2020). CO2 emissions (kt) – China | Data. [online] data.worldbank.org. Available at: https://data.worldbank.org/indicator/EN.ATM.CO2E.KT?end= 2016&locations=CN&most_recent_value_desc=true&start=1960 [Accessed 6 October 2020].

	Copenhagen Agreement (2009)		Paris Agreement (2016)
2020 target(s)	Carbon Intensity: minus 40-50% below 2005 by 2020 Non-fossil share of energy supply: 15% in 2020 Forest Cover: +40 million ha by 2020 compared to 2005 Forest Stock: +1.3 billion m³ by 2020 compared to 2005 [26% above 2010 by 2030 excluding LULUCF for non-fossil target] [26-37% above 2010 by 2030 excluding LULUCF for carbon intensity targets]	2030 unconditional target(s)	Peak CO2 emissions latest 2030 Carbon Intensity: minus 60-65% below 2005 by 2030 Non-fossil share of energy supply: 20% in 2030 Forest Stock: +4.5 billion m³ by 2030 compared to 2005 [33-47% above 2010 by 2030 excluding LULUCF for peaking and non-fossil target] [36-37% above 2010 by 2030 excluding LULUCF for carbon intensity targets]
Condition(s)	None	Coverage Land Use, Land-Use Change & Forestry (LULUCF)	Economy-wide Unclear how LULUCF is included

Figure 3.8: China's Copenhagen and Paris Environmental Commitments.
Source: Author, October 2020. Data source: The Climate Action Tracker (2020). China | Climate Action Tracker. [online] Climateactiontracker.org. Available at: https://climateactiontracker.org/countries/china/ [Accessed 6 October 2020].

Conference of Parties (COP) to the United Nations Framework Convention on Climate Change (UNFCCC).

Globally, the production and the consumption of energy is the most significant generator of greenhouse gas emissions, accounting for approximately 73 per cent of the total. The sector chiefly comprises transportation, electricity and heat, buildings, manufacturing and construction.[25] It accounted for 82.8 per cent (or 9,848 million tons of CO_2 equivalent) of China's total in 2016 (Figure 3.9). Industrial processes were responsible for 9.4 per cent (1,122 million tons), agriculture for 6.1 per cent (731 million tons), and waste for 1.6 per cent (186 million tons).[26] This is the reason why the nation has been aggressively constructing non-fossil fuel generation facilities, especially nuclear, solar, and wind power plants in addition to continuing to build up hydroelectric installed capacity, albeit this resource is increasingly scarce as its exploitation had been taking place for several decades.

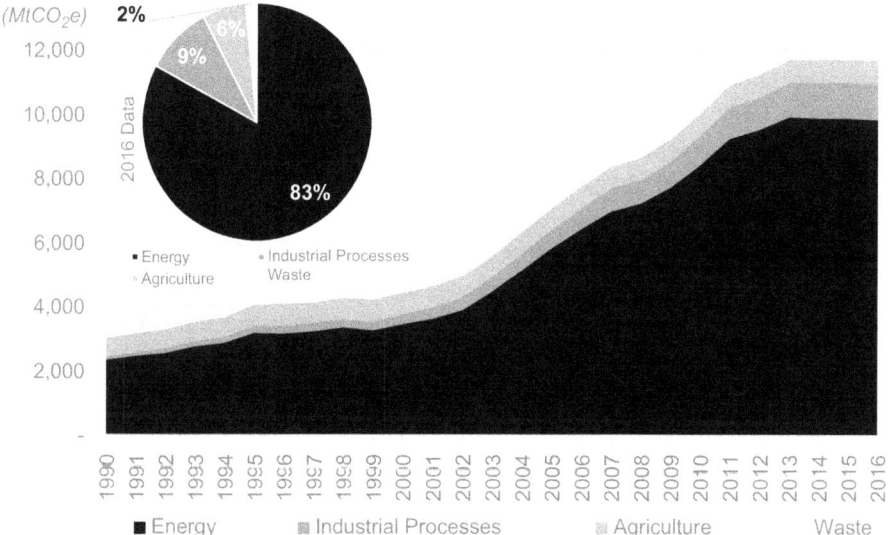

Figure 3.9: China's Historical Greenhouse Gases Emissions.
Source: World Resources Institute (2019). CAIT Climate Data Explorer. [online] Wri.org. Available at: https://www.wri.org/blog/2020/02/greenhouse-gas-emissions-by-country-sector [Accessed 7 October 2020].

The Chinese leadership concluded at least a decade ago that producing more than four in five units of electricity from highly polluting coal had been creating irreversible damage to the environment, people's health, and the economy in general. The government gradually constructed and executed over a period of many years dozens of different policies around the reduction of the contribution of the fuel in its energy mix. It did so successfully for the most part, in my view.

The changes in the energy mix that the Chinese electric power generation sector has seen in the past 10 to 20 years appear to be quite formidable to people familiar with the energy industry. I addressed a couple of times already the fact that changes to a nation's energy mix seldom happen rapidly. By rapidly, here I mean about ten years. So, the shift in China is incredible. Towards the end of the 2000s, the Middle Kingdom hardly had any renewable energy output. National electric power statistics did not even mention non-fossil fuel generation amounts, apart from hydro and nuclear power, on a regular basis, because it was too small. The whole primary energy supply was overly dominated by thermal power, which almost totally comprised coal-fired generation. This is because the nation had, and still has, vast amounts of cheap coal. Power generation from the fuel was the cheapest form of producing electricity apart from electricity from hydro-electric plants, itself a more limited source of generation. Limited in terms of how much new capacity can be added given that the bulk of cheap accessible sites have largely become exhausted; China did manage to raise hydro capacity to 356 gigawatts in 2019 from 52 gigawatts in 1995 but in the past three years the rate of growth has been decreasing, and the trend is unlikely to change in the future. Limited because of a lack in the flexibility of where a new hydro plant can actually be built. And limited because rainfall is not guaranteed, and a lack of rain causes some shortages in hydropower.

Going back to coal, the pollution generated by the fuel was a necessary evil. As a developing nation, it faced the tough choice of either having sufficient electricity to grow the economy or not. Using alternatives other than coal also posed a major issue. Apart from hydropower, alternative sources of electricity generation were either too expensive for the economy and the population to absorb, such as solar and wind power, or the technology was not sufficiently mature, as in the case of nuclear power. As the Chinese economy expanded massively in the 2000s and 2010s, it needed abundant and cheap power sources. During these two decades, on average the manufacturing industry took up between two-thirds and three-quarters of demand. Oil-fired generation (as well as solar and wind until recently) was prohibitively expensive – at least one to two-folds more expensive than coal generation. Also, because China added so many new coal plants in the past couple of decades, their construction became highly effective, efficient, and cost competitive. I think most engineering experts would concur when I say that China has been building coal plants cheaper, faster, and better than anyone else on earth.

Between January 2004 and November 2020, the nation added an average of about 99 gigawatts per year. This is roughly equivalent to the total installed capacity of such nations as South Korea, Spain, and the UK, and about a third that of India and Japan's. The additions included annual additions in thermal power capacity of about 52 gigawatts on average; thermal power comprises coal-fired power generation but also includes oil-fired, gas-fired, waste heat, residual pressure and residual gas, waste incineration and biomass power generation (Figure 3.10).[27] Importantly, since

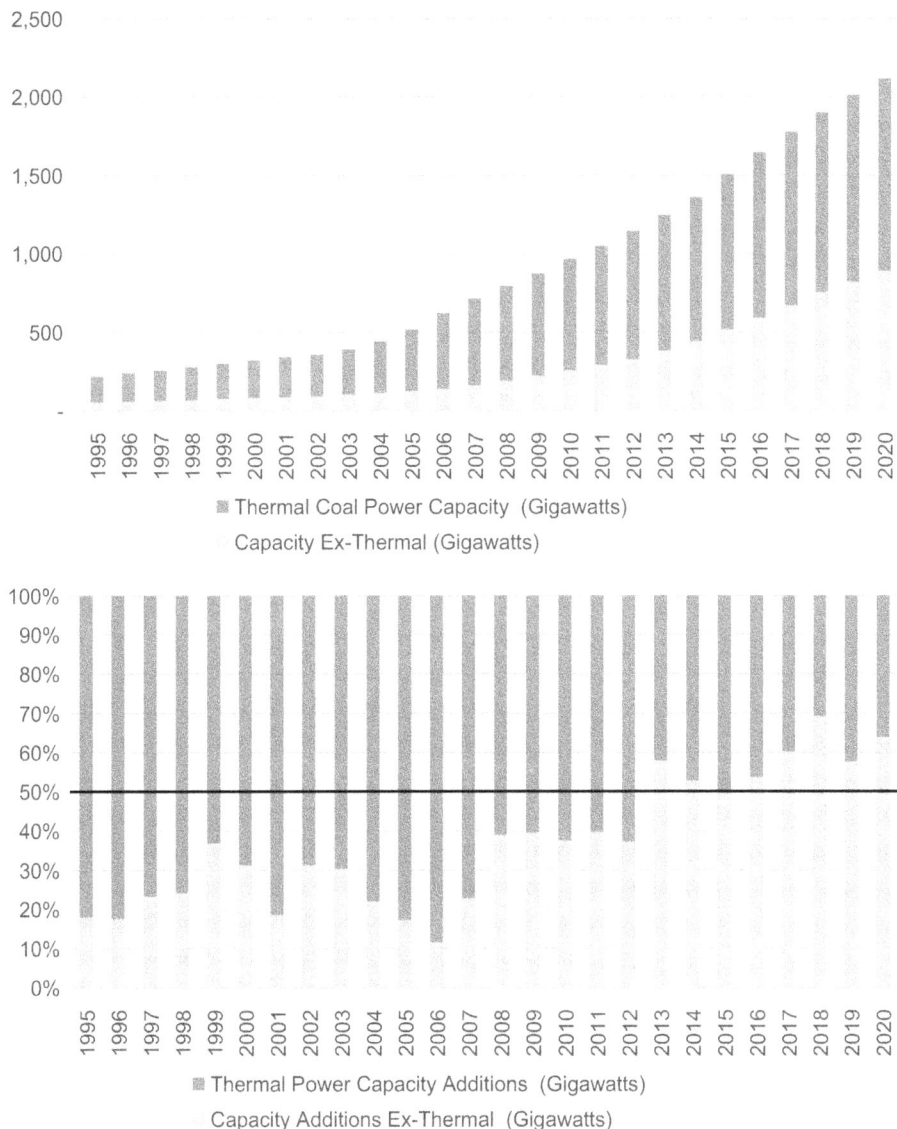

Figure 3.10: China Installed Capacity Additions (January 1995 to November 2020).
Source: Author, October 2020. Data source: National Bureau of Statistics (1994–2020). National Bureau of Statistics. [online] Stats.gov.cn. Available at: http://www.stats.gov.cn/ [Accessed 8 October 2020] (Chinese); China Electricity Council (1994–2020). China Electricity Council. [online] www.cec.org.cn. Available at: https://www.cec.org.cn/ [Accessed 8 October 2020]; National Energy Administration (2008–2019). National Energy Administration. [online] www.nea.gov.cn. Available at: http://www.nea.gov.cn/ [Accessed 8 October 2020] (Chinese).

about 2013, the electricity sector has been adding more non-thermal power plants than thermal ones, except for 2015 when the ratio was about 50:50.

The leadership finally realised sometime in the late 2000s that the price that the country and its people was paying – emissions – was just too high. Arguably, planners also realised that promoting local content would also translate into the creation of new industries and jobs. It managed to achieve that in the solar and wind industries, which now boast world-class manufacturers. Nuclear energy still has some ways to go as the localisation process takes much longer, but I am highly confident that Chinese manufacturers will also become a dominant force in this area as well. So, it embarked in promoting zero-carbon power plants, especially nuclear, solar and wind.

China started to get its head around the technology of the three non-fossil fuel sources in the 2000s and by the second half of the decade decided to sharply increase the individual capacities of the three. The clean energy boom was led with a sharp increase in wind capacity in 2006. The rise was followed by solar power in 2009, and then nuclear in 2014. I will look briefly at the development track record of all three technologies and then turn to their growth prospects over the next five years (the fourteenth Five-Year Plan) through 2025 and also the next 30 years. But first, a small footnote about generation capacities of different technologies, in general and in China in particular.

One kilowatt operating 24 hours a day and 365 days a year would have a maximum output of 8,760 kilowatts hours. The actual yields a plant can achieve differs depending on many factors. Some are the technology, the energy demand, the location, and maintenance requirements among the many factors. Nuclear generation capacity's utilisation rate was about 80 per cent and 84 per cent in the five years through December 2020 in China, I calculated (Figure 3.11); for 2020 I annualised the rates for the first eight months of the year. For thermal facilities, which include gas-fired generation as mentioned earlier, the rate was 47 per cent to 50 per cent. Hydro's was 41 per cent to 43 per cent, wind energy's was 20 per cent per cent to 24 per cent, and solar power's was 13 per cent per cent to 15 per cent. This in practice means that when the total capacity of a region or nation is increased by 10 per cent it does not mean that the actual capacity is necessarily increased by 10 per cent. It will, first, depend on the technology added. In China, based on the 2019 data, nuclear energy produced 1.7 times more electricity than thermal plants, 2 times more than hydro plants, 3.6 times more than wind power, and 6.3 times more than solar energy.

China begun its nuclear power development in the 1980s with help of foreign technology firms from Europe, Japan, and the US. The development thereafter was slow given cost and safety concerns, in my view. By the 2000s, plans were made to hasten the development of nuclear energy, but the plans were hindered by the March 2011 Fukushima Daiichi Nuclear Power Plant meltdown following the Tohoku earthquake and tsunami. Planners decide to slow the speed of nuclear energy expansion, relooking at all of the planned reactors on the drawing board, and halted the construction of those in progress pending detailed safety checks. These were completed in the second half of 2011 and then the development restarted.

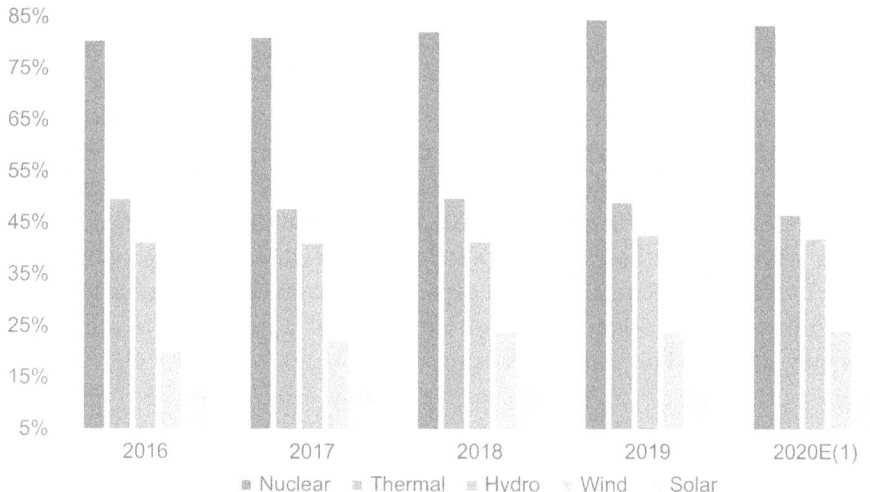

Figure 3.11: Capacity Utilisation Rates of Different Power Sources in China.
Source: Author, October 2020. Data sourced from: National Bureau of Statistics (2017–2020). National Bureau of Statistics. [online] Stats.gov.cn. Available at: http://www.stats.gov.cn/ [Accessed 9 October 2020] (Chinese); China Electricity Council (2017–2020). China Electricity Council. [online] www.cec.org.cn. Available at: https://www.cec.org.cn/ [Accessed 9 October 2020]; National Energy Administration (2017–2020). National Energy Administration. [online] www.nea.gov.cn. Available at: http://www.nea.gov.cn/ [Accessed 9 October 2020].

The big jump in capacity was in 2014; typically, nuclear plants take about five or six years to build. Capacity-wise, nuclear generation rose 2.2-fold to 20 gigawatts by 2014 versus 2009 and then increased 2.5-fold to 49 gigawatts. This capacity generated about 350 terawatt-hours in 2019; that's more than what any other Asian power market produces, except for India (1,559 terawatt-hours), Japan (1,036 terawatt-hours), and South Korea (585 terawatt-hours), and more than any country in Eastern or Western Europe, except France (555 terawatt-hours) and Germany (612 terawatt-hours). The average rate in the increase of nuclear energy production capacity during the thirteenth Five-Year Plan (2016–2020) was about 20 per cent but nuclear energy only accounted for 2.4 per cent of capacity as of August 2020 and for just 5 per cent of generation in the eight months through August 2020. On a personal note, I am well aware that nuclear is a highly contentious subject. I must admit that I strongly support the nuclear power development in China for as long as it is tightly, monitored by independent international bodies. With the advent of new generation reactors, the third generation, this form of power production will become increasingly safer.

While nuclear generation started to sharply contribute to the total energy mix from about 2014, that of solar power begun in 2009 or so. In my view, it is the steep fall in the price of solar power that promoted its exponential growth, a key subject I will discuss Section 3.1.2. Just like most major electricity markets the expansion has been with all three common types of systems. These include solar rooftop installations (panels atop of homes or factories), ground-mounted utility-scale solar farms (large arrays of solar panels fixed to the ground), and floating solar energy farms (panels floating on a water reservoir or lake), such as the massive 40 megawatt Sungrow Huainan Solar Farm atop of an artificial lake over a former coal mine in Anhui province, which comprises more than 160,000 panels over 800,000 square metres.[28] The nation's capacity increased to almost 205,500 megawatts in 2019 from just 3 megawatts in 1997 and less than 200 megawatts in 2007 (Figure 3.12),[29] accounting for about 10 per cent of the total national capacity compared to a negligible level ten years earlier. Given the low average yields from the capacity, at the generation end it only accounted for about 3 per cent.

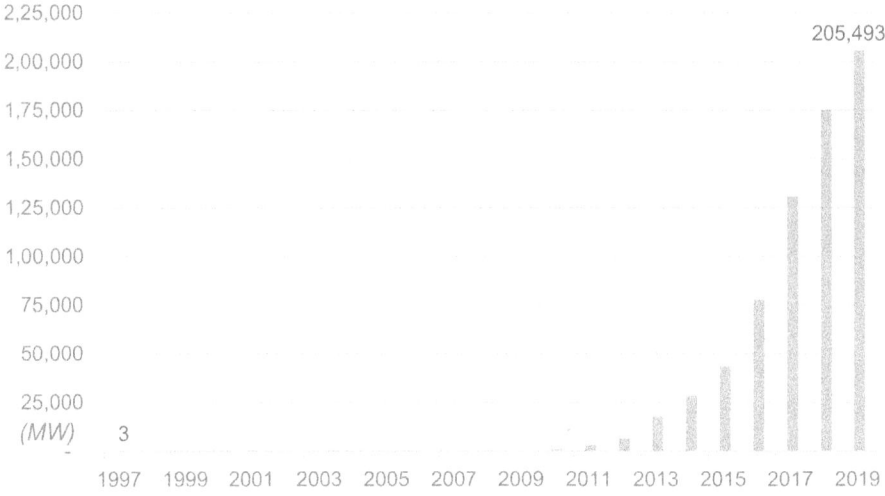

Figure 3.12: China Solar Power Generation Capacity.
Source: Author, October 2020. Data sourced from: bp p.l.c. (2020). Statistical Review of World Energy. [online] BP Global. Available at: https://www.bp.com/en/global/corporate/energy-economics/statistical-review-of-world-energy.html [Accessed 9 October 2020].

Wind was the earliest of the three technologies to take off in China. I would like to think that wind power is the golden child of China's clean energy, and it has proven to be an incredible success story for the country. The first wind turbine in China was imported from Denmark in the mid-1980s. The generation capacity remained tiny though until the mid-2000s when it finally exceeded 1 gigawatt. Planners were fully aware of the country's rich wind resource in Northern China (especially Xinjiang and Inner Mongolia, which account for 23 per cent of capacity and 24 per cent of output) but the cost per kilowatt-hour was pretty much prohibitive, being at least 1,200 yuan per megawatt-hour ($145 using the then rate) compared to as low as 290 yuan per kilowatt-hour ($43 using the 2020 rate) for coal power. Developers were mostly domestic, though there are some wind farms that are owned by operators from abroad, too. The development has also gone beyond on-shore sites with rich wind resources to sites with lower winds regimes as well as near-shore and off-shore locations thanks to technological developments. To give an idea of the growth, there was only 149 megawatts of wind power in 1997. This increased to 4.2 gigawatts by 2007, to 164 gigawatts in 2017, and to more than 210 in 2019 (Figure 3.13), or an average of about 19 gigawatts per year in the 2010s. The wind capacity as of June 2020 was about 217 gigawatts, of which about nearly 97 per cent was onshore and a little under 3 per cent was offshore or 6.99 gigawatts, of which 1.06 gigawatts was only added in the first half of the year.

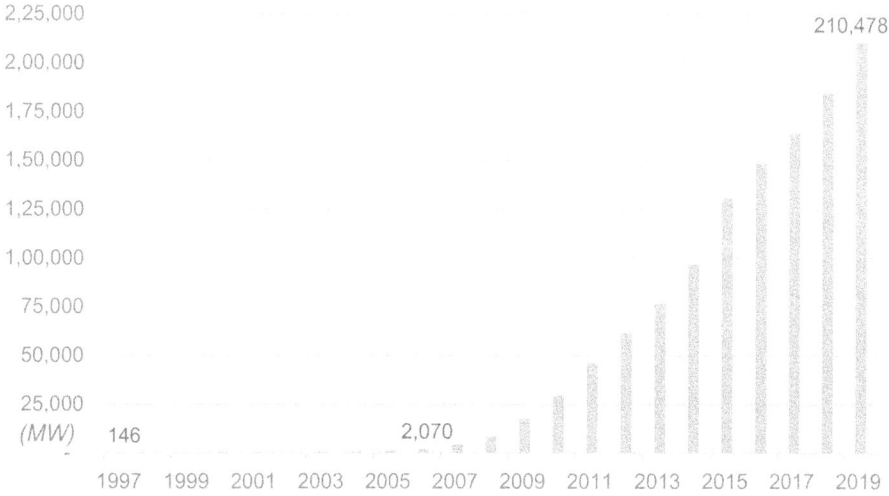

Figure 3.13: China Wind Power Generation Capacity.
Source: Author, October 2020. Data sourced from: bp p.l.c. (2020). Statistical Review of World Energy. [online] BP Global. Available at: https://www.bp.com/en/global/corporate/energy-economics/statistical-review-of-world-energy.html [Accessed 9 October 2020].

One issue that wind and solar energy have faced is curtailment. This explains why the rate of growth in generation capacity of wind and solar was faster than that of the rate in output. Curtailment means that the output had restricted connection to the grids. For wind, for example, between 2011 and 2016 the annual average rate was fairly volatile, ranging between 8 per cent (in 2014) and 17 per cent (2016). Aggressive action on the part of government authorities realised a steady decline to 4 per cent in 2019 (Figure 3.14). And I expect that the number in the short term will be kept at a low single-digit level before falling below 1 per cent. The wind curtailment was caused at different times by one or more factors, including network bottlenecks or transmission contracting, overcapacity as well as system imbalances. The bottleneck problem is interesting because it highlights a misunderstood fact about the Chinese electric power system, its planning, and its regulation. While

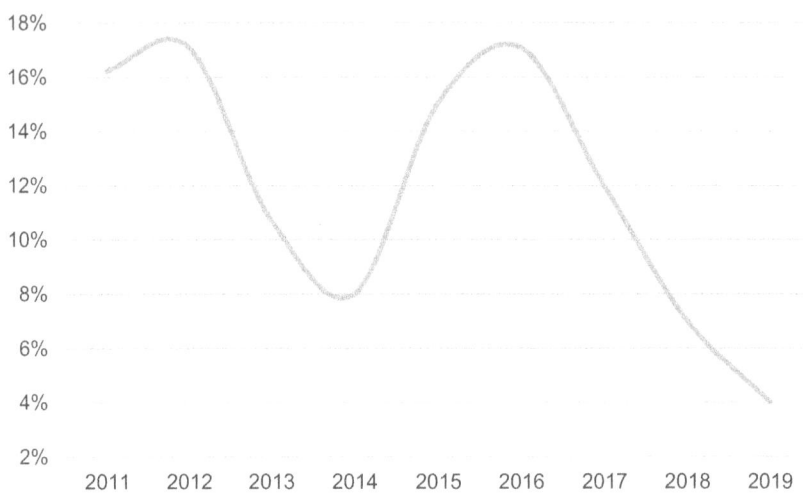

Figure 3.14: China Wind Power Average Curtailment Rate.
Source: Author, October 2020. Data sourced from: China Electricity Council. [online] www.cec.org.cn. Available at: https://www.cec.org.cn/ [Accessed 9 October 2020] (Chinese); National Energy Administration (2017–2020). National Energy Administration. [online] www.nea.gov.cn. Available at: http://www.nea.gov.cn/ [Accessed 9 October 2020].

the curtailment of solar and wind output happened for a whole bunch of reasons, when it comes to the power network issue, at times in some regions there was simply not enough transmission capacity. Developers constructed facilities at a much faster pace than the electric power grids could build transmission capacity (i.e., not enough power lines to transport the solar and wind power from the solar or wind farm). But at times, it was the grid that basically found whatever excuse it could not to buy power from renewable energy resources, based on my experience.

Another two challenges that solar and wind have faced is profitability. Broadly speaking developers will be looking to make a return of equity on their investments of somewhere from 7 to 10 per cent. Effective returns have probably been a little lower than what they had expected not just because the projects got no compensation for curtailment but also because of delays in the subsidy portion of the electric power tariff they receive. To clarify the curtailment impact on investment returns, let's imagine that my wind farm can produce in a certain year 100 units of power based on wind and maintenance schedule factors but the grid will only take up 90 units; my project will receive no compensation whatsoever for the units that could have been sold but were not. On the subsidy issue, if my wind farm has an approved tariff of 600 yuan per megawatt-hour ($90) but the average tariff for the grid where my wind farm is located is 400 yuan per megawatt-hour ($60), I should receive the later amount from the local grid and the difference of 200 yuan ($30) from the central government as a subsidy. Up to at least October 2020, this was not happening, with these subsidy payments being delayed by several months or several quarters. The delayed payment would also affect the investment return for my wind farm. Thankfully, both of these issues have been at the top of authorities' priority list for the energy sector and are being resolved. The subsidy will no longer be an issue at all in future given that they will be removed from 2021 and new wind farm's tariffs will have to be at the same level of that of their respective local grids.

Lastly, I will discuss two areas. One is outlook for clean energy in terms of high-level national targets for capacity additions over the next five years (the fourteenth Five-Year Plan, which is through 2025) and broad projections for the next 30 years through 2050. Another area is that of clean energy technologies other than hydro, nuclear, solar, and wind. These include, but are not limited to, energy storage, geothermal, hydrogen, and marine or ocean energy. Many of these technologies have a bright future but in China the bottom-line cost is what will allow a particular energy to prosper or not.

By any measure, the energy mix shift that China has already achieved in the past 20 years has been formidable. These, however, actually pale compared to the changes taking place in the next 30 years. The momentum to transform itself from a fossil fuel-heavy energy complex to a fossil fuels-light one is gaining speed, partly led by the drive to decarbonise. In 2020, many articles discussed and provided predictions on the likely targets for the fourteenth Five-Year Plan in terms of net capacity additions. Planners will be fine tuning in 2021 and may possibly even adjust targets during the Plan's period. One of several forecasts is that of the Beijing-based Global Energy Interconnection Development and Cooperation Organization (GEIDCO). I personally find the assumptions by GEIDCO quite realistic. Also, the organisation has a unique profile and background, boasting of a very strong leadership, including not one but two former chairmen of the all-powerful SGCC, among several luminaries. I will be using GEIDCO's assumptions in my analysis. These estimates should allow one to gain a better understanding of how much capacity China will add and how much capital will be invested over the next 30 years.

Researchers at GEIDCO expect the nation's total electric power generation capacity to increase threefold to more than 6,000 gigawatts by 2050 compared to a little over 2,000 gigawatt-hours as at the end of 2019. The fastest growing types of fuel sources will be thermal, solar PV, wind power, pumped storage, and nuclear energy (Table 3.8 and Figure 3.15). The energy mix will change in favour of non-fossil-fuels-based energy. Specifically, the ratio of fossil fuels (thermal coal and gas) electric power generation to non-fossil fuels in 2025, 2035, and 2050 is forecasted to be 42:58, 25:75, and 11:89, respectively (Table 3.8 and Figure 3.15). In Table 3.8, I added a column with my views on the possible upside or downside with the forecasts.

Chinese statistics historically have not often offered a break-down of thermal power capacity or production by type of thermal energy. In the discussions above for the purposes of showing the realm of growth I have used thermal power as a proxy for coal-fired power, knowing that the bulk of the thermal resource has been coal. But when it comes to forecasts, domestic newspaper articles, think tanks, and

Table 3.8: China Installed Capacity Projections (Adapted).

(*Gigawatts*)	2025	2035	2050	Potential
Total	2,950	4,370	6,010	
Hydro	392	486	571	⇔ Neutral
Pumped Storage	68	137	174	⇔ Neutral
Coal	1,101	911	403	⇩ Downside
Gas	152	192	229	⇩ Downside
Nuclear	72	125	177	⇧ Upside
Wind	536	1,107	1,967	⇧ Upside
Solar	551	1,270	2,248	⇧ Upside
Solar Thermal	9	41	110	⇧ Upside
Biomass	65	97	131	⇔ Neutral

Source: Global Energy Interconnection Development and Cooperation Organization (2020). New Development Concept on China's Energy Reform and Transformation Research Report. [online] Global Energy Interconnection Development and Cooperation Organization. Available at: https://www.geidco.org/ Accessed 13 October 2020]. 'Potential' column, Author, October 2020.

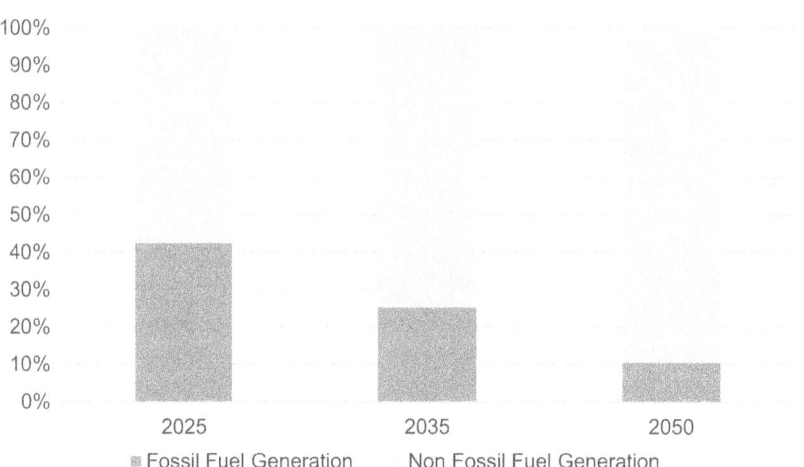

Figure 3.15: China Installed Capacity Projections (Adapted).
Source: Author, October 2020. Data sourced from: Global Energy Interconnection Development and Cooperation Organization (2020). New Development Concept on China's Energy Reform and Transformation Research Report. [online] Global Energy Interconnection Development and Cooperation Organization. Available at: https://www.geidco.org/ Accessed 13 October 2020]. 'Potential' column, Author, October 2020.

others have started to provide more transparency as to the thermal power numbers for 2019 and beyond in the past couple of years. This is probably due to the fact that gas-fired generation has been growing substantially in recent years and that this fuel source is roughly half the carbon footprint of coal, as well as due to the growth in biomass, which in some forms can also be low carbon.

China had approximately 1,040 gigawatts of coal-fired generation, representing more than 87 per cent of thermal power in 2019. The GEIDCO research indicates that the nation may add a net amount of about 60 gigawatts or so in the five years through December 2025. Then the number will fall by 190 gigawatts by 2035 and by a further 510 gigawatts by 2050 to 403 gigawatts. Solar plus wind capacity will jump to 4,325 gigawatts by 2050 from about 414 gigawatts at the end of 2019 (Table 3.8 and Figure 3.15). I have estimated that this may cost upwards of 30 trillion yuan ($4.5 trillion) excluding inflationary rises. Given that, historically, transmission and distribution costs have been in the ballpark of one-for-one with installed capacity costs, the bill could be roughly speaking 60 trillion yuan ($9 trillion) at a minimum, or 2 trillion yuan ($299 billion) per year, in the next 30 years. In terms of output, the GEIDCO research indicates that clean energy will contribute more than 50 per cent of electric power generation in China by 2050. The output share of biomass, hydro, nuclear, solar, and wind was about 25 per cent in 2019 and is set to rise to 23 per cent in 2025, to 41 per cent in 2035, and to 53 per cent in 2050. Obviously, a great number of things can happen over the next 30 years and these numbers will be revised and updated over time.

I strongly believe that by 2050 we will end up with a lot less coal and gas-fired capacity. Much less than the 640 gigawatts that GEIDCO is assuming for coal and gas combined. Coal is highly polluting, and China wants to decarbonise, so having 400 gigawatts of coal-fired generation in 2050 seems to me to be very much a bit of a stretch assumption. Gas generation is also polluting, it is a limited resource, and its volatile price raises unwarranted uncertainties. I think that nuclear, solar, and wind capacity is highly likely to be much higher than current estimates. This would allow for lower coal and gas numbers. In addition to higher nuclear, solar, and wind additions, I expect that China will add significant amounts of generating capacity from new technologies, once they are more mature and cost competitive.

The GEIDCO research estimates may not have factored into their current forecasts of energy mix through 2050 technologies other than the ones I mentioned. I am very confident that they are aware that the numbers could radically change if one of the newer clean energy solutions gains momentum in terms of cost and scale. For example, if a renewable energy solution, such as floating hybrid energy platforms, is able to overcome the current technological challenges, and if then it can be deployed at scale and at a competitive price, it could be one of the many that could substitute coal or gas-fired generation in two or three decades in China.

I categorise this new tech broadly in three buckets: new generation or offshoots of existing mainstream technologies (e.g., floating solar or floating offshore wind turbines), technologies at the prototype or early commercialisation stage (e.g., organic thin-film solar cell and wave energy converters), and substitutes or add-ons (e.g., storage and hydrogen). The development of these technologies is of course not solely specific to China. Development is taking place in many geographies all around the world. However, given the enormous scale of the nation's generation capacity requirements in the next 30 years, China is the leading region where this newer tech could rapidly reduce costs and become competitive thanks to the gigantic economies of scale that can be had in China given its enormous appetite for low to zero carbon energy. To put it into context, for wind alone, China will be adding roughly more than 1,400 gigawatts in the next 30 years; it will probably be higher but let us just stick to this more conservative number for now. This amount is greater than the total installed capacity of any country in the world, including the US, the second largest power market in the world, which had a total capacity of about 1,100 gigawatts as at the end of 2019.

When talking about these three buckets, the first and second are straightforward to understand given that they are simply about producing energy. But I would like to highlight two related solutions from the third bucket, the substitutes or add-ons bucket, in the China context, namely energy storage and hydrogen. Energy storage is intimately related to renewables because of its storage capability. Environmentally friendly hydrogen, green hydrogen, can be used to produce energy as well as a storage solution and is a very viable substitute to fossil fuels-based energy.

Energy storage is critical in any electric power system that hosts an increasing number or high number of renewables. The grid needs to be fed a stable amount of power supply at a rate similar to demand. Disruptions in this fine balance can cause a great variety of problems, including black outs. Solar and wind power are intermittent, as they are subject to the availability of the wind or solar resource. The electric power grid does have some mechanisms to balance things out, but it is limited. Storing the output from renewables sharply increases elasticity and dependability of the transmission and distribution network. There are a great variety of energy storage mechanisms all at different stages of maturity. The largest form currently in China is pumped storage but the fastest growing one is electrochemical storage. The former stores energy by pumping water up to a reservoir when demand is low and releasing it to produce electricity when demand picks up, often pumped up at night and released during the day. The latter is chemicals-based storage in batteries that will produce as much energy as their chemical components, such as lithium, allow.

China's enormous solar and wind capacity additions in the next 30 years will mean that it will need a formidable amount of energy storage to balance the electric power network. The nation has already been steadily adding pumped storage capacity, a very well-established technology. It will be commissioning greater amounts of capacity in the coming three decades. Pumped storage accounted for approximately 93.4 per cent of total energy storage capacity in 2019. Capacity should reach an estimated 40 gigawatts by the end of 2020 according to the Chinese Energy Storage

Association (CESA). And it may rise to 68 gigawatts and to 174 gigawatts by 2025 and 2050 respectively, based on GEIDCO estimates.

While pumped storage is the dominating form of energy storage, the fastest growing has been, and will remain in the coming few years, electrochemical (Figures 3.16 and 3.17) storage. It accounted for about 5.3 per cent of the total in 2019 according to CESA. The breakdown by type was Li-ion batteries (80.6 per cent), Lead-Acid batteries (17.8 per cent), Flow batteries (1.2 per cent), Super-Capacitors (0.4 per cent), and other (0.10 per cent). This technology added only 3 megawatts in the 2000s but another 1,700 megawatts in the 2010s. It may reach more than 2,700 megawatts in 2020, according to CESA. GEIDCO expects electrochemical storage capacity to jump to 40 gigawatts, 240 gigawatts, and 610 gigawatts in 2025, 2035, and 2050 respectively (Figure 3.17). I think that although these numbers may seem unreachable, the amount by 2035 and especially that by 2050 may actually be much greater. And, expansion of electrochemical storage may possibly be even faster in the 2030s and 2040s if some of the technologies currently at the prototype or early stage of development level prove themselves, are commercialised, and are competitively priced. Also, they must be more environmentally friendly at the usage-end (i.e., zero emissions) and at the value chain end (i.e., not using fossil fuels as part of the production process).

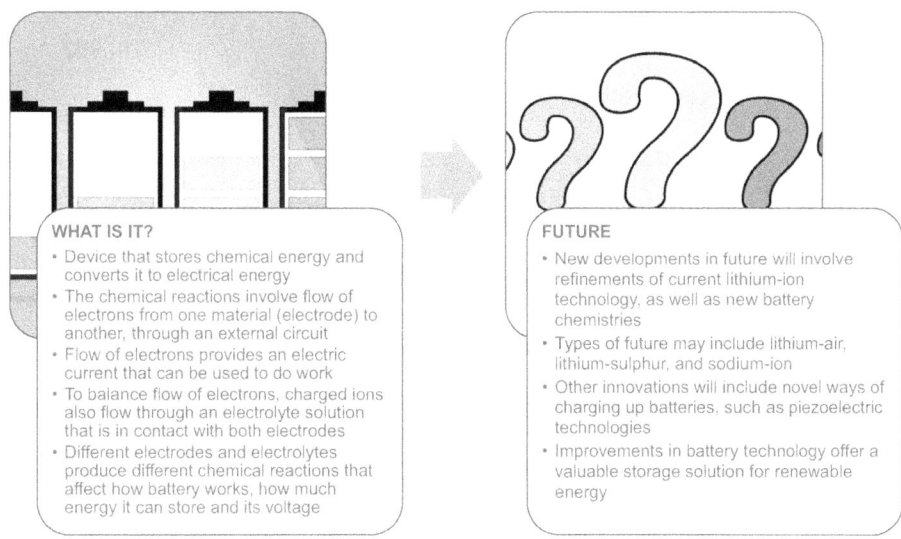

Figure 3.16: Electrochemical Batteries.
Source: Author, October 2020. Data source: Australian Academy of Science (2016b). How a Battery Works. [online] www.science.org.au. Available at: https://www.science.org.au/curious/technology-future/batteries [Accessed 13 October 2020]; Australian Academy of Science (2016). Batteries of the Future. [online] www.science.org.au. Available at: https://www.science.org.au/curious/technology-future/batteries-future [Accessed 13 October 2020]. Images by Mudassar Iqbal from Pixabay.

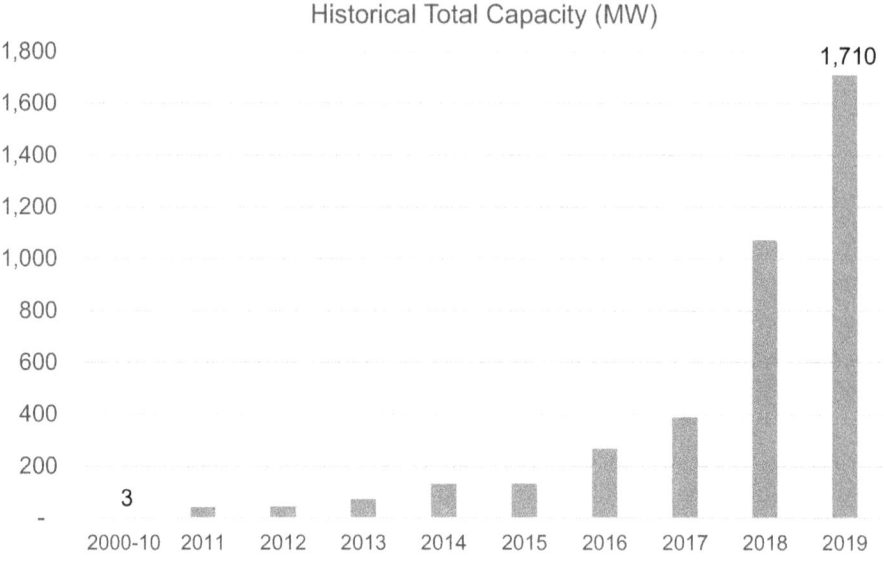

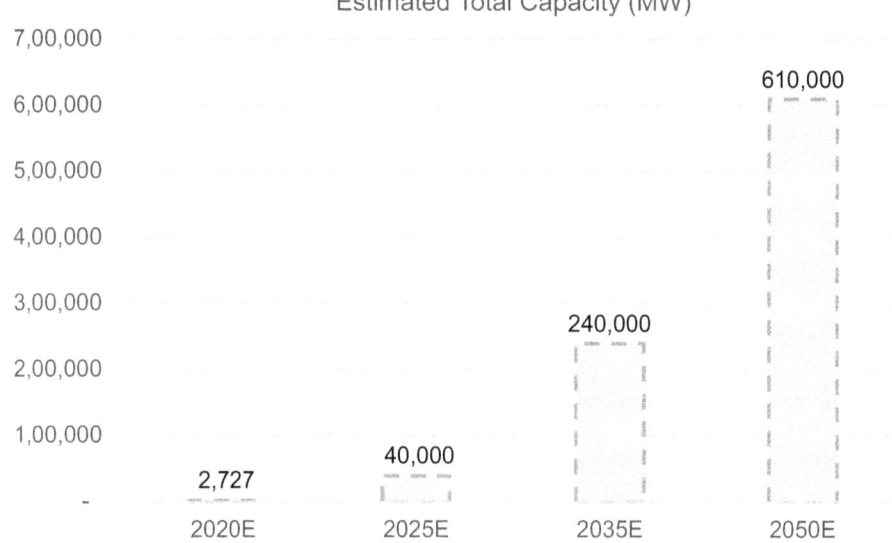

Figure 3.17: China's Historical and Estimated Electrochemical Storage Capacity.
Source: Author, October 2020. Data sourced from: Global Energy Interconnection Development and Cooperation Organization (2020). New Development Concept on China's Energy Reform and Transformation Research Report. [online] Global Energy Interconnection Development and Cooperation Organization. Available at: https://www.geidco.org/ Accessed 13 October 2020].

Hydrogen has been around a very long time. Potential uses were discovered more than 250 years ago. Today, it is used for a variety of industrial processes, ranging from the production of fertiliser, carbon steels, and semiconductors, to the processing of intermediate oil products. These processes, however, emit great amounts of CO_2. This is why this type of hydrogen is referred to as grey hydrogen (Table 3.9). In many countries all around the world, including China, there is an enormous amount of R&D been undertaken to develop green hydrogen solutions, which will include the storing of energy. An excellent comprehensive study by the International Energy Association published in 2019 discusses this topic from many different angles. It highlights that there is currently an unprecedented momentum on hydrogen all around the world with a remarkable breadth of stakeholders' interest widely spread among corporates and governments alike. The authors also underline that this is not the first time that hydrogen grabbed headlines as the 'new oil'. There have been at least three separate hype cycles. But they all ended with disappointment, and hydrogen never fully took off. The wide use of hydrogen, especially green hydrogen, faces several challenges. They include high production costs, a limited infrastructure, safety, and the lack of homogeneous regulatory standards; in other words, a widely officially recognised definition of what is grey and what is green hydrogen. The IEA thinks that this time around it may be different. I would tend to agree with this precept. Three positive momentum factors for hydrogen in general, and green hydrogen in particular, include the amount of interest and investments by stakeholders, government supporting policies driven by energy security interests, and hydrogen's unique capability to help some sectors that are hard to decarbonise, including airline transport, chemicals, shipping, and steel. A fourth factor is China. It should be regarded as a major driver because of the nation's extraordinary energy needs, the massive energy mix shift towards low carbon generation, its strong will and need to decarbonise to net-zero by 2060, green hydrogen storage capability, and the nation's capability to drive down costs given the enormous economies of scale it can attain. Similar to Australia's and the EU's publicly declared ambitions, it is certain China also wants to become a hydrogen superpower.

So, the global race is on towards cheap green hydrogen that uses electricity to break down water into oxygen and hydrogen, or electrolysis, using solar, wind, or other renewable energy. Essentially, any region with rich wind or solar as well as resources could build major facilities. We discussed earlier Australia's ambitions and the Asian Renewable Energy Hub project in Western Australia (see 3.1.1.1 Australia – Vast Energy Resources, a Curse in Disguise?). China is highly likely to follow this route as well. While the Asian Renewable Energy Hub is partly focused on exports, China would primarily consume the green hydrogen domestically as it currently needs, and will need even more in the next 30 years, as much storage capacity as it can build. I believe that in the coming years there will be some massive projects announced but first it must drive down the price per unit of energy – something that we discuss together with other energy costs and prices in the next section. The country is still fine tuning its overall plans, strategies, and regulations.

Table 3.9: Hydrogen Production Pathways Comparison (World Energy Council).

	Grey Hydrogen	Blue Hydrogen	Green Hydrogen
Main Production Routes	– Steam Methane Reforming (SMR) – Coal Gasification	– SMR + CSS – Coal Gasification + CSS	– Electrolysis Using Renewables
CO_2 Emissions	HIGH	LOW	ZERO
Current Cost	LOW	HIGH	HIGH
Social Acceptance	LOW	MID	HIGH

Source: Chart originally from World Energy Council. Source: Yin, P., Brauer, M., Cohen, A.J., Wang, H., Li, J., Burnett, R.T., Stanaway, J.D., Causey, K., Larson, S., Godwin, W., Frostad, J., Marks, A., Wang, L., Zhou, M. and Murray, C.J.L. (2020). 'The effect of air pollution on deaths, disease burden, and life expectancy across China and its provinces, 1990–2017: An analysis for the Global Burden of Disease Study 2017', The Lancet Planetary Health [online] vol. 4, no. 9, pp. e386–e398. Available at: https://www.thelancet.com/journals/lanplh/article/PIIS2542-5196(20)30161-3/fulltext#:~:text=We per cent20estimated per cent20that per cent201 per centC2 per centB724 [Accessed 6 October 2020].

There is some evidence that the acceleration will be at warp speed, by energy industry standards. For example, the China Hydrogen Alliance compiled some numbers regarding FCEVs; Asia's FCEVs potential was discussed in Chapter 2 (see 2.1.2 Higher Demand from Shift to Electric Vehicles). The nation only had 66 hydrogen refuelling stations as of February 2020. The Alliance wrote that the national target is to raise the number to 300, 1,500, and 10,000 by the end of 2025, 2030, and 2050, respectively. It also highlighted that by 2050, the nation should reach an annual production capacity of 5.5 million fuel cell systems, of more than 5 million FCEVs, and more than 20,000 fixed power generation equipment. By then, China's 60 million tons of hydrogen could account to 10 per cent of China's energy system (Table 3.10).

What has not yet been clearly stated is where the hydrogen will actually be coming from? Quoted by local media in 2019 and in 2020, experts and observers have been increasingly referring to green hydrogen, so I think that the momentum is fairly clear. They also concur that green hydrogen supply will increase in the future. Regulators have sporadically indicated that a national strategy or blueprint should be completed in the next year or two. Independent research group Energy Iceberg compiled an excellent sample of some of the current hydrogen-related projects by Chinese energy companies in China (Table 3.11) – proof that China is definitively far from being asleep at the wheel when it comes to green hydrogen.

Table 3.10: China's Hydrogen and FCEVs Industry Overall Targets.

Industry Target	2019	2020–2025	2026–2035	2036–2050
Hydrogen Energy Share	2.7%	4.0%	5.9%	10.0%
Value of Industry (CNY Billion)	300	1,000	5,000	12,000
Value of Industry (US$ Billion)	45	149	744	1,786
Refuelling Stations	23	200	1,500	10,000
Fuel Cell Electric Vehicles (Production/Year)	200	50,000	1,300,000	50,00,000
Fixed Energy/Power Facilities (Production/Year)	200	1,000	5,000	20,000
Fuel Cell Systems (Production/Year)	10,000	60,000	1,500,000	55,00,000

Source: Future Industry Research Institute (2020). Market Status and Development Prospect of China's Hydrogen Energy Industry in 2020. [online] www.sohu.com. Available at: https://www.sohu.com/a/404948594_99922905 [Accessed 14 October 2020].

Table 3.11: Sample of Hydrogen-Related Projects by Chinese Energy Companies in China.

Group	Investment(s)
China Huaneng	Agreement with Baicheng government, Jilin to develop a 2 gigawatts wind farm with power-to-hydrogen system. Also, in talks with the Tianjin government to explore the potential of building a power-to-hydrogen project that would adopt carbon capture devices.
China Datang	Will invest in a 6 megawatt solar-based hydrogen production demo project in Datong, Shanxi Province.
China Huadian	Will jointly develop a 100 megawatt solar project with FCEV manufacturer Weichai Power in Weifang, Shandong. The power generated will be used for green hydrogen production.
China General Nuclear	Agreement in October 2020 with Ningxia East to establish renewable power-to-hydrogen projects in the region.
China National Offshore Oil	Issued research and development tender for a tech partner to explore the potential for building an offshore electrolysis system for its offshore wind farms.
PowerChina	Plans to build a 200 megawatt 'poverty-alleviation' solar farm that will integrate with fish farming and electrolysis devices.
State Power Investment	Agreement with Siemens Energy to explore green-hydrogen cooperation; also purchased Siemens' skid-mounted proton exchange membrane (PEM) electrolysis system 'Silyzer 200'. Also, agreement with Hainan province to invest in hydrogen production facilities based on offshore wind projects in South China sea.

Note: List has been amended for clarity. Source: Energy Iceberg (2020). China's Green Hydrogen Effort in 2020: Gearing Up for Commercialization. [online] Energy Iceberg. Available at: https://energyiceberg.com/china-renewable-green-hydrogen/ [Accessed 15 October 2020].

3.1.2 Asian Clean Energy Prices' Fall is Cliff-Hanger for Coal, Gas, Oil

The energy transition from brown to green power in Asia in the next 30 years means a massive change in the energy mix in favour of clean energy. Financial investments will be concentrated on low and zero carbon solutions, including energy storage and new technologies. The massive quantities of clean power that will be added will lead to the average price per unit to fall even further thanks to the colossal economies of scale. This applies not just to China, the undisputed leader, but to Asia as a whole. In many regions, such as Australia, the average energy production price of solar and wind power is already equal or lower to that of fossil fuels. Many forecasters agree with these trends. Their estimates only differ as to the actual speed of the change. The continued progressive reduction in the average production cost of clean energies will have many wide implications in the broader energy sector in general, and affect the future consumption and price of fossil fuels-based electricity generation in particular.

The first area of discussion is looking at the energy prices trends of the principal electric power generation sources. There are a great number of institutions that produce financial models of the long-term costs of a particular energy. Here I will use data from three such institutions, the intergovernmental organisation International Renewable Energy Agency (IRENA), financial institution Lazard Ltd. (formerly Lazard Frères & Co.), and research and advisory provider BloombergNEF (formerly New Energy Finance). The estimates produced by these and other entities is called the levelised cost of energy (LCOE). The LCOE is a relatively straightforward approach because it factors in all of the costs incurred in the production of the energy during the lifetime of a project, including capital expenditures, borrowing, operations and maintenance, and other expenses. This is then divided by total volume produced by the project during its lifetime. There are some elements that require some assumptions that may sharply diverge from institution to institution, including interest rates, discount rates, or the equity investment to debt split. My intention here is not to offer a detailed evaluation of the LCOE nor is it to compare and contrast the approaches of the different institutions. Rather, it is to give a broad understanding as to the different current and future prices of energy for the various generation resources to better gauge the trend over the long term.

IRENA produces an annual study on the cost of generation of various clean energy resources. The data in the latest publication underlines the massive fall that the key renewables have experienced in the past decade (Figure 3.18). The weighted average global price per unit of energy produced by solar PV was about $68 per megawatt-hour. This was 82 per cent below the 2010 level or a compound annual decline rate of negative 17 per cent, I calculated. That for onshore wind was $53 per megawatt-hour, down 38 per cent in total and negative 5 per cent per year. For offshore wind it was $115 per megawatt-hour, down a total of 29 per cent or negative 4 per cent annually. IRENA also compiled numbers for some other clean energies, including concentrating solar power, which was down 47 per cent to $182 per megawatt-

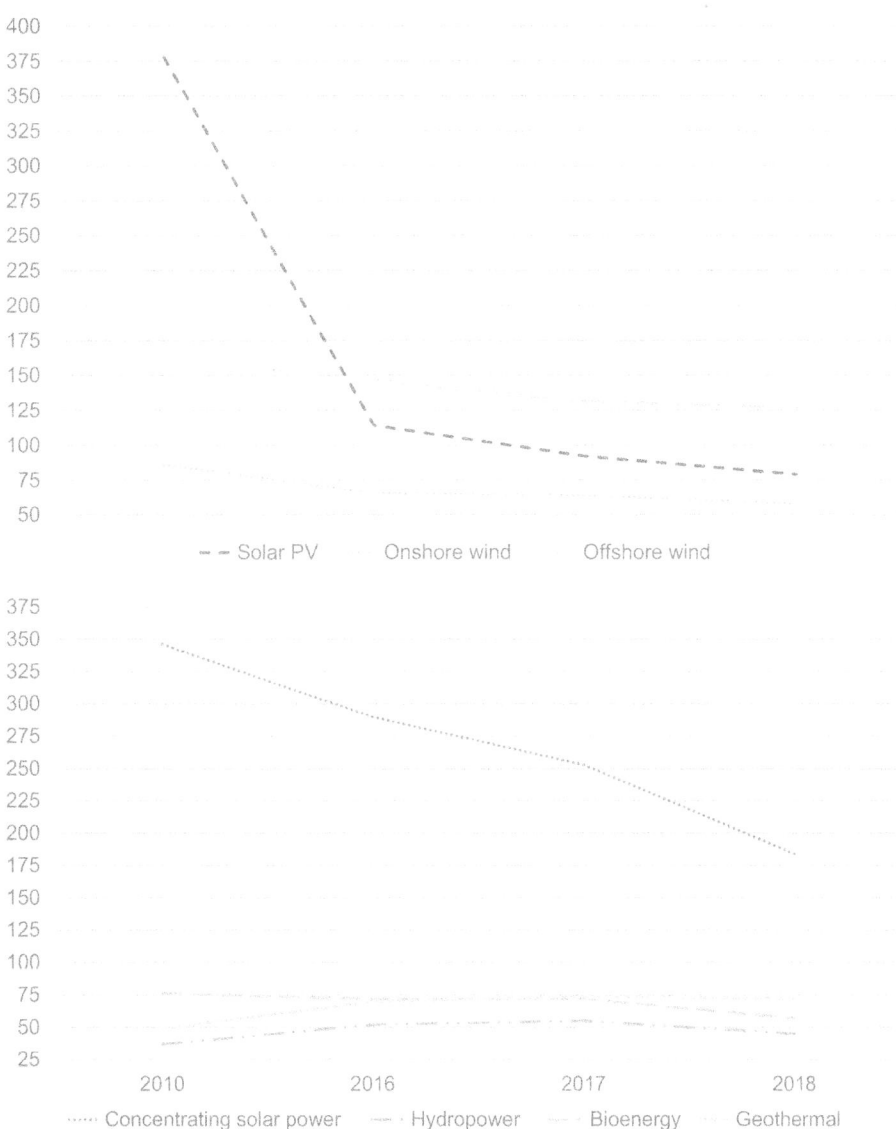

Figure 3.18: IRENA's Renewable Power Generation Costs ($/MWh) in 2019.
Source: IRENA (2020). Renewable Power Generation Costs in 2019. [online] /publications/2020/Jun/Renewable-Power-Costs-in-2019, Abu Dhabi: International Renewable Energy Agency, pp. 27–36. Available at: https://www.irena.org/publications/2020/Jun/Renewable-Power-Costs-in-2019 [Accessed 19 October 2020].

hour over the period, and bioenergy, which declined 13 per cent to $66 per megawatt-hour. The agency also included data for hydropower and geothermal, about $47 and $73 per megawatt-hour respectively. The price of the two actually increased over the ten-year period. What is different about these two generation technologies is that they are already quite mature (140 years for hydro and 110 years for geothermal) and it is particularly hard to compare two hydro or geothermal facilities apples to apples because of the geographic location and the geology of the project.

Lazard has been regularly publishing its own LCOE calculations. The latest iteration was for 2019 and confirms that solar PV (utility-scale ground-mounted) and onshore wind can now fully compete with coal and natural gas fired generation (Figure 3.19). Lazard put the solar PV range at $23 to $75 per megawatt-hour and onshore wind at $23 to $83 per megawatt-hour. These compare to the production costs for coal and gas power of $40 to $200 per megawatt-hour and $33 to $100 per megawatt-hour, respectively. There are, however, a couple of points that I should underline. For the two mainstream renewable technologies, solar and onshore wind, to compete head on with the two fossil fuels, energy storage is needed, and this would raise the final energy price per unit. Conversely, if a carbon price based on a project's emissions is applied to the two fossil fuels, it would make their production costs significantly higher.

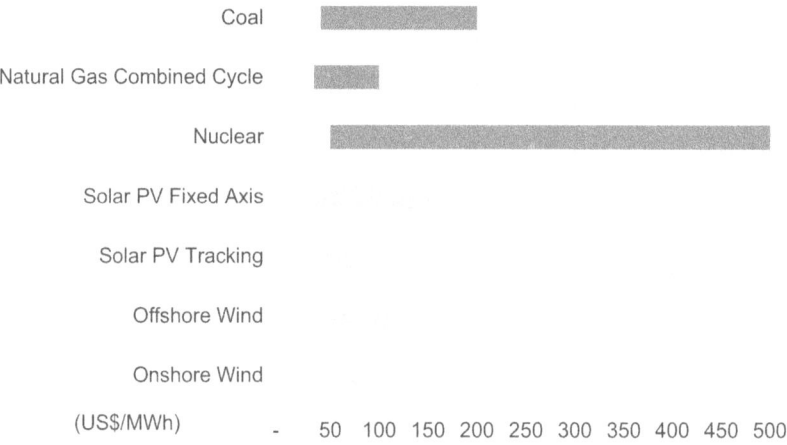

Figure 3.19: Lazard's Cost of Energy Comparison – Unsubsidised Analysis.
Source: Lazard (2019). Lazard's Latest Annual Levelized Cost of Energy Analysis – LCOE 13.0. [online] lazard.com, p. 2. Available at: https://www.lazard.com/perspective/lcoe2019 [Accessed 19 October 2020].

A World Economic Forum report, involving a great variety of experts from various agencies and companies, quoted some LCOE estimates from BloombergNEF. The numbers slightly differ from those of Lazard, but the conclusion remains the same, that solar and onshore wind today produce energy at a cost equal to or lower to that of coal and gas generation. The BloombergNEF numbers (Figure 3.20) break down solar power's energy cost with a higher degree of granularity, especially showing the difference between solar PV for residential rooftops compared to that for commercial and

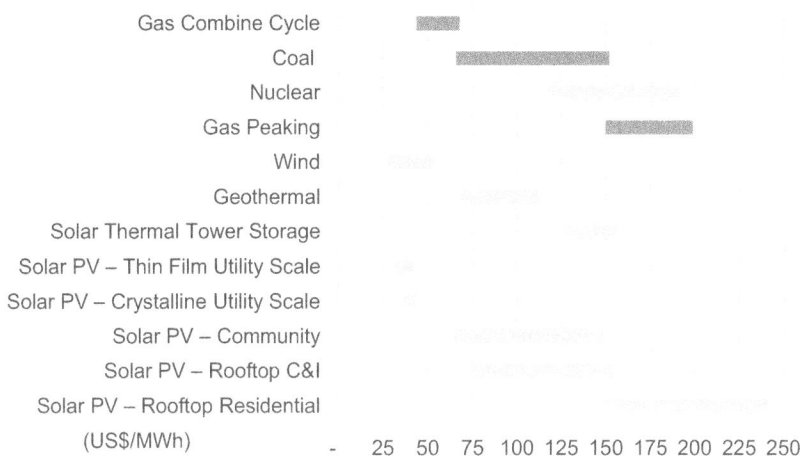

Figure 3.20: BloombergNEF Global LCOE Ranges, Second Half 2020 Updates.
Source: World Economic Forum (2020). Energy Technologies 2030: Wind and Solar PV Will Keep Taking the Lead. [online] World Energy Forum Global Future Council on Energy Technologies, p.1. Available at: http://www3.weforum.org/docs/WEF_Wind_and_Solar_2030.pdf [Accessed 19 October 2020]. Sourced from BloombergNEF (BNEF) 1H 2020 LCOE Update. https://www.bnef.com/.

industrial users or showing that for combined cycles gas plants versus those used for peaking, for example.

The decline in the cost of the mainstream renewable energies is set to continue. While different institutions have different estimates as to the realm of the fall over the next five, 10, or 20 years, all agree that their LCOE will drop further. The World Economic Forum report's group of experts, for example, believe that efficiency gains and falling equipment manufacturing costs should lower unit costs by 50 per cent for solar PV, 25 per cent for onshore wind, and 50 per cent for offshore wind energy by 2030. The paper also contends that the cost of onshore and offshore output should merge to about $30 per megawatt-hour by 2030.[30] For fossil fuels, on the other hand, I have not come across any observers arguing for sharp falls in the generation costs of coal or gas power. To the contrary, many jurisdictions will introduce some carbon penalties, such as a tax or price on carbon for emissions. It will make coal and gas power even more expensive. Plus, coal power will struggle to find attractively priced debt financing, if they can find it at all. So then why produce electricity from polluting coal-fired power plants at all? Why then not just simply replace coal plants with solar and/or wind facilities with some energy storage. Unfortunately, shutting them down before the end of their economic life will incur a cost, and the owner would have to bear these costs, called stranded costs. It is a good guess that over the next 20 to 40 years in Asia, depending on the specific electricity markets, many plants may be shut a bit earlier than the end of their economic life and they definitely would not see their life extended – something that can be done with some engineering works.

So, I have established that the price at which solar and onshore wind farms sell their output production is actually highly competitive with that of fossil fuels. Further, their prices will fall below those of fossil fuels level in the coming years. But what are the implications for coal and gas power in Asia in the coming years? My precept is that these trends will sharply cut the attraction of coal-fired generation and are also highly likely to weaken the attractiveness of gas-fired generation in many markets in the region.

It is a matter of when rather than how. My base case is that in 2040 at the very latest – and perhaps as soon as 2030 – none of the Asian markets will have new coal plants. This is because we have electricity markets at different stages of development and have some governments who are dead set, and sometimes have both. Japan and South Korea until recently were still planning on new coal-fired generation. Many of these plans have now gone sideways because of societal pressures. In some of the resources-constrained frontier markets such as Cambodia or Myanmar we could still see new plants coming through based on their perception of economic need, i.e., coal is a cheap source of generation and the fuel is plentiful, and logistically it is easier to build a large 1 or 2 gigawatt coal power plant than the same amount of solar generation. If key development financial institutions, such as the Asian Development Bank and the Asian Infrastructure Development Bank, really pull together and provide more proactive help to those markets still taking the coal option, then perhaps we will not have to wait until 2040 for the last new coal plant to be built. It could happen much earlier. What is certain is that demand for thermal coal is peaking in Asia. And the absolute amount of thermal coal will start declining in the next three to five years. This is because this is when it will peak in China, the biggest user. As such, unless producers somehow curtail supply, the price of thermal coal will also peak soon. Looking at one available thermal coal spot price benchmark in the Asia region, the Newcastle Freight-On-Board 6,000 kilocalories per kilo Net-As-Received, the 20-year monthly high of $180 will be an impossible dream in the future while the low of $22 may become a reality (Figure 3.21). The price is about $56.30 per tonne now (as of 18 October 2020) and the market consensus is for the price to rise to $69.10 per tonne in 2023, based on forecasts collated by Blomberg. I find these forecasts somewhat overly bullish.

The future evolution of gas prices in Asia is trickier than that of thermal coal. Gas generation offers several positive attributes. The availability of the fuel is increasing, pricing mechanisms are progressively becoming more buyer friendly, plants can be called on more rapidly than coal (which works well with renewables), and gas plants have a carbon footprint that is half or less than that of coal. My thesis, though, is that ultimately demand will sharply fall because natural gas-based power generation is still polluting, and the prices will remain volatile. This is more likely to happen in the second half of the 30-year period that I have been looking at. There is simply too much gas generation capacity, and thus demand, being built around the region, especially in China and East Asia. The expected consumption increase is expected by a variety of institutions, including BP, the International Energy Agency, Oxford Economics, McKinsey, Shell, and many others. To give a sense of this, I

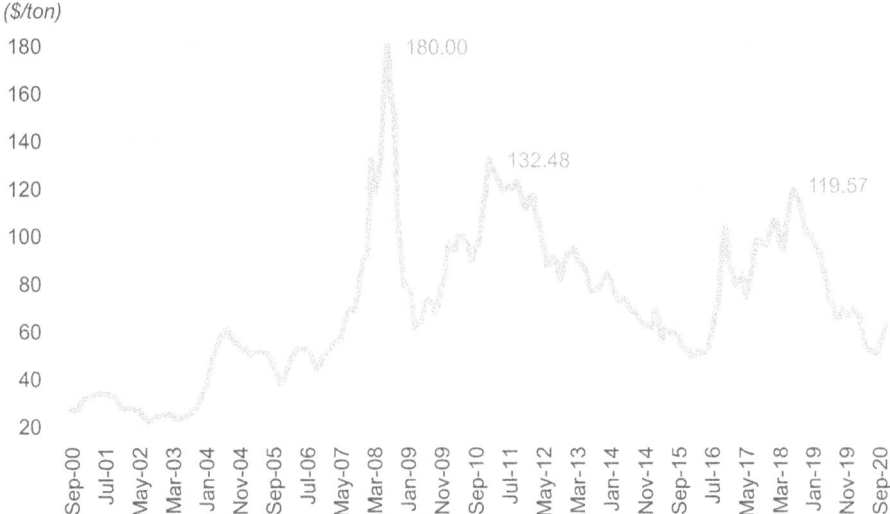

Figure 3.21: Newcastle Freight-On-Board 6,000 Kilocalories per Kilo Net-As-Received Monthly Price – $ per Metric Tonne.
Source: Index Mundi (2020). Coal, Australian Thermal Coal – Monthly Price – Commodity Prices – Price Charts, Data, and News – IndexMundi. [online] www.indexmundi.com. Available at: https://www.indexmundi.com/commodities/?commodity=coal-australian&months=240 [Accessed 17 January 2021]. Note: Coal (Australia), thermal GAR, f.o.b. piers, Newcastle/Port Kembla from 2002 onwards, 6,300 kcal/kg (11,340 btu/lb), less than 0.8%, sulfur 13% ash; previously 6,667 kcal/kg (12,000 btu/lb), less than 1.0% sulfur, 14% ash.

will just mention a couple of views. The International Energy Agency expected shorter term demand to be impacted by the COVID-19 induced economic slowdown. Still, it estimates that global demand will on average grow at a rate of 1.5 per cent between 2019 and 2025.[31] And, this increase will actually be centred in the Asia region. Among many longer term estimates we can use one from oil major Shell. It forecasted that global demand for liquified natural gas will reach 700 million tonnes by 2040 compared to 359 million tonnes in 2019.[32] With this kind of sturdy demand, prices are unlike to weaken in the short term. For this we will need to have to wait until the outlook for long-term demand falls. Something I think we will start seeing from 2035 onwards because of the emissions element and also the price volatility, as evidenced by Japan's spot price for liquified natural gas price statistics from the Ministry of Economy, Trade and Industry (Figure 3.22).[33]

Another element that could in part or significantly displace gas for power generation is green hydrogen. Currently, green hydrogen cannot compete commercially with gas. It is still too expensive. But many experts are highly confident that it could be produced at significantly lower prices in a few years. Currently, green hydrogen is around a level of 3.50 to 5.00 Euros ($4.13–$5.91) per kilo according to the International Energy Agency. To be commercially competitive it needs to be at or below

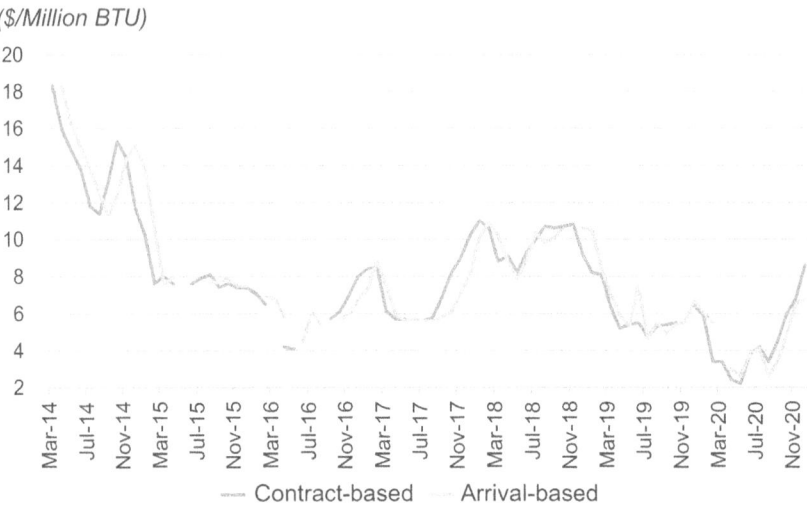

Figure 3.22: Japan Spot LNG Price Statistics.
Source: METI (2020). Spot LNG Price Statistics | METI. [online] www.meti.go.jp. Available at: https://www.meti.go.jp/english/statistics/sho/slng/index.html [Accessed 17 January 2020].

$2 per kilo. Several studies argue that this versatile energy carrier can achieve the $2 per kilo mark thanks to bigger and better equipment (i.e., electrolysers) and cheap renewable energy. Research and consultancy firm IHS Market forecasts that the cost will fall 30 per cent by 2025. BloombergNEF thinks this can be reached by 2030 and that it could further fall to $0.8 to $1.60 in many geographies by 2050. Both the European Commission and the Australian government also forecast that the two regions will be able to produce green hydrogen at below $2 per kilo in future.[34]

I think that we cannot pinpoint for now an exact time for green hydrogen to get to $2 per kilo or below in Asia's key electricity markets in large scale. I concur with the many experts and their many sophisticated financial models that it is highly likely to happen before 2050. When emissions-free green hydrogen does arrive at scale, thermal coal should already have become a dwindling fuel source, say in the 2040s. This means that it is the consumption of natural gas that will be negatively impacted.

3.2 New Opportunities through the Digital Door

The last smartphone that I bought, a Samsung (sorry Apple fans), was a couple of years ago and the price had hardly changed compared to the previous model that I had, which was also a couple of years old. While the price had hardly changed, the speed and memory had vastly improved. This is true for most digital technologies. They are getting better and cheaper and the prediction and computational tools faster. Whether one is tech-savvy or not or whether one is keenly interested in tech or not, only a hermit would have not noticed the massive improvement and falling cost of new tech. And, this is very much true for digital tech solutions in the energy industry as well.

The energy world is increasingly adopting digital technologies and solutions. It is an energy digitalisation revolution. New technologies in the energy industry drive cost savings throughout the value chain of the production and supply of the energy, they facilitate raising environmental sustainability, and they can effectively raise the efficiency of energy-related financial transactions. The advent of digital technologies and solutions for energy is part of the Fourth Industrial Revolution, a new industrial revolution involving smart tech in industry and manufacturing processes. It is called Industry 4.0, which, funnily enough, is a term that was allegedly coined by a German government entity. Something ironic given that usually governments are behind the tech curve. The digital technological solutions for energy in part help to address challenges faced by the industry, including the intermittent supply nature of renewables – no power when there is no wind or sun, help with optimising the transition of the energy mix away from thermal coal, oil, and eventually natural gas. They also help to improve the operations of new economy electric power networks, including mini-grids and distributed energy resources, and that of energy storage. These new tech solutions are also essential for the management and processing of the incredibly formidable amounts of energy-related data whose exponential growth has been witnessed in recent years.

The massive shift in the energy mix that we discussed in the last chapter will happen at the same time as the digitalisation revolution. In the energy industry, the principal manifestation of this will be through smart energy supply and usage. The 'smart' part will have at least three pillars: AI, blockchain solutions, and the IoT. In these areas, China will most likely be one of the leaders in the Asia region and the world. I believe that China's leadership is coming from several sources or circumstances. The nation is already a tech giant and a tech adoption leader. The tech advancements are driven by both private players and the government. The ongoing energy market reforms are sharply increasing competition and, at the same time, innovation. And the economies of scale from the gigantic size of the deployment of new tech will drive costs down.

Let me first quantify why the digital solutions revolution is a major event. First, this new tech will allow for the significant cost savings in the energy supply value chain, notably benefiting providers as well as consumers. Second, this new tech will make the investments in smaller projects, including mini-grids and distributed energy, significantly more economically and logistically feasible (avoiding the construction of expensive transmission lines). Third, this new tech will enhance energy efficiency and conservation, and drive more sustainable and environmentally friendly energy supply. Fourth, new tech creates vastly more flexible and more cost-effective financing solutions, as well as cost savings in such areas as feasibility studies, legal fees, and other development costs. It introduces new financial transactions applications, including additional financing channels. All of these developments translate into a vast amount of new business and investment opportunities for existing industry participants and new entrants to the market as well.

In this section, I will first try to profile what we are referring to when we are talking about digital technologies for energy. I will provide some detail as to the different forms of digital technologies for energy as well as the potential magnitude of costs optimisation. Then I will provide several real-world examples and case studies of the applications of the technology. Finally, I will evaluate why China will be one of the key actors in this area in Asia as well as in the rest of the world.

3.2.1 New Digital Tech and the New Energy World

The COVID-19 pandemic has made high-level energy discussions more easily accessible thanks to technology as a lot of international meetings went online, for example. I attended many of these events in 2020 and all of them, without exception, mentioned the word 'digitalisation' – and a few variants – throughout the event. Speakers featured ministers, international agencies, CEOs, prominent financial institutions, and others. They all had something to say about digitalisation. This is not just because digital technologies is an integral part of the energy business. That is only part of the reason. The other part is that it plays a crucial role in the world's quest for decarbonisation. In this section, first we will try to get our head around what the tools used by energy companies actually are and explain how they are increasingly attractive to these companies. Thereafter, I will look at a couple of the many positives that they bring to the energy business; cost and environmental sustainability related benefits.

Computational technologies have always been an indispensable part of energy generation and supply. More than a decade ago I remember visiting the control room of a nuclear power plant in Southern China. I was shocked by the size and complexity of the control room. I joked with the plant manager that one may need a PhD to work out what is what in the control room given all of the buttons, displays, and computers. Without smiling, he said that all operators had at least one PhD. This says a lot about the symbiotic relationship between digital technologies and energy production and supply. In fact, energy companies worldwide have been looking at new digital solutions for many years.

3.2.1.1 Getting One's Head Around Digital Energy Tech: How it Works

There are large numbers of digital technologies and solutions. A general understanding of what they are and how energy companies may leverage them will help us in the identification of present and future business or investments possibilities. Let us first try to identify what the tools used by energy companies are and explain how they are increasingly attractive to these companies.

The energy industry's progressive adoption of digitalisation is not new. It is part of the Fourth Industrial Revolution or Industry 4.0, which is impacting all sectors and industries, albeit to different degrees. For market incumbent energy generators or suppliers, it goes well beyond just upgrading their information technology hardware and software. For new market entrants it means convincing risk-averse energy companies, and often conservative consumers, that their solutions are better and will save them money or time. The energy industry's Fourth Industrial Revolution, Energy 4.0, means an acceleration in the deployment of energy tech solutions within the existing energy infrastructure as well as at the periphery of the grid that includes introducing new tools for big data and cloud operations. In essence the bulk of the digital technologies, be it proven ones or those still under development, have an intimate relationship with the various segments of the electric power value chain. Also, many of the Energy 4.0 solutions actually work in parallel or together. After an analysis of related literature, one author put the digital technologies and solutions in three broad baskets (Table 3.12). There are applications related to the balancing of the demand and supply of the electricity (system balance), applications for process optimisation where the solutions help to raise the efficiency and effectiveness of internal

Table 3.12: Digital Applications Categories Based on Findings in Recent Literature (Adapted).

System Balance	Smart Grid and Optimised Operations	**Smart Grid** – Real-time condition monitoring and load, voltage, and frequency control via remotely controllable transformers, generators, and consumers **Optimised generation** – Optimised operation of generation assets based on real-time monitoring, remote control, and direct communication with grid and market
	Smart Market and Flexibility Integration	**Integration of flexibilities** – 'Flexibilisation' of existing generation units and consumers – Creation of new flexibilities via storage units and sector coupling (e.g., batteries, e-mobility) – Marketing of flexibilities of generators, consumers, and storage units (e.g., virtual power plants, operating reserve, demand-side management, demand response) **Smart market** – Smart contracts – High frequency microtransactions – Peer-to-peer trades and peer-to-peer trading platforms – Variable tariffs for (small) generators or consumers

Table 3.12 (continued)

Process Optimisation	Anomaly Identification and Predictions	– Optimisation of forecasting for generation, demand, grid conditions, and prices. Minimisation of balancing energy – Optimised anomaly detection, failure localisation, and remote fixing – Predictive maintenance – Digital twin for optimisation of operations and maintenance and lifetime extension – Optimised strategic development of new business models, avoidance of risks, and optimised investment decisions – Analysis of customer interactions (Net Promoter Score, reasons for leaving) and prediction of customer wishes and reactions
	Process Efficiency	– Digital document management system – Digitally supported work preparation and documentation (availability of documents and instructions, virtual reality training, augmented reality support) – Robotic process automation (automation of reparative tasks), e.g., automated regulatory reporting – Connected internal processes with digital processes of customers and suppliers
Customer Orientation	Smart Home	**Smart home management and energy management services** – Desegregation and visualisation of energy consumption in real time – Energy-efficiency optimisation of appliances such as lighting, smart devices – Optimisation of tariff and consumption by owner of solar PV system, energy storage, e-mobility charging, etc. – Ambient assisted living and security systems
	Communication Channels	– Digitalised customer communication channels and applications of bots (i.e., computing autonomous programs)
	Trust and Transparency	– Automated billing based on real consumption – Certification of origin of electricity – Supplying the secure connection to the smart meter to third parties (e.g., for contracts submissions)

Source: Weigel, P. (n.d.). Digital Applications in the Energy Sector – A Review. [online] encyclopedia.pub. Available at: https://encyclopedia.pub/427 [Accessed 26 October 2020].

processes, and customer-service-focused applications, providing the company with additional benefits, including higher revenues, by helping to secure more customers or by providing additional services users.[35]

I created a schematic chart (Figure 3.23) showing some of the elements in the electric power value chain and many digital tech applications. Power elements include fossil fuel power plants, renewable energy facilities, distributed energy resources, and grid networks as well as power-related solutions, including energy storage and energy efficiency and conservation. I also listed some digital technologies that are tied to these power sector elements, including a variety of hardware and software solutions such as blockchain, big data and data management, or drones and remote sensing. Digital Twins and drones are two examples that can illustrate this. The Electric Digital Twins is the virtual representation of some electric power infrastructure, such as a transmission line. It can model the performance of the transmission line and exchange network or model data internally or externally efficiently and securely. Other solutions such as AI and big data processing solutions could help in the analysis of the virtual model representation. Benefits include time savings on the model's design and maintenance, higher accuracy, avoidance of duplication, and cutting information technology integration costs according to industrial products producer giant Siemens.[36] Drones, called Unmanned Air Vehicles (UAVs) in the industry, can be used to scrutinise electric power transmission networks for evaluating the condition of the high voltage power lines, and pylons can be affected or can suffer damage from storms, lightning strikes, and rust and corrosion, for example. Using UAVs to inspect and maintain transmission equipment will save operations and maintenance costs, as the operator will not have to rely on expensive helicopters and ground crews, also increasing safety and reducing hazardous work. Also, the UAVs carry high-definition cameras so they can get extremely close to the lines for inspection – they are also designed to not be affected by the overhead power lines magnetic field. They cheaply and effectively collect high-quality data, including three-dimensional (3D) data, which the operator can crunch to forecast, for example, maintenance schedules or to optimise designs.

Figure 3.23: Summary of Energy Industry Intersection with Energy 4.0.
Source: Author, October 2020.

In Energy 4.0, there are hardly any digital technologies or solutions, such as blockchain, cloud computing, or virtual and augmented reality, that are broadly or intimately not exposed to the energy sector (Figure 3.24). Many studies and insights have been published in the past two or three years about what market incumbents need to do for the transition to Energy 4.0. These reports, including several from top-tier global consultancies, including Deloitte, EY, and PwC, pretty much all point in the same direction – to aggressively invest time and money in energy tech solutions. Apart from Energy 4.0 optimising internal operations and the usage of assets, the primary focus must also be on Know-Your-Client (KYC) related data. The aim should be to have as much granular real-time information as possible. One of several studies I read was from the Economist Intelligence Unit[37] about the IoT and business applications. For the energy industry, the Economist Intelligence Unit placed the various applications into five broad categories. One is 'connecting applications', such as smart home appliances. Another is 'collecting applications', such as the gathering of user consumption data. A third is 'monitoring applications', including the real-time monitoring of the health of appliances at home or power-related equipment, such as a transformer, controlled by the energy company. Another is 'monetising applications' for generating value from the assets' data, such as a better distribution of capital expenditure for R&D Finally, is 'optimising applications' related to optimising assets such as alert systems for machinery or for equipment upgrades.

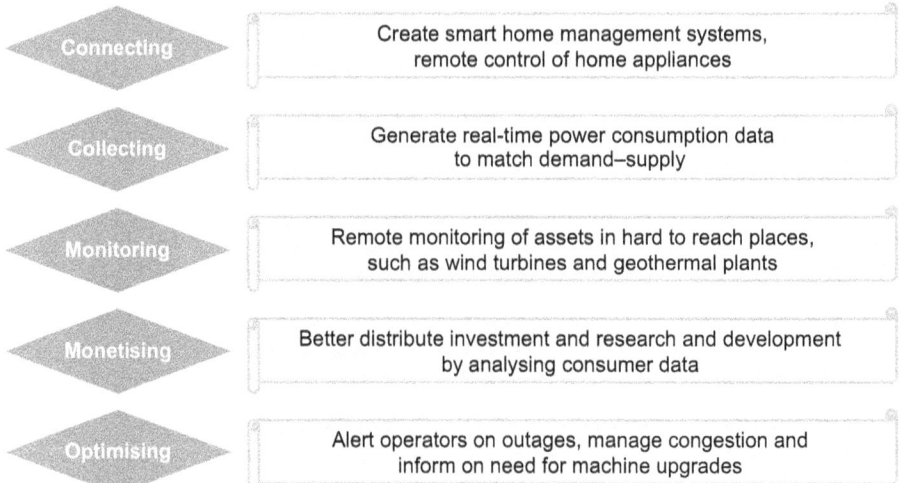

Figure 3.24: Internet of Things Applications for Energy Industry (Adapted).
Source: The Economist Intelligence Unit Limited (2020). The Internet of Things Applications for Business – Exploring the Transformative Potential of IoT. The Economist Intelligence Unit Limited, p. 11.

Many of the insights available have a particular focus on the smart home, i.e., the residential customer, in terms of the energy company relationship with the end user. I would argue that a more substantial one is the relationship with industrial and commercial users simply because they are much larger consumers of energy. It is not just smart homes that will have smart meters, sensors, and smart appliances and equipment hooked up to the cloud, linking the user directly to the energy company. Many commercial and industrial users in many jurisdictions, such as China, are already doing so. One key difference perhaps is that the company–residential user link must be totally transparent and its handling very simple and easy. This is because an average customer will not want to spend any time interacting with their home energy systems. They would expect full automation and a seamless process. On the other hand, a commercial or industrial customer will almost certainly want to dynamically manage its energy systems. They may appoint one or more individuals internally completely dedicated to this task or they could hire an external service provider, such as an energy management systems consultant, to do that for them.

In terms of a customer's personalised, data-driven experience, a Deloitte report[38] lists sensors, smart meters, and analytics as tools to facilitate the viewing of real-time and historic consumption, receiving usage and outage alerts, bill payment, and applying or cancelling new services offered by the energy provider. At the home energy management level, smart appliances and AI tools would be the user's virtual assistant, helping with managing energy usage as well as dynamically optimising the usage by price, time, green attributes, and the like. Energy 4.0 can help a home that also has some energy-generation capacity – being chiefly rooftop solar systems – to trade electricity with the energy company or others in real time (i.e., peer-to-peer transactions), e-mobility charging and energy storage solutions, for example. This list is in no way exhaustive but should provide a good understanding on the great number of options that will open up with Energy 4.0 to all users.

The highlights of the Deloitte report are aligned with the experience of Enel, the Italian utility. In my view it is one of the most proactive and advanced Energy 4.0 companies among the large-size utilities in the world. It has fully endorsed and is investing heavily in the digitalisation transition as part of its long-term strategy. For example, it had set aside almost 12 billion Euros in capital expenditure towards continued digitalisation and grid automation for 2020 to 2022, it had said in late 2019.[39] About two years earlier, Enel participated in an event organised by Amazon Web Services about an IoT proof-of-concept pilot project the two had completed in just six months. They had ensured that the project would be future proof, in other words able to be updated with the latest applications and solutions, just like when one upgrades a smartphone app or operating system. The Enel representative said that the company's objective was to open energy access to more people, to new technologies, to new ways for people to manage energy, and to new uses of energy. He added that the idea was to use the IoT to get closer to the customer, get data real time so that Enel can act on the data and produce insights to enable customers to use energy in a different way. Enel used Amazon Web Services' IoT to build a connected home solution that provides services for home automation, energy management, security, safety, and assisted living. He added that for industrial customers, the energy-management systems would allow the customer to use electricity in a more efficient way as well as saving expenses. Also, for Enel's own facilities the aim was to improve the performance of processes in areas such as predictive maintenance, materials management, logistics, health, and safety.[40]

Data and analytics are at the very core of the energy business with many uses. Some would include the management of materials supply and inventories, maintenance prediction and scheduling, plant optimisation, demand forecasting, risk assessment, and billing and payments. Another important consideration is understating that it can create new income for the companies or help the company to retain customers in a brutally competitive environment, such as that in Australia or the UK. In recent years energy companies have more aggressively sought new digital technologies, sometimes even investing in early stage technology companies so as to be ahead of the curve and competition. A major part of the reason is because many of the companies are trying to provide their customers with better services as well as services and products other than their traditional core ones. We mentioned several examples in Asia such as gas supplier Tokyo Gas and telecom services provider KDDI when discussing the changing nature of the industry players (see 2.3.4 Japan's Tokyo Gas, KDDI Cautiously Progressive Approach).

When a market completely opens up to competition and the end user has a real choice as to the preferred company for his or her electricity, gas, or water, the provider is rapidly forced to offer better services or more products, and ideally also include more value-added ones. So, they adopt digital solutions to their work flows and sharply modernise their information technologies infrastructure and capabilities.[41] This was highlighted, for example, in a report by the global consultancy McKinsey & Company.[42] It looked at some of the energy companies that are trying to transform themselves into digital enterprises as they turn themselves into comprehensive energy solutions providers as opposed to just energy suppliers, again a trend discussed in detail previously (see 2.3 Changing Nature of Industry Players).

In a book serving as a guide to prepare the managements of energy utilities for Energy 4.0, author, consultant, and ex-colleague Wayne Pales strongly emphasises the importance for energy companies to critically address the collection and usage of internal and external data. He importantly mentions what he coins the six truths facing the industry:
1. new energy technology is getting cheaper, physically smaller, and more efficient
2. everything is becoming connected
3. growth of data is increasing insights into companies and consumers
4. energy consumers are striving for greater choice and control
5. energy consumers expect simplicity
6. reducing the impact we have on our planet is an increasingly important part of investment decisions[43]

So, in a nutshell, industry participants are going to be collecting more and more data and it is imperative for them to process it effectively and efficiently, and use it strategically so as to be able to be competitive in increasingly complex energy markets.

3.2.1.2 New Digital Energy Tech, Many New Benefits

Energy 4.0 provides a great number of benefits as one can easily appreciate from Section 3.2.1.1 above. At the revenue level, it can enhance the competitiveness of an energy company or provide tools to successfully retain customers. At the capital expenditure level, it can optimise the spending, such as getting a more accurate view of what equipment needs to be replaced and the ideal timing. A crucial area of course is operating costs. It can help to reduce expenses overall through the better management of operations. Here I will present a series of examples detailing the potential financial savings. The savings amount will differ case by case, but the general trajectory can be easily understood. The variance is due to different assessments approaches and the different experiences of the companies doing the evaluation. Still, it has to be appreciated that much of this is very new. In years to come, the tech will get even better and cheaper, so the financial benefits will be even greater.

The Deloitte report mentioned earlier provides a nice, easy to understand table (Figure 3.25) showing where the benefits of Energy 4.0 can actually be identified. They include the impact on revenues, on costs, on news entrants, and on new products and services, and whether there are opportunities to create value through digital innovation. When it comes to new digital technologies their rapid entrance is resulting in sharp cuts 'for key building blocks such as computing power, data storage, and internet bandwidth'.[44] The lower costs are allowing for a quicker, and cheaper, introduction of sensors and AI. On the new energy generation and supply technologies, Deloitte argues that the fall in the LCOE of many clean technologies is a new business opportunity for established energy companies. But also, that they have to take advantage of these opportunities so as to remain competitive with 'start-ups, entrepreneurs, and companies from adjacent industries who may compete with incumbents'.[45] Regarding new practices and business models, in essence the incumbent energy companies need to adapt or perish. The old business model, often based on a monopoly or oligopoly environments, no longer works in this new energy transition world. Flexibility, an open mind, and fast reactions will be key (Figure 3.25).

A study published by International Energy Agency in 2017[46] also found several areas of cost savings from Energy 4.0. The agency calculated that over the period from 2016 through 2040, the savings from adopting Energy 4.0 solutions for power generation and electric power networks worldwide could amount to almost $84 billion, in 2016 dollars. A life extension for power plants of about five years accounted for 41 per cent of the total. A five-year life extension for networks would amount to 24 per cent. Five per cent in lower operations and maintenance expenses would account for 23 per cent. And 5 per cent in lower total network

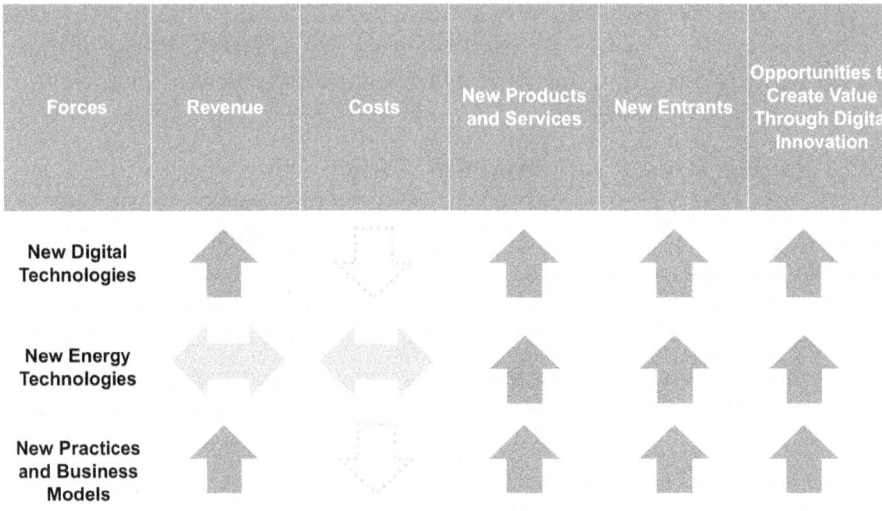

Figure 3.25: Disruptive Forces' Potential Impact on Businesses in the Power Sector (Adapted). Source: Wei, J., Sanborn, S., and Slaughter, A. (2019). Digital Transformation and the Utility of the Future | Deloitte Insights. [online] www2.deloitte.com. Available at: https://www2.deloitte.com/us/en/insights/industry/power-and-utilities/digital-transformation-utility-of-the-future.html [Accessed 22 October 2020].

losses and in higher electricity output per unit of fuel amounted for 7 per cent and 5 per cent respectively (Figure 3.26).

A third analysis is by another global consultancy, McKinsey & Co, and was published in 2018.[47] This study also evaluated the ways that Energy 4.0 can raise revenues and reduce operating costs for energy companies (Table 3.13). McKinsey concluded that Energy 4.0 adoption, including process automation, digital enablement, and advanced analytics would reduce expenses in the energy company's value chain of between 11 and 26 percent. Breaking this down it was 11 per cent savings at the generation end, 26 per cent for transmission and distribution and related network costs, 25 per cent for customer and retail, and a 21 per cent reduction in expenses at the company's corporate centres.

I sought some real-world examples and identified a few that provide some granularity as to the financial impact of adopting some Energy 4.0 solutions. They come from a services and appliances provider, an industrial equipment maker, a hardware to software tech company, and lastly from an electronics manufacturer.

Schneider Electric is a French-based multinational focused on providing services related to energy and automation digital solutions for efficiency and sustainability through combining energy technologies, real-time automation, software and service (Table 3.14). The company has fully recognised that the world is heading towards a digitised future. They published a study in early 2019 about the impact of Energy 4.0

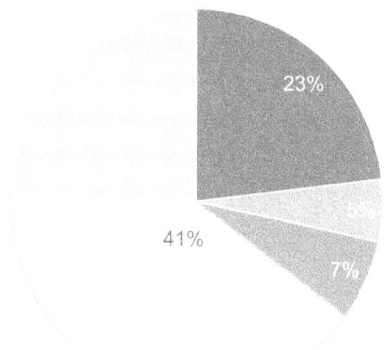

- 5% lower O&M costs
- Efficiency: 5% more electricity output per unit of fuel
- Efficiency: 5% lower total network losses
- 5 year life extension for power plants
- 5 year life extension for networks

Figure 3.26: Worldwide Cost Savings from Enhanced Digitalisation in Power Plants and Electricity Networks over 2016–2040 (Adapted).
Source: Turk, D. and Cozzi, L. (2017). Digitalization & Energy. [online] iea.org. Paris, France: International Energy Agency. Available at: https://www.iea.org/reports/digitalisation-and-energy [Accessed 24 October 2020].

Table 3.13: Illustrative Potential Worldwide Cost Savings from Enhanced Digitalisation in Power Plants and Electricity Networks over 2016–2040 (Adapted).

(Percentage)	Generation	Transmission and Distribution	Customer and Retail	Corporate Center
Process Automation	2.1	3.3	2.7	8.9
Digital Enablement	3.4	10.3	12.7	3.5
Advanced Analytics	5.5	12.4	9.1	8.9

Source: Booth, A., de Jong, E., and Peters, P. (2018). The Digital Utility: New Challenges, Capabilities, and Opportunities. [online] mckinsey.com, p. 10. Available at: https://www.mckinsey.com/~/media/McKinsey/Industries/Electric%20Power%20and%20Natural%20Gas/Our%20Insights/The%20Digital%20Utility/The%20Digital%20Utility.ashx [Accessed 24 October 2020].

on 230 of its customers, assessing 330 data points.[48] It concluded that the digital transition undertaken by these customers generated an increase of up to 50 per cent in productivity, 50 per cent in carbon footprint optimisation, 75 per cent in maintenance cost optimisation, 85 per cent in energy-consumption savings, and 80 per cent in engineering cost and time optimisation.

Swiss-Swedish industrial giant ABB has a large portfolio of products, systems, services, and solutions for the electric power and other industries. It has been one of

Table 3.14: Schneider Electric Survey of 12 Key Business Benefits of Digital Transformation.

BENEFIT	UP TO	AVERAGE
Capital Expenditure		
Engineering costs and time optimisation	80%	35%
Commissioning costs and time optimisation	60%	29%
Investment costs optimisation	50%	23%
Operating Expenses		
Energy consumption savings	85%	24%
Energy costs savings	80%	28%
Productivity	50%	24%
Equipment availability and uptime	50%	22%
Maintenance costs optimisation	75%	28%
Sustainability, Speed, and Performance		
CO^2 footprint optimisation	50%	20%
Time to market optimisation	20%	11%
Decrease in occupant comfort-related incidents	33%	24%
Return on investment	0.75 years	5.30 years

Source: Schneider Electric (2019). 2019 Global Digital Transformation Benefits Report: Schneider Electric. [online] www.se.com. Available at: https://www.se.com/ww/en/download/document/998-20387771_DTBR/ [Accessed 25 October 2020].

the leading companies globally in Energy 4.0, in my view, looking at digital solutions to raise the efficiency and reduce costs for its customers. One example of this is with switchgears, a market worth over $100 billion a year. Switchgears are a critical component for power supply. They are typically in substations and serve to supply the electricity from high-voltage power (6 to 36 kilovolt) from generation plants and utility networks to low-voltage distribution networks. ABB has been promoting digitalising switchgears, which can help the facility owner save up to 30 per cent thanks to operating costs optimisation, such as by massively improving the monitoring of the switchgears' circuit breakers, power monitors, motor and feeder controls and protection devices, which decreases down-time.[49] The company also provides digital technologies to lessen energy and maintenance expenses for industrial sites and buildings by as much as 30 per cent[50] – something that we had discussed with the case study from Aden in China (see 2.1.3.4 Making a Business Out of Saving Energy). Another example is from ABB's Industrial Automation business and a new project virtualisation solution – i.e., Electrical Digital Twins – for large capital projects, common in the energy sector. Using digital technologies, this solution promises to lessen projects' cost and schedule overruns through shrinking automation related capital costs by up to 40 per cent, and lower delivery schedules by up to 30 per cent, and start-up hours by as much as 40 per cent, including up to 85 per cent less project installation, commissioning and testing engineering hours.[51]

Tech giant IBM commissioned a research firm, Forrester Consulting, to evaluate the economic benefits of IBM clients employing its blockchain platform and services solution.[52] Admittedly, IBM paid for the study for product promotion, but there were valuable insights for our analysis on the cost benefits of Energy 4.0. Forrester well summarised the type of business situations that would find blockchain solutions useful. It argued that circumstances include when multiple parties need to access the same data or when they need assurance that the data is valid and tampering free, for example. One of the conclusions is that the blockchain solution introduced new revenue streams to some businesses through, for example, simply charging blockchain customers a fee for every transaction. Another is a reduction in both capital and operating expenses as the blockchain solution would allow the business to maximise the use of its assets through opening access to the resources, including, for example, a car fleet or energy distribution. Yet another is cost savings by more efficient billing and disputes caused by inconsistent documentation, where legal and financial resources dedicated to conflict resolution may be reduced by 70 to 80 per cent. The study also highlights that the potential payback in a low case scenario can be 36 months and may generate a return on investment of 43 per cent, and in a high-case scenario it would be 10 months and 590 per cent.[53]

The fourth and last real-world example aimed at providing some granularity as to the potential financial impact of Energy 4.0 is from another large corporation – China headquartered Huawei Technologies, a major global information and communications technology manufacturer. The company has quite a few businesses directly or indirectly in the energy arena, such as solar photovoltaic and related equipment. Huawei has published[54] that it forecasts there will be at least 6.5 million 5G telecommunications base stations and 2.8 billion 5G users globally by 2025. The company said that its Huawei 5G base stations will be as much as 30 times more energy efficient than 4G ones, although the electricity consumption will be similar, in other words more output for the same amount of energy used. Importantly, at a global level Huawei also forecasts innovation will lead the information and communications technology industry's energy conservation solutions to lower average annual carbon emission per connection to 15 kilos in 2025, 80 per cent lower than that (75 kilos) in 2015.

Energy 4.0 and the new tech solutions will be of progressively greater importance to global decarbonisation and environmental sustainability. At a high level, we can confidently state that Energy 4.0 is present at every block of the energy value chain as one can gather from the various technological solutions described in Sections 3.2.1 and 3.2.1.1. Be it proven technologies or those still developing. For example, blockchain, big data and data management, mobile connectivity, and other solutions can all help to verify and optimise electric power production from a fossil fuel power plant, reducing the plant's emissions. Importantly, it will help for the management by the grids of variable renewable energy (solar and wind), and so will be crucial for the transition from brown to green.

I am of course not forgetting about the heated environmental debate about tech. The argument of those concerned over the massive increase in the physical number of tech apparatus, such as smart tablets, is that the production and usage of these technologies actually increases emissions. They argue, for example, information and communication technology hardware is associated with the extraction of scarce metals or toxic materials in the production process. Another argument is that the massive increase in the number in data centres or EVs actually increases energy demand, which will mean the need for additional generation facilities. These arguments are of course true. However, I am not overly concerned. This is because there is a clear global awareness of this problem and because I believe Energy 4.0 is helping to create a better, more sustainable world, and that the additional side effects are being, and will continue to be, addressed and mitigated, and ultimately resolved. I guess it is an argument not too different from the environmental Kuznets curve,[55] famous in debates two to three decades ago, and still going; the 'original Kuznets curve postulates that income inequality first rises and then decreases in the process of economic development. Similarly, the EKC suggests that environmental impacts rise and then decrease in growing economies.'[56] I am quite certain that the debate will continue for some time.

3.2.2 The Real Life of Digital Energy Tech: Some Use Cases

In Sections 3.2.1.1 and 3.2.1.2, I hopefully achieved three things. One is providing an understanding of some of the major forms of digital energy technologies and solutions. Another is examples of interactions between the technologies and the various bits of the energy generation and supply value chain. And finally, some empirical mini case studies on the financial and environmental benefits the energy industry can harvest from Energy 4.0. Before discussing China's present and future leading role in the last section of this chapter, I will first present some further colour on the current and future applications and solutions from three angles. They include experiences from one of Asia's largest utilities, from a small nation in the region, and from a variety of new entrants and start-ups, particularly some revolving around AI, blockchain, and the IoT. All of these experiences will show that new digital energy tech and solutions for energy are not conceptual, they are a reality already, albeit some are not as yet economically viable or are yet to be widely accepted.

3.2.2.1 South Korea Has a Surprising Tech Innovator
This should not be surprising to many people: state-owned utilities in Asia and elsewhere in the world are typically only tasked to run the energy system effectively, efficiently, and safely. And, for some, profitability may also be a key performance indicator, although not always. On the very surface, these precepts are little

different when it comes to Korea Electric Power Corporation (KEPCO) relative to its peers. I covered the company as a financial analyst for about a quarter of a century. What was surprising to me is the drive the company has to consistently strive to do better. In part this may be due to the company culture. It may also be related to the fact that it operates in a fairly unique energy market. Features including the fact that it is a market that is not interconnected with any neighbouring networks, a market where the government curtails tariffs adjustments causing some financial stress on KEPCO, and a market where the government has decided to take the decarbonisation route. Also, it is a market where the government does offer massive support for innovation domestically so that the technology or knowhow can be exported.

First, let us put KEPCO into context in terms of its profile. KEPCO is a vertically integrated electric power utility chiefly with operations in South Korea; its domestic versus overseas business ratio was 94.4:5.6 in the six months through June 2020. The corporation is 58 per cent owned by the government and state-owned enterprises, while 42 per cent of the shares are listed on stock markets, as of the end of 2019. The corporation generated about 67 per cent, and transmitted, distributed, and sold about 100 per cent of the nation's electricity, more than 520 terawatt-hours in total. It had assets worth 197.6 trillion won ($174.3 billion) and sales of 59.2 trillion won ($52.2 billion). It also had 45 projects in 25 countries including generation, transmission, and distribution, as well as smart grid projects.

What is good about KEPCO's innovation is the amount of resources it has put into digital tech and solutions so as to be more efficient and be able to export its knowhow abroad, too. This was well before decarbonisation hit the headlines at home and abroad post the 2015 Paris Agreement. KEPCO has been spending about 720 to 735 trillion won ($635 to $647 million) annually in the three years through 2019, representing roughly 1.2 per cent of its revenue, on R&D (Figure 3.27); I should point out that this is only the amount recognised in its financial statements and does not include expenses it may be writing off or capitalising (i.e., included in its overall capital expenditure). As such, the actual R&D amount could be even larger. While the annual spending goes into a great variety of baskets and that on digital technologies is not split out, the amount should be in the tens of millions at a minimum. I should note that in my opinion spending would be even higher as if KEPCO's financial position were stronger. Its revenues would be higher if the government had raised end-user tariffs, as it is supposed to, given the higher fuel costs that the company incurred during the 2017 to 2019 period I am looking at here. In fact, the company posted negative income after tax numbers in 2018, 2019, and 2017 fell 82 per cent.

At the end of 2018, the company reshaped its long-term business plans as it adopted a pro-Paris Agreement strategy. It said it would target to have 20 per cent of its output from renewable energy sources by 2030 as well as looking at energy conservation and efficiency solutions, which of course are tech intensive. It devised a four-pronged approach (Table 3.15). One is focusing R&D and other investment efforts on

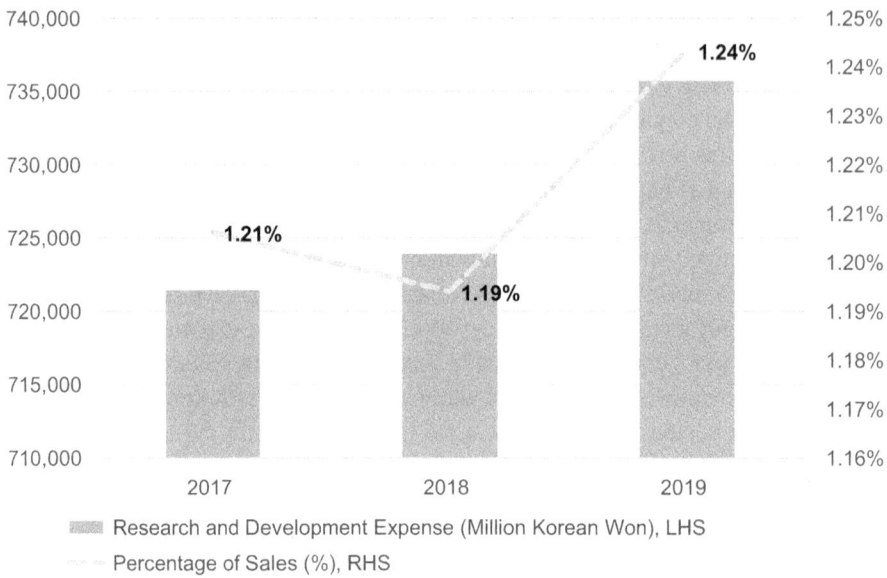

Figure 3.27: KEPCO's Research and Development Expenditure.
Source: Author, October 2020. Data sourced from: Korea Electric Power Corporation (2020a). R&D Investment | KEPCO. [online] home.kepco.co.kr. Available at: http://home.kepco.co.kr/kepco/EN/ntcob/ntcobView.do?pageIndex=1&boardSeq=21047780&boardCd=BRD_000510&menuCd=EN04010901&parnScrpSeq=0&searchCondition=title&searchKeyword= [Accessed 28 October 2020].

the Fourth Industrial Revolution, including energy platforms and digital power systems; it does not call this Energy 4.0 but KEPCO 4.0. Then it is a focus on a climate change response, including energy-efficiency improvements and clean convergence power generation – referring to using fossil fuels but reducing their emissions such as capturing the carbon. Another is an energy transition focus, which comprises grid-energy storage systems, and new and renewable energy. Lastly, it is a focus on the power grid enhancement, which means building a super grid (e.g., extremely long transmission networks linked with other countries), and on an active distribution network geared towards optimising the demand–supply balance. All this is captured in Table 3.15 which I adapted from KEPCO but purposely added a column about my perceived amount of exposure of digital energy technologies and solutions to each individual goal to show the likely level of the realm of tech investments. I believe that half of the eight objectives will chiefly revolve around digital energy technologies and solutions while for the other half they will play a significant role.

KEPCO's list of strategic technologies was published in its 2019 sustainability report. In the 2018 edition, it had actually already detailed the digital energy related technologies that it was focusing on. It broke these down into how it would develop the technologies in six areas and explained how it would go about securing these technologies. The six areas revolve around sensors, IoT, the cloud, big data, AI, and robots

(Table 3.16). KEPCO actually has implemented these areas in various settings, many through a wholly owned subsidiary, KEPCO KDN, which is solely involved in information and communication technologies, including smart grids and what it calls AICBM (AI, IoT, Cloud, Big Data, Mobile).

Table 3.15: KEPCO 2030 Technical Strategies and Developing Eight Core Strategic Technologies (Adapted).

Fourth Industrial Revolution	Energy Platforms	Big Data Platform; AI Platform; Business Platform	DigiTech: HEAVY
	Digital Power Systems	Digital Power Plant; Intelligent Substation; Asset Management System	DigiTech: HEAVY
Climate Change Response	Energy Efficiency Improvement	Smart Energy City; Demand Response and Behind-The-Meter Solution; High-Efficiency / Eco-Friendly Devices	DigiTech: HEAVY
	Clean Convergence Power Gen	Capturing Carbon and Conversion into Resources; Pure Oxygen Power Gen; Supercritical CO_2 Power Gen	DigiTech: MEDIUM
Energy Transition	Grid-Energy Storage System	EESS Operation Tech; ESS Engineering; Hydrogen Energy (Power-2-Gas)	DigiTech: MEDIUM
	New and Renewable Energy	Next-Gen Wind Power; Renewable Energy Engineering; Renewable Energy with Other Industries	DigiTech: MEDIUM
Power Grid Enhancement	Super Grid	Power System Connections among Countries; HVDC Transmission Tech; HVDC Transformation Tech	DigiTech: MEDIUM
	Active Distribution Network	Active Distribution Network Tech; Distributed Gen Connection; DC Distribution Tech	DigiTech: HEAVY

Source: Author, October 2020. Data sourced from: Korea Electric Power Corporation (2020). Clean Energy Smart KEPCO: Sustainability Report 2019. [online] Naju-si, Jeollanam-do, Republic of Korea: KEPCO Corporate Strategy Team, Corporate Planning Department, p. 38. Available at: http://home.kepco.co.kr/kepco/EN/D/C/KEDCPP004.do [Accessed 28 October 2020].

KEPCO is not developing its Energy 4.0 just for self-use. It wants Energy 4.0 to become an important future income stream. One example is microgrids. It completed two island microgrids demonstration projects in South Korea, which won a prestigious international smart grid award in 2018. A related one is smart cities. The company is playing a leading role in building up a template that can be used nationwide. There are more than 80 smart city projects in the country at various stages of development, as of early 2019, so KEPCO can easily secure more demonstration sites to further develop its integrated energy management and other technologies. KEPCO targets growing these new business areas together with EV charging infrastructure and

Table 3.16: Major Technologies of KEPCO 4.0 (Adapted).

	Sensor	IoT	Cloud	Big Data	AI	Robot
Development of Core technologies	Develop 33 sensors for power generation, transmission, substations, and distribution	Acquire the international standard for the in-house developed protocol	Transform the entire work process in to a cloud system	Establisha KEPCO big data platform	Develop basic AI technologies including status diagnostics	Develop unmanned robots and drones to monitor and inspect electric power facilities
Strategies to Secure Technologies	Develop the energy harvesting technology to secure sources for the sensors	Build infrastructure for IoT	Identify new business models base on the cloud	Build infrastructure to store and analyze big data	Develop AI machine learning technologies	Become the first public corporation to develop AI-based chat bots

Source: Korea Electric Power Corporation (2019). Clean Energy Smart KEPCO: Sustainability Report 2018. [online] Naju-si, Jeollanam-do, Republic of Korea: KEPCO Corporate Strategy Team, Corporate Planning Department, p. 38. Available at: http://home.kepco.co.kr/kepco/EN/D/C/KEDCPP004.do [Accessed 28 October 2020].

management systems, for example. Its near-term target was to realise about 0.4 trillion Korean won ($0.35 billion) in annual revenues by 2020 or so. It wanted to raise the amount to 1.3 trillion Korean won ($1.14 billion) by 2022 and then to 7.5 trillion Korean won ($6.6 billion) by 2030 (Figure 3.28).[57]

3.2.2.2 Asia's Little Big Digital Energy Technologies Giant

Singapore may be a small place but it thinks big. Singapore has been highly proactive with its energy sector. In the past few decades it shifted away from predominantly oil generation, in favour of natural gas, and now increasingly solar power. Now the city state wants to make another bold move, a sustainable energy future. One, it wants to sharply increase the amount of solar-installed capacity to 2 gigawatts by 2030. Two, it is exploring ways to be part of a regional grid, which would include neighbouring countries such as Indonesia and Malaysia. Three, it is evaluating the potential of deploying emerging solutions including hydrogen and carbon capture, utilisation and storage (CCUS). And finally, it will continue to support the use of natural gas, given the nation's limited renewable energy resource. The backbone for the development of all of these solutions is digitalisation and that of a smart city.

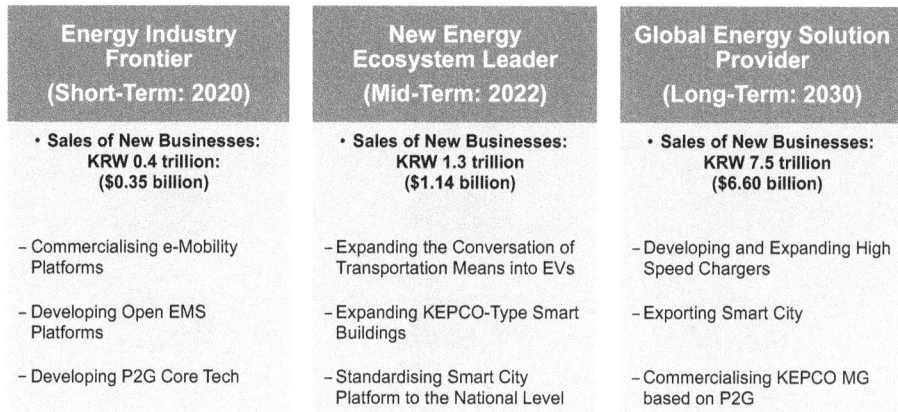

Figure 3.28: KEPCO's New Industries Expansion (Adapted).
Source: Korea Electric Power Corporation (2020). Clean Energy Smart KEPCO: Sustainability Report 2019. [online] Naju-si, Jeollanam-do, Republic of Korea: KEPCO Corporate Strategy Team, Corporate Planning Department, p. 36. Available at: http://home.kepco.co.kr/kepco/EN/D/C/KEDCPP004.do [Accessed 28 October 2020].

The public and private sectors will continue to work together to build out a smart grids that are coupled with remote monitoring will balance the supply from gas power plants, solar PV generation, and energy storage with the demand from industry, buildings, smart homes, and e-mobility (Figure 3.29).

Leveraging digitalisation, especially through data analytics, process automation, and remote monitoring, will be key for these interactions to work successfully. Singapore already ranks as the top smart city in the world (Table 3.17) and government statutory agency the Energy Market Authority is leading the charge with the help of industry. It said that it had co-created more than 40 innovative solutions as of 2020. And I am sure there will be many more to come. Examples of these solutions include automated pipeline monitoring using drones, more energy efficient data centres, power plant digitalisation solutions to promote higher reliability, and smart grid management systems in ports.[58] Another, S$10 million ($ 0.83 million), innovation example is a partnership between the Energy Market Authority and Keppel O&M, a private company, aimed at deploying a 7.5 MW/7.5MWh lithium-ion battery floating energy storage system, storing enough electricity to power more than 600 four-room apartments.

3.2.2.3 The New Breed of New Energy Digitalisation Market Entrants

One of the fastest growing areas in the energy digitalisation world are companies, particularly start-ups, trying to leverage, blockchain, AI, and the IoT. Established companies, including KEPCO and TEPCO and many others, have been researching and developing solutions in this area. But also, there are many newer entrants also

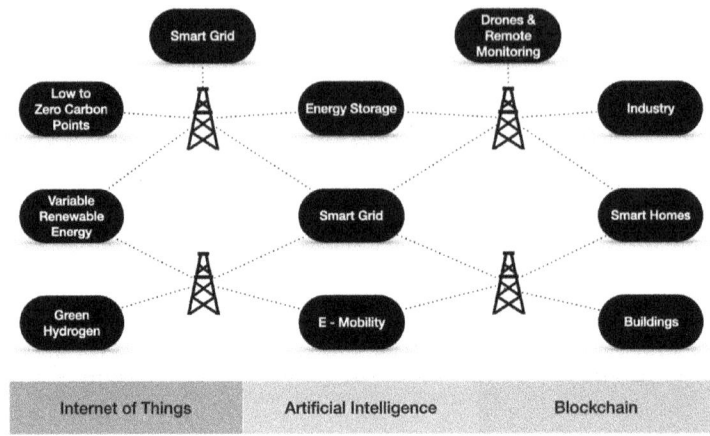

Figure 3.29: Schematic View and Some Features of Smart Grid.
Source: Author, October 2020. Source of Text: National Smart Grid Mission (2020). Smart Grid | National Smart Grid Mission, Ministry of Power, Government of India. [online] Nsgm.gov.in. Available at: https://www.nsgm.gov.in/en/content/smart-grid [Accessed 29 October 2020].

Table 3.17: IMD Smart City Index 2019 and 2020.

	2019 Ranking		2020 Ranking
#1	Singapore	#1	Singapore
#2	Zurich	#2	Helsinki
#3	Oslo	#3	Zurich
#4	Geneva	#4	Auckland
#5	Copenhagen	#5	Oslo
#6	Auckland	#6	Copenhagen
#7	Taipei	#7	Geneva
#8	Helsinki	#8	Taipei
#9	Bilbao	#9	Amsterdam
#10	Dusseldorf	#10	New York

Source: IMD (2020). Smart City Index 2020. [online] IMD Business School. Available at: https://www.imd.org/smart-city-observatory/smart-city-index/ [Accessed 29 October 2020].

looking for ways to monetise this form of energy digitalisation. I will broadly explain the solution and its applications and then present some use cases, among the dozens currently all around the world, particularly looking at some of the slightly more established new breed of companies.

The three technologies complement one another comfortably. The IoT is the connectivity piece that will be responsible for collecting the data as well as providing data. Blockchain is the key bridge or infrastructure combining IoT and AI and setting up the rules of engagement. AI crunches the data as well as optimising the processes and the rules. The three can work with autonomous agents including sensors, EVs, IoT devices and other equipment (Figure 3.30).[59] I would think that IoT and AI or machine learning are relatively easy to understand for the non-expert given that they are an integral part of our new digital everyday life. For IoT, there are a great number of smart devices such as smart refrigerators, fitness trackers, and smart home security systems. For AI, it is the technology used for opening one's phone with face ID, searching information on the Internet, or the recommendation engines found in social media. Blockchain, however, is a little bit more complex to understand even at a high level; at least it was for me when I first came across it five years ago or so.

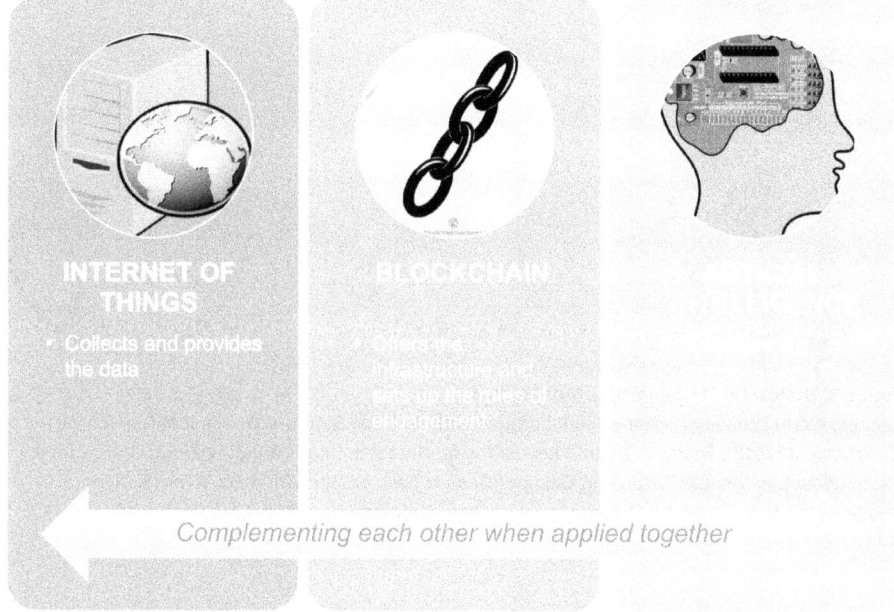

Figure 3.30: Convergence of Blockchain, IoT, and AI.
Source: Author, October 2020. In part sourced from: Sandner, P. (2020). Blockchain, IoT and AI – a Perfect Fit. [online] Medium. Available at: https://medium.com/@philippsandner/blockchain-iot-and-ai-a-perfect-fit-c863c0761b6 [Accessed 30 October 2020]. Images from: Publicdomainvectors.org.

At a high level, the definition of blockchain technology is a digital record of transactions where 'blocks' (i.e., individual records) are linked together in a 'chain' (i.e., a single list. So how is the transaction recorded? First, a transaction is requested. This could be from an individual or it could even be automated. The requested transaction is then broadly broadcasted to a peer-to-peer network, which is made up of computers, called nodes. The transaction is then verified by all of the participants of the blockchain. After the verification is complete the transaction is combined with other transactions, which creates a new block of data for the ledger. The new blockchain is added to the existing blockchain and becomes permanent and unalterable. Then the transaction is complete (Figure 3.31).

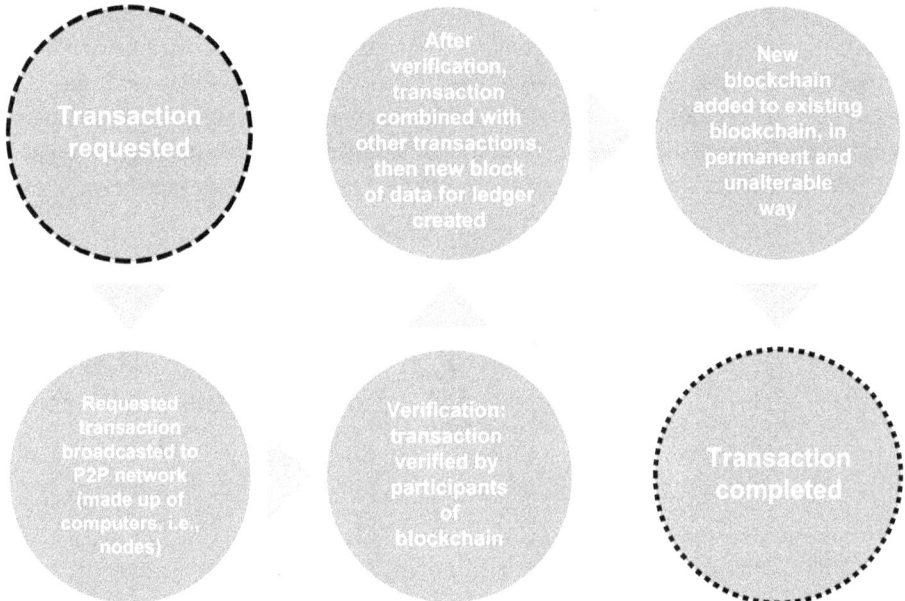

Figure 3.31: How Blockchain Works (Adapted).
Source: World Economic Forum in collaboration with PwC and Stanford Woods Institute for the Environment (2018). Fourth Industrial Revolution for the Earth Series Building Block(chain)s for a Better Planet. [online] weforum.org, Cologny/Geneva Switzerland: World Economic Forum, p. 10. Available at: http://www3.weforum.org/docs/WEF_Building-Blockchains.pdf [Accessed 30 October 2020].

The International Renewable Energy Agency is one of the many entities that has published studies on the convergence of the three technologies. It looked specifically at blockchain in the energy industry. It identified that there were almost 190 companies working in this area, over 71 projects, and about $466 million in investment as of September 2018; needless to say, the growth has continued in the past couple of years. The agency identified 30 key innovations where AI, IoT, and blockchain play a significant role. These include enabling technologies such as e-charging, business models such as peer-to-peer electricity trading, market designs such as net billing schemes, and system operations such as the advanced forecasting of variable renewable generation (Table 3.18).

Table 3.18: Irena Innovation Landscape Brief: Blockchain (Adapted).

ENABLING TECHNOLOGIES	BUSINESS MODELS	MARKET DESIGN	SYSTEM OPERATION
Utility scale batteries	Aggregators	Increasing time granularity in electricity markets	Future role of distribution system operators
Behind-the-meter batteries	Peer-to-peer electricity trading	Increasing space granularity in electricity markets	Co-operation between transmission and distribution system operators
Electric-vehicle smart charging	Energy-as-a-service	Innovative ancillary services	Advanced forecasting of variable renewable power generation
Renewable power-to-heat	Community-ownership models	Re-designing capacity markets	Innovative operation of pumped hydropower storage
Renewable power-to-hydrogen	Pay-as-you-go models	Regional markets	Virtual power lines
Internet of Things		Time-of-use tariffs	Dynamic line rating
Artificial intelligence and big data		Market integration of distributed energy resources	
Blockchain		Net billing schemes	
Renewable mini-grids			
Supergrids			
Flexibility in conventional power plants			

Source: Ratka, S., Boshel, F., and Anisi, A. (2019). Innovation Landscape Brief: Blockchain. [online] irena.org, Abu Dhabi: International Renewable Energy Agency, p. 4. Available at: https://www.irena.org/-/media/Files/IRENA/Agency/Publication/2019/Sep/IRENA_Blockchain_2019.pdf [Accessed 1 November 2020].

All of the above-mentioned developments apply very much worldwide, including the Asia region. I undertook an evaluation of AI, Blockchain, and IoT of pilot projects in individual Asian electricity markets. Out of 22, I only found four that did not have any publicly available information regarding possible trials. The four markets mostly happened to be frontier markets, at an early stage of economic development. They include Cambodia, Laos, Mongolia, and Myanmar. Unsurprisingly the most vibrant markets in terms of the number of pilots are Asia's more developed economies including Australia, Japan, New Zealand, Singapore, South Korea. Surprisingly, there is little publicly available information on developments in Hong Kong and Macau, despite the massive amount of innovation in this front in China. The other markets where I found a large number of references are Bangladesh and Thailand (Table 3.19).

Table 3.19: Snapshot of AI/Blockchain/IoT/Pilot Projects in Individual Asian Power Markets.

Power Market	Pilot(s)	Power Market	Pilot(s)
Australia	✓	Mongolia	N/A
Bangladesh	✓	Myanmar	N/A
Cambodia	N/A	New Zealand	✓
China	✓	Pakistan	✓
Hong Kong	✓	Philippines	✓
India	✓	Singapore	✓
Indonesia	✓	South Korea	✓
Japan	✓	Sri Lanka	✓
Laos	N/A	Taiwan	✓
Macau	✓	Thailand	✓
Malaysia	✓	Vietnam	✓

Source: Author, October 2020.

Asia's leading utilities are all involved in digital innovation projects. The names that I have come across the most are Australia's AGL Energy, Energy Australia and Origin Energy, China's SGCC, Hong Kong's CLP Holdings, Japan's Kansai Electric Power and TEPCO, South Korea's KEPCO, and Thailand's Electricity Generating Authority of Thailand. Yet there are a great number of start-ups working hard on developing their own Energy 4.0 technologies and solutions. Apart from those in China, which are discussed in Section 3.2.3, some of the higher profile names native to Asia are Australia's Power Ledger and Singapore's Electrify.

Power Ledger is a blockchain technology developer with a focus on energy and environmental commodities trading. The Perth-headquartered tech developer was founded in mid-2016. Less than five years on, it has managed to build partnerships with 24 companies, have 21,000 users of its technology, and has launched pilot projects in Australia, India, Japan, Malaysia, and Thailand as well as in Europe and

North America, as of June 2020.[60] There is limited publicly available information regarding its capital funding. In 2017, it secured A$8 million ($5.6 million) in government grants and partners' funding and raised A$34 million ($23.8 million) through a utility token offering; a utility token is different from a cryptocurrency or a security token, as utility tokens only offer the owner the right to a service or product in the same way as a pre-paid voucher or a cinema or concert ticket, and in the Power Ledger context it is a means of payment on the platform developed by the issuing company. We can understand what Power Ledger offers through two examples.

In late 2018, Power Ledger launched a pilot project in Bangkok. It partnered with one of the largest domestic real estate developers, Sansiri, and a Thai renewable energy projects developer, BCPG, chiefly with solar and wind assets in Thailand and as well as some hydro projects in Laos and solar projects in Japan. The project involves a 635 kilowatts solar rooftop development at Sansiri's upmarket 20-acre (about 81,000 square meters) mixed-use development in the centre of Bangkok called Town Sukhumvit 77 (T77). A shopping centre, a large school, an apartment building, and a dental clinic were selected for the peer-to-peer energy trading proof-of-concept project. The generation capacity is divided into 54 kilowatts for the shopping centre, 413 kilowatts for the school, 168 kilowatts for the apartment, plus another separate 150 kilowatts in addition to an energy storage facility operated by BCPG. The electricity will be consumed by each of the buildings but as some will consume more than they produce and others produce more than they consume, the excess can be traded through a BCPG computer and smartphone trading application. If there still is excess produced, it can be sold to the energy storage or to the grid, which is operated by a state-owned utility Metropolitan Electricity Authority (MEA). The trial has shown that the power traded reduced electric power bills from MEA by about 10 per cent and that excess output from solar panels was 10 per cent, said a BCPG representative. I should point out that currently Thailand is not an open market, so this pilot was in a sandbox (i.e., isolated) environment. All parties have been seemingly quite happy with the results, including the government. BCPG and Sansiri are keen to replicate the T77 pilot at other Sansiri property developments in the country. In 2019, BCPG established a new company, Thai Digital Energy Development, with a state-controlled company tasked to engage in energy investment related businesses. Power Ledger signed an exclusive partnership with Thai Digital Energy Development for developing peer-to-peer energy and environmental commodities, including renewable energy certificates and carbon credits, trading solutions in Thailand in May 2020. This will start with a 12 megawatt project at the campus of a university in the Northern Thai city of Chiang Mai.[61]

Electrify is another company with a focus on energy and environmental commodities trading through blockchain technology. The Singapore-headquartered company, which calls itself an energy innovator, was founded in mid-2017. In 2018, it raised $30 million through a token offering. It has not managed to build as large a footprint as Power Ledger. It currently is running a pilot project in Singapore, although it has been actively looking for opportunities in the rest of Southeast Asia. Electrify operates its 'Energy Marketplace' application for producing retail contracts after evaluating the consumer's usage data so they can choose the most suitable provider. The company says that it has transacted more than 60 gigawatt-hours and saved 500 commercial and industrial companies a combined S$1.5 million ($1.1 million). It signed an agreement with a large local generation company, Senoko Energy, and a subsidiary of France's energy giant ENGIE in 2020. It will launch a commercial pilot-project for peer-to-peer (P2P) energy trading in Singapore. The year-long pilot, through June 2021, will allow Senoko Energy's residential and business clients to deploy peer-to-peer energy trading.[62]

Power Ledger and Electrify are only two examples but they do have a more developed profile, including the partnerships they have built, compared to peers. There is actually a lot of activity on the part of start-ups working with Energy 4.0 technologies and solutions, such as Australia's Solara, as well as several from outside Asia looking for business development opportunities in the region. These companies include, but are not limited to, the US's LO3 Energy and Lithuania's WePower, which are both active in Australia and Japan. As well as others such as Electron from the UK and Swytch from the US. The activities by all of these companies and the Asian countries they are involved in is nowhere as buoyant as that in China.

I should add two trends that I believe will continue in the coming years. The first is a struggle for incumbent vertically integrated utilities in the region to leverage digitalisation with internal resources. These companies include Hong Kong's CLP Holdings and Hongkong Electric, Japan's TEPCO and Kansai Electric Power, Indonesia's PT Perusahaan Listrik Negara (Persero), Malaysia's Tenaga Nasional, Taiwan's Taiwan Power Company, and many others. Probably a rare exception is KEPCO. While many of these Asian integrated utilities have made some efforts on the Energy 4.0 front, few have had high-profile breakthroughs. This is typical of large, conservative companies where change happens at a snail's pace, if at all. They are most likely, in my opinion, going to follow the path of Thailand's Electricity Generating Authority of Thailand, namely seeking partnerships with the innovators. It is also possible that some of them may actually decide to acquire a stake in an innovator similar to what we have seen in Europe. Another possible trend is that some of these innovators may merge or one could acquire another. One thing for sure is that the rise in the number of open competitive markets in the region as well as the increase in micro-grids and decentralised energy resources will further open the markets for companies with AI, blockchain, IoT-based solutions.

3.2.3 China's Fast-Track Energy Tech Digitalisation

China has made tremendous broad-based advances on the digital technology front in general, including in the energy industry front. In this last section, I will first highlight some of the general advances emphasising use cases with the thought in mind that many people outside China may perhaps have some scepticism regarding the nation's digital progress. And I will argue that intensifying geopolitical pressures, which we can refer to as the tech cold war, is actually pushing the government and private enterprises in China to speed up energy digitalisation innovation. Then I will dive into the nation's development plans for its smart grids as well as into related digital innovation from some clean energy equipment manufacturers through some real-world examples.

3.2.3.1 Augmenting Reality of Digital Leadership

In late October 2020, suddenly about $3 trillion disappeared from circulation. The amount is bigger than the estimated 2020 GDP of the UK, the world's fifth largest. The amount ended up in brokerages accounts as bids for the largest ever planned initial public offering in the world, that of Ant Group, which was supposed to list on the Shanghai and Hong Kong stock exchanges on 5 November 2020, after having raised about $35–39 billion, according to various media reports. The listing was suspended but here I want to discuss Ant's achievements as an example of China's digital technology prowess. Ant is part of one of the largest e-commerce empires in the world, Alibaba. It launched an application called Alipay in 2004 as a way to build a trust system for online transactions between buyers and sellers. Today, more than one billion people, including myself, use the app for a great variety of financial transactions. The Alipay app supplies more than 1,000 daily life services, such as car hailing and supermarket purchases, and two million mini programmes, which include mobility and municipal services, and many others; as a user I can testify that the Ant's financial transactions platform is very easy and convenient to use. Ant's digital infrastructure processed almost 120 trillion yuan ($17.9 trillion) worth of transactions in the 12 months through June 2020 and at its peak it could handle almost 460,000 transactions per second (Table 3.20).[63] A gargantuan task!

In terms of innovation, Ant has expanded in all major financial services segments. It initially started with digital payments through Alipay, which comprises more than 80 million active merchants on a monthly basis. Its CreditTech provides unsecured credit available to consumers and small businesses online. Its InvestmentTech offers tailor-made investment products, with the consumer's financial capability and risk appetite determined by its algorithms. Its InsureTech, in collaboration with some insurance companies, also delivers tailor-made solutions. All of these solutions are all thanks to the digital technologies infrastructure that it has built, which comprises 'AI, security and blockchain that provide speed, reliability, scalability and security that

Table 3.20: Size and Scale Summary of Ant Group (Adapted).

1,000+ million	Alipay app Annual Active Users [1]	80+ million	Alipay app Monthly Active Merchants [2]
711 million	Alipay app Monthly Active Users [1]	729 million	Alipay app Digital Finance Annual Active Users [3]
2,000+	Partner Financial Institutions [4]	200+	Countries and Regions with Online Payment Services [5]
RMB 118 trillion ($17.6 trillion)	Digital Payments Total Payment Volume in Mainland China [6]	RMB 1.7 / 0.4 trillion ($254 / 60 billion)	CreditTech Consumer / SMB Credit Balance [7]
RMB 4.1 trillion ($613 billion)	InvestmentTech AUM [8]	RMB 52 billion ($7.8 billion)	InsureTech Insurance Premiums and Contributions [9]

Source: Ant Group Co., Ltd. (2020). Ant Group Co., Ltd. H Share IPO Prospectus. [online] Ant Group Co., Ltd., Hong Kong S.A.R.: Ant Group Co., Ltd., p. 9. Available at: https://www1.hkexnews.hk/listedco/listconews/sehk/2020/1026/2020102600165.pdf [Accessed 2 November 2020]. Sourced from the Hong Kong Exchanges and Clearing Limited. Notes: (1) During the twelve months ended August 17, 2020. (2) In the month ended June 30, 2020. (3) Users who transacted in one or more digital finance services on the Alipay platform during the twelve months ended June 30, 2020. (4) Total number of partner financial institutions across our digital payments and digital finance services, including banks, asset managers, insurance institutions and other licensed financial institutions as of June 30, 2020. (5) Overseas countries and regions where we support online transactions through Alipay as of June 30, 2020. (6) During the twelve months ended June 30, 2020. (7) Balance of consumer and SMB credit enabled through our platform as of June 30, 2020, including balance of third-party partner financial institutions (including MYbank) and our licensed financial services subsidiaries (which accounted for approximately 2% of the balance of consumer and SMB credit enabled through our platform), as well as balance securitized. (8) Balance of AUM enabled through our platform as of June 30, 2020, including AUM of third-party partner financial institutions and our licensed financial services subsidiary (which accounted for approximately 33% of the AUM enabled through our platform). (9) Insurance premiums enabled through our platform as well as contributions by Xianghubao participants during the twelve months ended June 30, 2020. Insurance premiums include premiums of third-party partner financial institutions and our licensed financial services subsidiary (which accounted for approximately 9% of the insurance premiums and contributions enabled through our platform).

support the massive transaction volumes in our ecosystem', which enables Ant to 'sustain high performance at scale, while maintaining the delivery of differentiated services and superior user experience'.[64] The group believes that digital technologies innovation is its raison d'être and will continue to spend much of its time and money on further developing the ecosystem. I can highlight three more benchmarks underlying this focus. One is that two out of every three of its current 16,660 employees work on technology. Another is that 40 per cent of the planned initial public offering net proceeds, about $16 billion, would had gone towards product innovation and technologies development. Finally, the company had 26,308 patents and patent applications, of which 28 per cent had been granted and 59 per cent were filed overseas. Ant is a big proponent of blockchain for which 1,146 patents were approved or registered worldwide to complement its AI and IoT capabilities.[65]

I spend a bit of time talking about Ant not just because I am a satisfied customer of some of its products offerings but because it is a flag bearer of Chinese innovation. Those who use Alibaba's e-commerce consumer platform Taobao will testify on what a great, easy, smooth shopping experience it is and the vastly richer products and price range it offers versus Amazon or eBay. But there are many other examples of Chinese innovation beyond Alibaba's Ant, such as Tencent's WeChat, which is so much more than a messaging app. For Chinese speakers it is a new convenient, efficient, seamless way of conducting business. To participate in a webinar, for example, it is down to just a couple of clicks from invitation to attendance. Other examples are the world-leading AI capabilities of video-sharing social networking app TikTok and Unmanned Aerial Vehicles maker SZ DJI Technology. A final note is from consultancy EY. In an insight report it noted that China's share of global e-commerce will rise to 63 per cent by 2022, from just under 56 per cent by the end of 2020, with the US's share expected to decline to 15 per cent.[66]

I hope that I have established that China is one of the digital transformation leaders globally and also addressed doubts on the part of some sceptics. I now turn to why I belong to the school of thought that its leadership will nothing but advance over the next few years. This is down to momentum and geopolitics. The momentum is not from innovators such as Alibaba but also from a strong support, especially financial investment, from the Chinese government. The current state of the tech geopolitics simply reinforces that the tech war ignited by the Trump administration, though it probably has bipartisan support, has fuelled the Chinese government to further accelerate its encouragement of the transition to Energy 4.0.

The Trump administration adopted a policy that 'rests on restricting the flow of technology to China, restructuring global supply chains, and investing in emerging technologies at home. Even a new US administration is unlikely to stray from these fundamentals' believes Director of the Digital and Cyberspace Policy Program at the US Council on Foreign Relations Adam Segal. He adds, 'Beijing's counterstrategy, too, has crystallized. China is racing to develop semiconductors and other core technologies so as to reduce its vulnerability to supply chains that pass through the US. In pursuit of that goal, its leaders are mobilizing tech companies.'[67] So this will up China's Energy 4.0 momentum.

During a press briefing related to the fourteenth Five-Year Plan recently the Minister of Science and Technology pointed out that during the thirteenth Five-Year Plan China's ranking in the Global Innovation Index jumped to 14 from 29, R&D expenditure rose to 2.21 trillion yuan ($314 billion) from 1.42 trillion yuan ($212 billion), and basic research spending almost doubled to 133.6 billion yuan ($20 billion).[68] There are a few numbers flying around regarding the planned spending on 'new infrastructure', including 5G networks, AI, data centres, industrial Internet, inter-city transportation and rail system, new-energy vehicle charging stations, and ultra-high voltage power transmission[69] during the forthcoming Five-Year Plan. The forecasts that emanate from a great variety of sources including industry bodies and securities firms, ranged from about 10 trillion yuan ($1.5 trillion) to as much as three times that amount in the second quarter of 2020. The estimates had seemingly converged to a more frequently published 11 to 15 trillion yuan ($1.6–2.2 trillion) by September 2020.

A forecast by a domestic think tank, the China Electronic Information Industry Development Research Institute, published in March,[70] and another from US investment bank Goldman Sachs reported by Chinese Media[71] in September, put the numbers at 17 trillion yuan ($2.5 trillion) and 15 trillion yuan, respectively. Both forecasts provided breakdowns on the expected expenditure. Both estimates concur that approximately 34 per cent of the total will go to high-speed and urban transit rails related infrastructure. The balance will go towards building the nationwide 5G network, data centres ultra-high voltage power lines, IoT for industry, AI, and EV charging stations. I should note that the two do not actually concur on the actual breakdown apart from the rail-related spending (Figure 3.32).

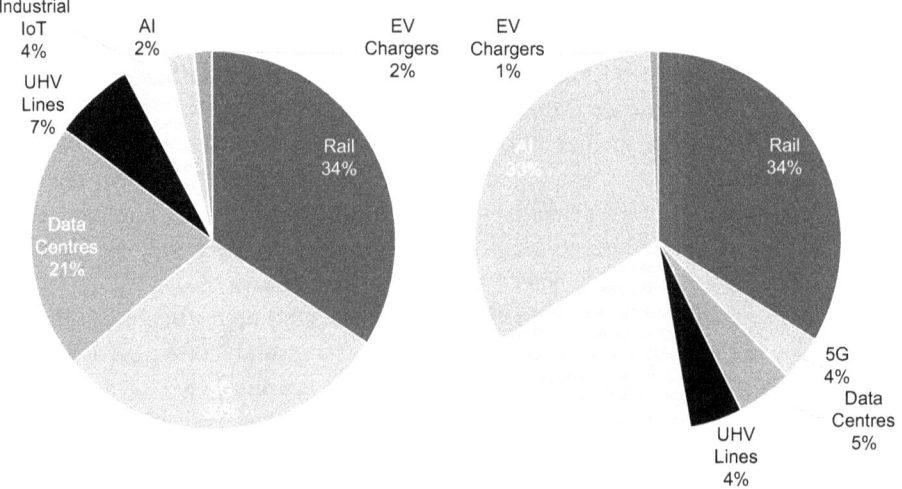

Figure 3.32: Two Different Breakdowns of New Energy Infrastructure Spending Estimated for 2020–2025. Source: China Science Daily, Science net, Science News Magazine (2020). SADI Think Tank Released the First White Paper on 'New Infrastructure' Development in the Industry. [online] news.sciencenet.cn. Available at: http://news.sciencenet.cn/htmlnews/2020/3/437386.shtm?id=437386 [Accessed 4 November 2020]. Estimates from China Electronic Information Industry Development Research Institute. (Original in Chinese.) And Lingyun Capital Research Center and Goldman Sachs (2020). China's New Infrastructure: Prospects for the Next Five Years. [online] www.sohu.com. Available at: https://www.sohu.com/a/425191576_733114 [Accessed 4 November 2020]. Article quotes: Linyunhui (issue 889) | China's New Infrastructure: Prospects for the Next Five years from Lingyun Capital Research Center and Goldman Sachs. (Original in Chinese.)

Whatever the amount, it is clear to me that this offers almost unlimited growth opportunities for domestic innovators and for those few companies from abroad that are already involved in China's Energy 4.0. Policy-wise, the leadership has laid out its modernisation objectives through 2025 and tech will play a fundamental role. At the fifth plenary session of the 19th Communist Party of China Central Committee in October 2020, headed by President Xi Jinping, it stated:

> China's economic and technological strength, and composite national strength will increase significantly. ... Making major breakthroughs in core technologies in key areas, China will become a global leader in innovation. New industrialization, [Information Technology] application, urbanization, and agricultural modernization will be basically achieved.[72]

Earlier, in April 2020, the National Development and Reform Commission, the nation's leading economic planner, had put 'new infrastructure' in three categories. I believe that these categories put Energy 4.0 under an important light: Energy 4.0 in China overlaps well with these new digital tech infrastructure categories.

> First, information infrastructure. It mainly refers to the infrastructure generated by the evolution of new generation information technology, such as communication network infrastructure represented by 5G, Internet of things, industrial IoT and satellite IoT, new technology infrastructure represented by AI, cloud computing and blockchain, and computing infrastructure represented by data centre and Intelligent Computing Centre.
>
> Second is to integrate infrastructure. It mainly refers to the in-depth application of IoT, big data, AI and other technologies to support the transformation and upgrading of traditional infrastructure, and then form the integrated infrastructure, such as intelligent transportation infrastructure, and intelligent energy infrastructure.
>
> Third, innovation infrastructure. It mainly refers to the public welfare infrastructure supporting scientific research, technology development and product development, such as major science and technology infrastructure, science and education infrastructure, and industrial technology innovation infrastructure.[73]

3.2.3.2 China's Smartening of Energy and its Earth-shaking Near-Term Future

In Section 3.2.3.1, I looked at the potential for China to be one of the leaders in Energy 4.0 in Asia and the rest at the world. I also presented some examples of the nation's digital innovation strengths and achievements, including from central industry players such as Ant. These also very much exist when it comes to digital innovation specifically in the energy industry. Energy 4.0 has strong policy backing and has a variety of strong market actors. I will first highlight some of the most important recent developments on the policy front. I will then present several use cases from some of these industry participants and gauge the role of AI, blockchain, and IoT in the grand plans.

Energy 4.0 in China is first supported by a global national agenda for innovation. This stems from an objective of being globally competitive and relevant when it comes to digital technologies development and their addressable domestic and international markets. a strategy that has been advocated by the central government for about a decade but has gained much momentum in the past couple of years. It also stems from a newfound urgency to drive the country towards being technologically as self-sufficient as possible given the global tech Cold War, which is increasingly affecting global supply chains. This I had discussed at the end of the previous section. Energy 4.0 is also supported by policies and guidelines specifically catered to the electric power industry.

Throughout the past decades there have been several government edicts specific to Energy 4.0. These have driven the Chinese grid networks to become smarter, but I personally had not found that there was a particular sense of urgency – especially when compared to the related priorities to shift the electricity mix away from thermal coal in favour of clean end renewable energy as well as emissions-reduction-related initiatives. At least that is my personal perception. In the past couple of years this to me has changed. The smartening of the grid and of the management of the energy have significantly risen on the priority agenda.

One example is a document issued in late September 2020. It came from not one but from four key government bodies. These include the Ministry of Science and Technology, the Ministry of Industry and Information Technology, the National Development and Reform Commission, and the Ministry of Finance, the latter two being among the most influential of bodies. The document was titled NDRC High Tech Document [2020] No. 1409 'Guidance on Broadening Investment in Strategic Emerging Industries, Cultivating and Strengthening the Growth Posts of New Growth Points'[74] – my translation. The title sounds quite grand and its objectives are even more ambitious. It specifically encourages clean energy growth bottlenecks. It calls for an acceleration in the development of new energy industries to help address the intermittency issue created by the planned massive addition of variable renewable energy like wind and solar in the next 30 years. Some of the solutions mentioned are the addition of new infrastructure networks, including smart grids, microgrids, distributed energy, new energy storage solutions, hydrogen production and hydrogenation facilities, and fuel cell systems. At the same time the document emphasised the importance of concurrently launching more energy conservation and environmental protection pilot projects as well as cultivating energy-saving, the water-saving and environmental protection equipment industry, and the seawater desalination industry, and accelerating the demonstration and application of advanced technologies and equipment.[75] All this to support the sustainable development of the environmental protection industry.

During a press conference, authorities were asked what kind of measures would be executed so as to optimise the investment environment for these new energies. The National Development and Reform Commission put forward what it called 'four policy guarantees'. It will reform the processes so as to improve the investment approval mechanisms. This includes a simplification of approvals, reducing processing time, and ensuring that major projects are implemented swiftly. Second, resources will be allocated in a market-oriented manner, especially land, other natural resources, and necessities like electricity and water. Third, the way projects are supervised will be overhauled and new regulatory mechanisms introduced. And, lastly, broad consultation and exchanges between investors, enterprises, local government bodies, local financial institutions, and the like will be implemented. All of these initiatives sound very positive, but the devil is in the detail and in the implementation.

Having witnessed policymaking and the subsequent implementation in China for more than three decades, I can testify that it is never smooth. But then to be fair neither is it smooth in such regions as Europe or North America. Especially, when it comes to complex issues and an overhaul in regulatory mechanisms. Still my high level of confidence in an acceleration in new energy digitalisation and other new energy investments succeeding is that China has a decent record in the transformation of its energy industry. Of course, at times change arrived at a slower pace than originally expected but the reality is that the country is vast, and one must not think of China as having one big, united, homogenous energy market. It contains several and they behave differently. Still, their development paths slowly but surely converge. One such example is the development of the solar and wind sector. It was definitely not a smooth and straight path but what the industry has achieved is formidable in terms of capacity built and lowering the energy output costs.

Having broadly provided a broad policy backscene, we can turn to real-world examples of China's Energy 4.0 development and its outlook. To provide a picture on the profile of some of the major actors of Energy 4.0 I will use state-owned SGCC, and private companies Huawei Technologies and Blue Aspirations. Before digging into these companies and their involvement in the development of China's Energy 4.0, I should emphasise that these companies are by no means the only ones. A recent report on the nation's smart cities,[76] for example, highlighted that there is a myriad of other companies involved in many aspects of the nation's digitalisation path, including AI, IoT, big data, cloud computing, and mobile infrastructure. These companies also include the two giant telecommunications providers, China Unicom and China Mobile (Table 3.21). In fact, China Mobile, for example is already generating good revenue growth from what it calls industrial digitalisation. Its revenues from this segment amounted to almost 43 billion yuan ($6.4 billion) in the first half of 2020, an increase of 5 per cent compared to the same period a year earlier.

I have mentioned the SGCC a few times in this book. It is a global giant by many measures (Table 3.22), including operating capacity, revenues, and assets. There are four things that are important to know about the group that affect its role in China's digitalisation. Apart from the fact that it has several grid investments and collaborative projects abroad, its size gives it a unique influential position in the global electric power industry, which has turned it into a trend influencer at home and abroad. Second, at home its core business is progressively transitioning to an open competition market from a single-buyer one. This has impacted its world view and future strategies. Third, for the past two decades – it was formed in 2002 – its primary capital investment target has been more on the side of transmission and distribution

Table 3.21: Leading Enterprises in Key Smart City Technology Domains (Adapted).

	Technologies			
	IoT and Mobile Infrastructure	Big Data	Cloud Computing	AI
Leading Enterprises	– Huawei	– Neusoft	– Sugon	– Alibaba
	– China Mobile	– Tencent	– Alibaba Cloud	– Baidu
	– Inspur	– Huawei	– Tencent Cloud	– iFlytek
	– China Unicorn	– Inspur	– Huawei	– Huawei
	– Tencent	– Beiming Software	– UCloud	– SenseTime
	– ZTE	– H3C	– China Telecom	– Megvii
	– H3C	– Sugon	– Amazon Web Services	– Intellifusion
	– Sugon	– Taiji	– Kingsoft	– CloudWalk
	– Alibaba Cloud	– Digital China	– Microsoft Azure	– Yitu
	– Hikvision	– Alibaba Cloud	– Baidu Cloud	– Hikvision
	– Dahua			– Dahua

Source: Atha, K., Callahan, J., Chen, J., Drun, J., Green, K., Lafferty, B., Mcreynolds, J., Mulvenon, J., Rosen, B., and Walz, E. (2020). China's Smart Cities Development. [online] USA: SOS International LLC, p. 38. Available at: https://www.uscc.gov/sites/default/files/2020-04/China_Smart_Cities_Development.pdf [Accessed 4 November 2020].

network assets relative to digital technologies. An example is its obsession with building extremely long, ultra-high voltage transmission lines that have proven expensive in terms of R&D and capital expenditure in long duration assets. In recent times, however, it has increasingly paid more attention to digitalisation infrastructure and technologies and has been diverting more capital into this area. Finally, planning authorities have been squeezing the group's revenues by either not adjusting tariffs when needed or even cutting them. In future its transmission and distribution network fees will be based on a fixed return, just like regulated grid networks in Australia or the UK, for example. This is likely to put even more pressure on profits. For this reason, being an active participant in the nation's Energy 4.0 is particularly important for the corporation because among the many benefits it can help with controlling or reducing costs.

The company has set itself some clear targets for its Energy 4.0 path. Such investment is nothing new for the group, but it has declared an official acceleration in its efforts. It published a series of its achievements for 2019 and stated its eight-track digitalisation strategy in a white paper entitled 'Internet of Things in

Table 3.22: China's Main Grids Financial Snapshot (Six Months to June 2020).

(6 Months to 30/06/2020)	State Grid Corp. of China		China Southern Power Grid	
	Billion Yuan	Billion $	Billion Yuan	Billion $
Revenue	1,186.1	177.3	254.3	38.0
Operating Expenses	1,128.9	168.7	240.0	35.9
Net Profit	5.7	0.8	3.0	0.5
Capex	176.0	26.3	n/a	n/a
Liabilities	2,399.4	358.6	597.1	89.3
Total Assets	4,246.0	634.7	979.3	146.4

Source: Hua Energy Net (2020). State Grid, China Southern Power Grid and Five Major Power Generation Companies: Who Makes the Most Money? Who Loses the Most Money? [online] baijiahao.baidu.com. Available at: https://baijiahao.baidu.com/s?id= 1676780776901936835&wfr=spider&for=pc [Accessed 5 November 2020]. (Chinese). China Bond (2020). Semi Annual Report on Corporate Bonds of China Southern Power Grid Co., Ltd. (2020). [online] chinabond.com.cn. Available at: https://pdf.dfcfw.com/pdf/H2_AN202008281403499589_1.pdf [Accessed 5 November 2020]. (Original in Chinese.) China Power Net (2020). Five Major Networks and Two Networks Gather in 2020. [online] Tencent. Available at: https://xw.qq.com/cmsid/20200901A0A3BY00 [Accessed 5 November 2020].

Electricity' in October 2019.[77] It plans to build up an energy ecosystem (e.g., managing demand-side management, electricity trading); set up smart services (e.g., digital solutions to help energy retailers); set up operations optimisation (e.g., automated grid despatch, digital twins); set up smart operations management (i.e., human resources and supply chains); build data centres at group and provincial levels (i.e., power supply and customer data mining); construct IoT networks comprising grid equipment, users, and the like; advance cloud computing for operations management (e.g., cloud-based despatch); and, finally, increase digital applications R&D.[78]

The group will be dedicating specific funds for its eight-track strategy. In 2016 to 2018, funds dedicated to digitalisation amounted to about 12 to 15 billion yuan ($1.8 to 2.2 billion). This amount is estimated by one forecaster, from Zhongtai Securities, to have doubled to 30 billion yuan ($4.5 billion) in 2019 and it expects that it will almost double again, to approximately 58.5 billion yuan ($8.7 billion) in 2021 (Figure 3.33). These numbers are useful to broadly gauge the amount of capital involved. But I think that the amounts involved may be much greater because many of the projects, including pilot projects, and much of the R&D is often undertaken jointly with partners, be it other state companies, private companies, institutes, or others. For example, recently Chinese tech giant ZTE Corp. announced that it had successfully completed a 5G smart grid end-to-end pilot partnering with China Mobile and China Southern Power Grid.[79] This is just one of dozens of examples of course.

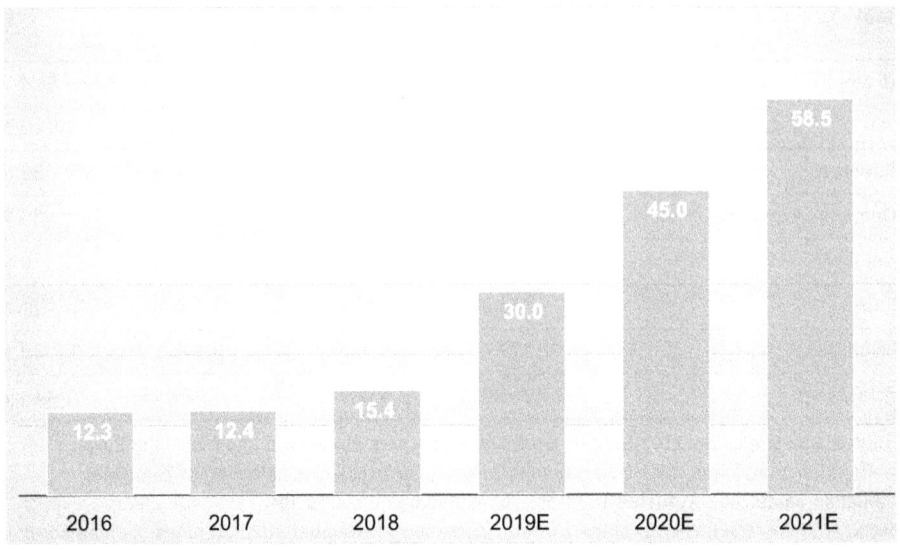

Figure 3.33: State Grid Corp. of China Spending on Digitalisation (Billion Yuan).
Source: Yu, Y. (2019). Chinese Grids' Transformation to Benefit Digital & Tech Companies Globally. [online] Energy Iceberg. Available at: https://energyiceberg.com/chinese-grids-transformation-to-benefit-digital-tech-companies-globally/ [Accessed 4 November 2020].

At the core of the SGCC's eight-track strategy and that of its smaller sister company China Southern Power Grid is a key piece of equipment: smart electricity meters. That's the key link between the electricity company and the consumer of the energy. One can guess that it is no easy task or small feat in a nation as vast as China to add new meters and replace old ones – typically they have a life of five to eight years according to Chinese experts. Here the two grid operators also hope to make the task more efficient and cost effective thanks to digitalisation. To put this into context, the two companies installed 224 million meters between 2009 and 2012, 282 million between 2013 and 2015, replacing 59 million, and between 2016 and 2020 203 million were installed, including 163 million replacements and 40 million new ones (Figure 3.34). Spending was about 19 billion yuan for the two companies in 2019 and it is expected to rise to 28 billion by 2018 according to the China Business Industry Research Institute.

To put this into context, the recorded total completed investment amount going towards power grid capital construction of the three Chinese grid operators on a combined basis was 485.6 billion yuan ($72.6 billion) in 2019, a decline of about 10 per cent compared to the previous year. The construction included the addition of 230.42 million kVA new capacity of 220 kV and above substation equipment (10,000 kVA) and 34,022 kilometres in new 220 kV and above transmission lines.[80]

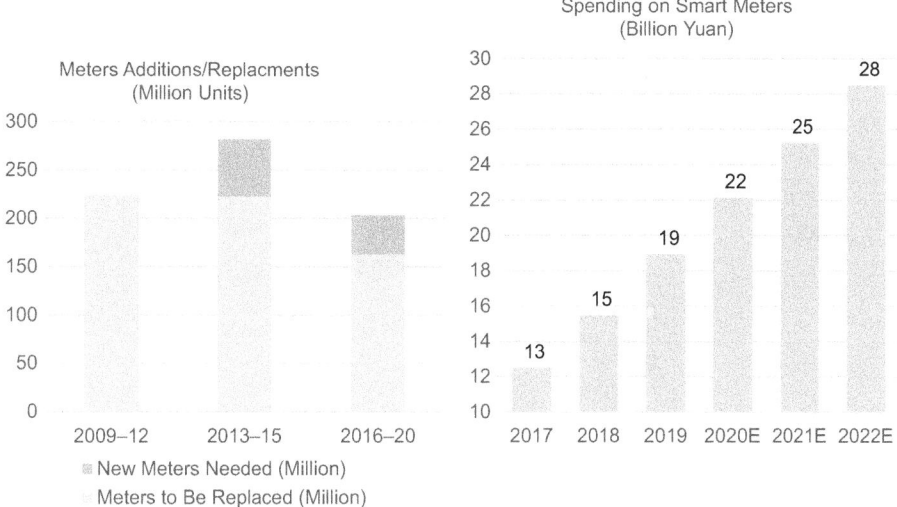

Figure 3.34: China Power Meters Additions and Replacements, and Smart Meters Spending. Source: China Business Industry Research Institute (2020). Analysis on Current Situation and Prospects of China's Ubiquitous Electricity IOT Industrial Chain in 2020. [online] baijiahao.baidu.com. Available at: https://baijiahao.baidu.com/s?id=1680967794659554149&wfr=spider&for=pc [Accessed 4 November 2020].

Moving from China's grids to some of the nation's flagbearers, I will profile Huawei Technologies first. Some sizing up first. Huawei is one of China's and the world's biggest technology companies. It is a private, employee-owned information and communications technology infrastructure and smart devices giant. It was founded in 1987, in the then sleepy Chinese city of Shenzhen, which borders Hong Kong. Today, Huawei boasts almost 200,000 employees, revenues of 858.8 billion yuan ($128.4 billion), net profits of 62.7 billion yuan ($9.4 billion), and total assets of 858.7 billion yuan ($128.4 billion) as of December 2019,[81] and it is ranked 49 in the global Fortune 500. The company has two principal business lines. One is consumer products, including smartphones and laptops. The other is business products, services, and enterprise solutions, including for the energy industry and smart cities (Figure 3.35).

To adequately gauge how the company will actively be involved in the development of Energy 4.0 in China, I will first list some its use cases and then move on to discussing some of the digital energy technologies and solutions that it can supply.

The first example is that, in trial at the premises of one of its domestic customers, the company found that its 5G power solution saved the customer as much as 4.1 megawatt-hours per site per year. It also tested this equipment with a European user and found that energy could be cut by 50 per cent. In another setting, a knowledge-based campus, it found that its intelligent analytics were able to reduce energy consumption by more than 15 per cent. Another example is that a router

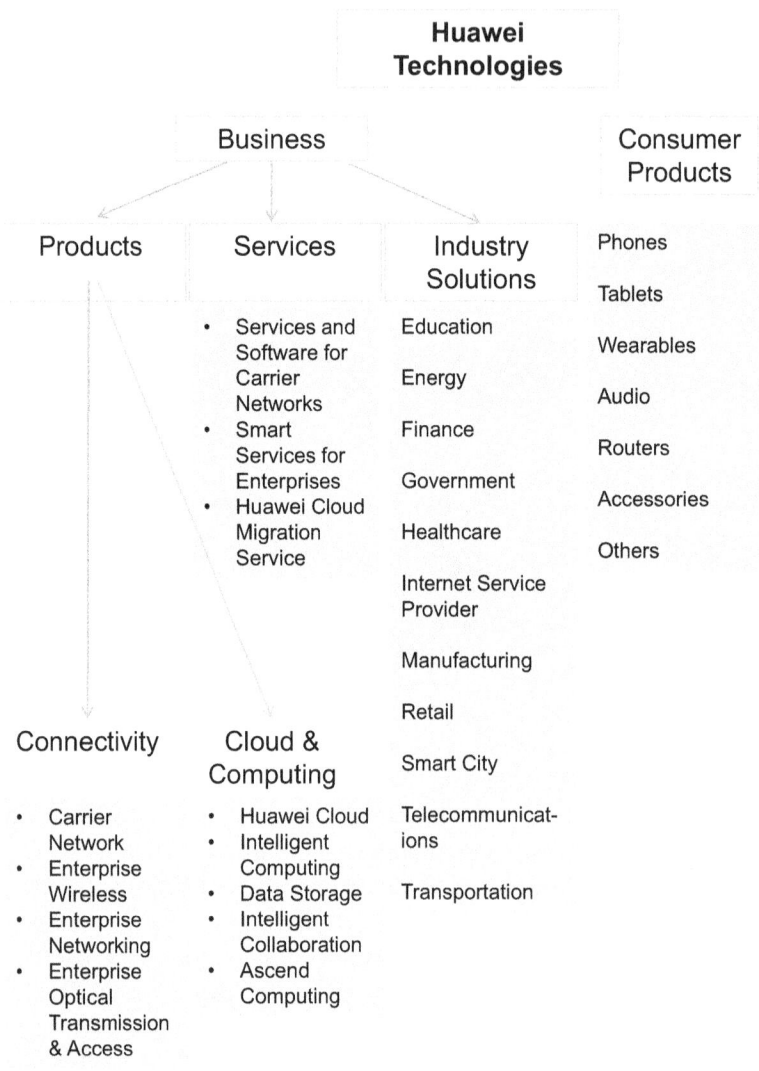

Figure 3.35: Snapshot of Huawei's Businesses.
Source: Author, October 2020. Data source: Huawei Technologies' website https://www.huawei.com/en/corporate-information.

model that it developed, the NetEngine 8000 X8, is able to save 90 megawatt-hours per year as the company found more efficient power supply and heat dissipation solutions. A final example is cooling for data centres. The AI-based solution was developed in conjunction with China Mobile at a site in Ningxia province. It expects savings of 6 gigawatt-hours per year, an especially high saving rate I find, which is partly thanks to its technology allowing for 100 per cent free cooling in the winter.[82]

The company already supplies its Energy 4.0 tech and solutions to more than 190 power utilities around the world. It wants to further expand its business domestically and abroad through a strategy that it calls 'platform plus ecosystem'. As one may guess from my many mentions of 5G, AI, blockchain, and IoT, this is what Huawei is looking at. In my own view, key advantages that Huawei has are its size and the size of its home market. Its size means it has sufficient financial and human resources to nurture various initiatives. Its home market size means that it has a vast choice of potential testing size and once the testing is done it can rapidly scale up the technologies and reduce costs. In a recent event, the Huawei Global Power Summit Online 2020, the company listed a great number of use cases. The first set I found interesting is its work with some large utilities. In China, Huawei is working with the SGCC on smart grids and IoT so the company can become a 'world-class energy Internet enterprise'. It is helping the Dubai Electricity and Water Authority to set up a digital arm comprising AI applications, digital services, energy storage, and renewable energy. It is working on digitalisation also with the Saudi Electricity Authority, especially grid operations optimisation, creating an asset leasing business, and executing digital transformation. Finally, it is also working on Energy 4.0 with Thai government-controlled utility the Provincial Electricity Authority, with the objectives to cut operations costs 80 per cent and generate 50 per cent of its revenue from digital technology.[83]

Having presented use cases from a gigantic state-owned enterprise, the SGCC, and from a leading non-state-owned technology company, Huawei, I thought that it would be good to use a start-up as the last example of Chinese Energy 4.0 innovation.

Blue Aspirations is the second of the two private companies examples. It is a young company. Though only founded in 2019, it is already having an impact on offshore wind development with its proprietary floating LiDAR system (FLS) technology (Figure 3.36). LiDAR stands for Light Detection and Ranging and is a remote sensing tool (it relies on the doppler shift of light pulses to measure the speed of aerosols in the wind). FLS are commonly used at offshore wind farms projects to measure and collect wind and meteorological data to gather insights into the profitability, profile, and optimal design specifications for a site. Blue Aspirations manufactures a FLS that can be deployed rapidly – without sacrificing quality – and embodies a design philosophy that incorporates redundancy at every level to eliminate the risk of single-point failures. These two features are chiefly what distinguishes their systems from competitors in the market. The company was formed by a group of engineers, software developers, and other professionals with a great deal of experience in the wind industry and telecommunications. The team's experience and agility led them to construct two FLS prototypes and test them offshore by October 2019 and by January 2020, they had already obtained a data accuracy verification from DNV GL, a leading global engineering consultancy and industrial certification company. The company then iteratively finetuned its prototypes to give rise to the commercial model in use today. Its first two commercial FLSs were deployed in August 2020 – I personally feel that this is quite a rapid progression and an impressive achievement for any young start-up.

Figure 3.36: Blue Aspirations ZX300M Model Floating Lidar System.
Source: Photo courtesy of Blue Aspirations.

Notes

1 Bocse, A.-M. and Gegenbauer, C. (2017). UK's Dash for Gas: Implications for the Role of Natural Gas in European Power Generation. London, UK: The European Centre for Energy and Resource Security (EUCERS), p. 7. [online] eucers.com. Available at: https://eucers.com/wp-content/uploads/2019/03/strategy-paper-14.pdf [Accessed 21 September 2020].
2 energypedia (2019). Nationally Determined Contributions (NDC) – energypedia.info. [online] energypedia.info. Available at: https://energypedia.info/wiki/Nationally_Determined_Contributions_(NDC) [Accessed 26 September 2020].
3 BP (2020). BP Energy Outlook 2020. [online] BP Global. Available at: https://www.bp.com/en/global/corporate/news-and-insights/press-releases/bp-energy-outlook-2020.html [Accessed 25 September 2020].
4 Department of the Environment and Energy, Australian Government (2019). Australian Energy Update 2019 | energy.gov.au. [online] Energy.gov.au. Available at: https://www.energy.gov.au/publications/australian-energy-update-2019 [Accessed 26 September 2020].
5 Thornhill, J. (2020). Australia's $200 Billion LNG Boom Waylaid by Covid and Cracks – BNN Bloomberg. [online] BNN Bloomberg. Available at: https://www.bnnbloomberg.ca/australia-s-200-billion-lng-boom-waylaid-by-covid-and-cracks-1.1490307 [Accessed 26 September 2020].
6 Department of the Environment and Energy, Australian Government (2019). Australian Energy Update 2019 | energy.gov.au. [online] Energy.gov.au. Available at: https://www.energy.gov.au/publications/australian-energy-update-2019 [Accessed 26 September 2020].
7 Institute for Energy Economics and Financial Analysis (2020). IEEFA Podcast Australia: Why Gas?: The Australian Government's New Gas Announcements are Flogging a Dead Horse. [online] IEEFA. Available at: https://ieefa.org/ [Accessed 27 September 2020].

8 Rystad Energy (2020). Rystad Energy: Australia's Gas and Coal Power Demand has Peaked. [online] World Coal. Available at: https://www.worldcoal.com/power/11092020/rystad-energy-australias-gas-and-coal-power-demand-has-peaked/ [Accessed 27 September 2020].

9 Institute for Energy Economics and Financial Analysis (2020). IEEFA Podcast Australia: Why Gas?: The Australian Government's New Gas Announcements are Flogging a Dead Horse. [online] IEEFA. Available at: https://ieefa.org/ [Accessed 27 September 2020].

10 Graham, P., Hayward, J., Foster, J., Story, O., and Havas, L. (2018). GenCost 2018 Updated Projections of Electricity Generation Technology Costs. [online] https://publications.csiro.au/publications/, Commonwealth Scientific and Industrial Research Organisation, pp. 28–31. Available at: https://publications.csiro.au/rpr/download?pid=csiro:EP189502&dsid=DS1 [Accessed 27 September 2020].

11 Australian Energy Market Operator (AEMO) (2020). 2020 Integrated System Plan (ISP): For the National Electricity Market. [online] aemo.com.au. Available at: https://aemo.com.au/energy-systems/major-publications/integrated-system-plan-isp/2020-integrated-system-plan-isp [Accessed 27 September 2020].

12 Australian Energy Market Operator (AEMO) (2020). 2020 Integrated System Plan (ISP): For the National Electricity Market. [online] aemo.com.au. Available at: https://aemo.com.au/energy-systems/major-publications/integrated-system-plan-isp/2020-integrated-system-plan-isp [Accessed 27 September 2020].

13 Australian Energy Market Operator (AEMO) (2020). 2020 Integrated System Plan (ISP): For the National Electricity Market. [online] aemo.com.au. Available at: https://aemo.com.au/energy-systems/major-publications/integrated-system-plan-isp/2020-integrated-system-plan-isp [Accessed 27 September 2020].

14 Asian Renewable Energy Hub (2020). Asian Renewable Energy Hub. [online] Asianrehub.com. Available at: https://asianrehub.com/ [Accessed 28 September 2020]; Ammonia Energy Association (2020). Green Ammonia at Oil and Gas Scale: the 15 GW Asian Renewable Energy Hub. [online] Ammonia Energy Association. Available at: https://www.ammoniaenergy.org/articles/green-ammonia-at-oil-and-gas-scale-the-15-gw-asian-renewable-energy-hub/ [Accessed 27 September 2020]; Law Business Research Ltd (2020). Renewable Energy Law Review. [online] p. 13. Available at: https://thelawreviews.co.uk//digital_assets/534c53ca-4ec8-45b2-8ec2-a8853f998a4e/The-Renewable-Energy-Law-Review—Edition-3.pdf [Accessed 27 September 2020].

15 Sun Cable Pte Ltd (2020). Power Generation | Singapore | Sun Cable. [online] Sun Cable Singapore. Available at: https://www.suncable.sg/ [Accessed 27 September 2020]; Maisch, M. (2020). Federal Government Awards Major Project Status to Sun Cable's 10 GW Solar Plan. [online] pv magazine Australia. Available at: https://www.pv-magazine-australia.com/2020/07/29/federal-government-awards-major-project-status-to-sun-cables-10-gw-solar-plan/ [Accessed 28 September 2020].

16 bp p.l.c. (2020). Statistical Review of World Energy. [online] BP Global. Available at: https://www.bp.com/en/global/corporate/energy-economics/statistical-review-of-world-energy.html [Accessed 29 September 2020].

17 Renewable Energy Institute (2020). Proposal for 2030 Energy Mix in Japan. [online] Renewable Energy Institute, Tokyo, Japan: Renewable Energy Institute, pp. 7–8,11. Available at: https://www.renewable-ei.org/pdfdownload/activities/REI_Summary_2030Proposal_EN.pdf [Accessed 30 September 2020].

18 Taiwan Power Company (2020). Taiwan Power Company 2019 Sustainability Report. [online] https://csr.taipower.com.tw/en/index.aspx, Taiwan Power Company, p.59. Available at: https://csr.taipower.com.tw/upload/132/2019110109130980581.pdf [Accessed 30 September 2020]; Global Wind Energy Council (2020). From 0 to 15GW by 2030: Four Reasons Why Taiwan is the Offshore Wind Market in Asia. [online] Global Wind Energy Council. Available at: https://gwec.net/from-0-to-15gw-by-2030-four-reasons-why-taiwan-is-the-offshore-wind-market-in-asia/ [Accessed 29 September 2020].

19 CGTN (2020). Full text: Xi Jinping's Speech at the General Debate of the 75th Session of the United Nations General Assembly. [online] news.cgtn.com. Available at: https://news.cgtn.com/news/2020-09-23/Full-text-Xi-Jinping-s-speech-at-General-Debate-of-UNGA-U07X2dn8Ag/index.html [Accessed 5 October 2020].

20 CGTN (2020). Full text: Xi Jinping's Speech at the General Debate of the 75th session of the United Nations General Assembly. [online] news.cgtn.com. Available at: https://news.cgtn.com/news/2020-09-23/Full-text-Xi-Jinping-s-speech-at-General-Debate-of-UNGA-U07X2dn8Ag/index.html [Accessed 5 October 2020].

21 Yin, P., Brauer, M., Cohen, A.J., Wang, H., Li, J., Burnett, R.T., Stanaway, J.D., Causey, K., Larson, S., Godwin, W., Frostad, J., Marks, A., Wang, L., Zhou, M. and Murray, C.J.L. (2020). 'The effect of air pollution on deaths, disease burden, and life expectancy across China and its provinces, 1990–2017: An analysis for the Global Burden of Disease Study 2017', The Lancet Planetary Health [online] vol. 4, no. 9, pp. e386–e398. Available at: https://www.thelancet.com/journals/lanplh/article/PIIS2542-5196(20)30161-3/fulltext#:~:text=We per cent20estimated per cent20that per cent201 per centC2 per centB724 [Accessed 6 October 2020].

22 Crippa, M., Oreggioni, G., Guizzardi, D., Muntean, M., Schaaf, E., Lo Vullo, E., Solazzo, E., Monforti-Ferrario, F., Olivier, J.G.J., and Vignati, E. (2019). Fossil CO2 and GHG Emissions of all World Countries – 2019 Report, EUR 29849 EN, Publications Office of the European Union, Luxembourg, ISBN 978-92-76-11100-9, doi:10.2760/687800, JRC117610. Available at: https://edgar.jrc.ec.europa.eu/overview.php?v=booklet2019 [Accessed 6 October 2020].

23 Carbon Dioxide Information Analysis Center, Environmental Sciences Division, Oak Ridge National Laboratory, Tennessee, United States (2020). CO2 emissions (kt) – China | Data. [online] data.worldbank.org. Available at: https://data.worldbank.org/indicator/EN.ATM.CO2E.KT?end=2016&locations=CN&most_recent_value_desc=true&start=1960 [Accessed 6 October 2020].

24 The Climate Action Tracker (2020). China | Climate Action Tracker. [online] Climateactiontracker.org. Available at: https://climateactiontracker.org/countries/china/ [Accessed 6 October 2020].

25 Ge, M. and Friedrich, J. (2020). 4 Charts Explain Greenhouse Gas Emissions by Countries and Sectors. [online] World Resources Institute. Available at: https://www.wri.org/blog/2020/02/greenhouse-gas-emissions-by-country-sector [Accessed 7 October 2020].

26 World Resources Institute (2019). CAIT Climate Data Explorer. [online] Wri.org. Available at: https://www.wri.org/blog/2020/02/greenhouse-gas-emissions-by-country-sector [Accessed 7 October 2020].

27 National Bureau of Statistics (1994–2020). National Bureau of Statistics. [online] Stats.gov.cn. Available at: http://www.stats.gov.cn/ [Accessed 8 October 2020] (Chinese); China Electricity Council (1994–2020). China Electricity Council. [online] www.cec.org.cn. Available at: https://www.cec.org.cn/ [Accessed 8 October 2020]; National Energy Administration (2008–2019). National Energy Administration. [online] www.nea.gov.cn. Available at: http://www.nea.gov.cn/ [Accessed 8 October 2020] (Chinese).

28 Planete Energies (2018). Huainan: Largest Floating Solar Farm in the World. [online] Planète Énergies. Available at: https://www.planete-energies.com/en/medias/close/huainan-largest-floating-solar-farm-world [Accessed 9 October 2020].

29 bp p.l.c. (2020). Statistical Review of World Energy. [online] BP Global. Available at: https://www.bp.com/en/global/corporate/energy-economics/statistical-review-of-world-energy.html [Accessed 9 October 2020].

30 World Economic Forum (2020). Energy Technologies 2030: Wind and Solar PV Will Keeptaking the Lead. [online] World Energy Forum Global Future Council on Energy Technologies, p. 1. Available at: http://www3.weforum.org/docs/WEF_Wind_and_Solar_2030.pdf [Accessed 19 October 2020]. Sourced from BloombergNEF (BNEF) 1H 2020 LCOE Update. https://www.bnef.com/.

31 International Energy Agency (2020). 2021–2025: Rebound and Beyond – Gas 2020 – Analysis. [online] IEA. Available at: https://www.iea.org/reports/gas-2020/2021-2025-rebound-and-beyond [Accessed 19 October 2020].
32 Shell (2020). Shell LNG Outlook 2020. [online] www.shell.com. Available at: https://www.shell.com/energy-and-innovation/natural-gas/liquefied-natural-gas-lng/lng-outlook-2020.html#iframe=L3dlYmFwcHMvTE5HX291dGxvb2sv [Accessed 19 October 2020].
33 METI (2020). Spot LNG Price Statistics | METI. [online] www.meti.go.jp. Available at: https://www.meti.go.jp/english/statistics/sho/slng/index.html [Accessed 19 October 2020].
34 Willuhn, M. (2020). Green Hydrogen to Reach Price Parity with Grey Hydrogen in 2030. [online] pv magazine International. Available at: https://www.pv-magazine.com/2020/07/16/green-hydrogen-to-reach-price-parity-with-grey-hydrogen-in-2030/ [Accessed 20 October 2020]; Edwardes-Evans, H. (2020). Green Hydrogen Costs 'Can Hit $2/kg Benchmark' by 2030: BNEF | S&P Global Platts. [online] www.spglobal.com. Available at: https://www.spglobal.com/platts/en/market-insights/latest-news/coal/033020-green-hydrogen-costs-can-hit-2kg-benchmark-by-2030-bnef [Accessed 20 October 2020].
35 Weigel, P. (n.d.). Digital Applications in the Energy Sector – A Review. [online] encyclopedia.pub. Available at: https://encyclopedia.pub/427 [Accessed 26 October 2020].
36 Siemens (2018). Siemens Electrical Digital Twin – A Single Source of Truth to Unlock the Potential within a Modern Utility's Data Landscape. [online] siemens.com, Erlangen, Germany, p. 6. Available at: https://assets.new.siemens.com/siemens/assets/api/uuid:66c9013092b493265e091c154a33f9dd38d36c20/electricaldigitaltwin-brochure-final-intl-version-singlepages-no.pdf [Accessed 27 October 2020]. Article Number: EMDG-B10153-00-7600–Electrical Digital Twin Brochure.
37 The Economist Intelligence Unit Limited (2020). The Internet of Things Applications for Business – Exploring the Transformative Potential of IoT. The Economist Intelligence Unit Limited, p. 11.
38 Wei, J., Sanborn, S., and Slaughter, A. (2019). Digital Transformation and the Utility of the Future | Deloitte Insights. [online] www2.deloitte.com. Available at: https://www2.deloitte.com/us/en/insights/industry/power-and-utilities/digital-transformation-utility-of-the-future.html [Accessed 22 October 2020].
39 Enel (2019). Enel 2020–2022 Strategic Plan. [online] enel.com. Available at: https://www.enel.com/content/dam/enel-common/press/en/2019-November/Enel%20Strategic%20Plan%202020%202022%20ENG.pdf [Accessed 22 October 2020].
40 Amazon Web Services (2017). AWS re:Invent 2017: IoT @ Enel: A New Generation IoT Core Platform (IOT312). [online] YouTube. Available at: https://www.youtube.com/watch?v=ZhiBrx1lJEs [Accessed 22 October 2020].
41 McKinsey (2018). The Digital Utility: New Challenges, Capabilities, and Opportunities. [online] mckinsey.com. U.S.A. Available at: https://www.mckinsey.com/~/media/McKinsey/Industries/Electric%20Power%20and%20Natural%20Gas/Our%20Insights/The%20Digital%20Utility/The%20Digital%20Utility.ashx [Accessed 22 October 2020].
42 McKinsey (2018). The Digital Utility: New Challenges, Capabilities, and Opportunities. [online] mckinsey.com. U.S.A. Available at: https://www.mckinsey.com/~/media/McKinsey/Industries/Electric%20Power%20and%20Natural%20Gas/Our%20Insights/The%20Digital%20Utility/The%20Digital%20Utility.ashx [Accessed 22 October 2020].
43 Pales, W. (2018). *The Digital Utility: Using Energy Data to Increase Customer Value and Grow Your Business*. Melbourne, Victoria: Grammar Factory Pty. Ltd.
44 Wei, J., Sanborn, S., and Slaughter, A. (2019). Digital Transformation and the Utility of the Future | Deloitte Insights. [online] www2.deloitte.com. Available at: https://www2.deloitte.com/us/en/insights/industry/power-and-utilities/digital-transformation-utility-of-the-future.html [Accessed 22 October 2020].

45 Wei, J., Sanborn, S., and Slaughter, A. (2019). Digital Transformation and the Utility of the Future | Deloitte Insights. [online] www2.deloitte.com. Available at: https://www2.deloitte.com/us/en/insights/industry/power-and-utilities/digital-transformation-utility-of-the-future.html [Accessed 22 October 2020].

46 Turk, D. and Cozzi, L. (2017). Digitalization & Energy. [online] iea.org. Paris, France: International Energy Agency. Available at: https://www.iea.org/reports/digitalisation-and-energy [Accessed 24 October 2020].

47 Booth, A., de Jong, E., and Peters, P. (2018). The Digital Utility: New Challenges, Capabilities, and Opportunities. [online] mckinsey.com, p. 10. Available at: https://www.mckinsey.com/~/media/McKinsey/Industries/Electric%20Power%20and%20Natural%20Gas/Our%20Insights/The%20Digital%20Utility/The%20Digital%20Utility.ashx [Accessed 24 October 2020].

48 Schneider Electric (2019). 2019 Global Digital Transformation Benefits Report: Schneider Electric. [online] www.se.com. Available at: https://www.se.com/ww/en/download/document/998-20387771_DTBR/ [Accessed 25 October 2020].

49 ABB (2020). ABB Reveals How Digital Switchgear Can Lead to Major Cost Savings. [online] new.abb.com. Available at: https://new.abb.com/news/detail/23097/abb-integrates-advanced-digital-technologies-to-cut-energy-and-maintenance-costs-for-buildings [Accessed 24 October 2020]. Also see White Paper: 'Making the Switch to Digital Switchgear'.

50 ABB (2019). ABB Integrates Advanced Digital Technologies to Cut Energy and Maintenance Costs for Buildings. [online] new.abb.com. Available at: https://new.abb.com/news/detail/23097/abb-integrates-advanced-digital-technologies-to-cut-energy-and-maintenance-costs-for-buildings [Accessed 25 October 2020].

51 ABB (2020b). ABB to Lower Costs and Increase Value of Capital Projects with ABB Adaptive Execution Offering. [online] new.abb.com. Available at: https://new.abb.com/news/detail/69387/abb-to-lower-costs-and-increase-value-of-capital-projects-with-abb-adaptive-execution-offering [Accessed 25 October 2020].

52 Odell, S. and Fadzeyeva, J. (2018a). Emerging Technology Projection: The Total Economic Impact of IBM Blockchain Projected Cost Savings and Business Benefits Enabled by IBM Blockchain. [online] ibm.com. Forrester Research, Inc. Available at: https://www.ibm.com/downloads/cas/QJ4XA0MD [Accessed 24 October 2020].

53 Odell, S. and Fadzeyeva, J. (2018a). Emerging Technology Projection: The Total Economic Impact of IBM Blockchain Projected Cost Savings and Business Benefits Enabled by IBM Blockchain. [online] ibm.com, Forrester Research, Inc., p. 30. Available at: https://www.ibm.com/downloads/cas/QJ4XA0MD [Accessed 24 October 2020].

54 Huawei Technologies (2020). Leading Energy Digitalization for a Smart and Sustainable World. [online] huawei.com. Available at: https://www.huawei.com/en/events/ict2019/leading-energy-digitalization-for-a-smart-and-sustainable-world [Accessed 27 October 2020].

55 Santarius, T., Pohl, J., and Lange, S. (2020). 'Digitalization and the decoupling debate: Can ICT help to reduce environmental impacts while the economy keeps growing?', *Sustainability*, vol. 12, no. 7496, pp. 4–5.

56 Santarius, T., Pohl, J., and Lange, S. (2020). 'Digitalization and the decoupling debate: Can ICT help to reduce environmental impacts while the economy keeps growing?', *Sustainability*, vol. 12, no. 7496, pp. 4–5.

57 Korea Electric Power Corporation (2020). Clean Energy Smart KEPCO: Sustainability Report 2019. [online] Naju-si, Jeollanam-do, Republic of Korea: KEPCO Corporate Strategy Team, Corporate Planning Department, p. 36. Available at: http://home.kepco.co.kr/kepco/EN/D/C/KEDCPP004.do [Accessed 28 October 2020].

58 Energy Market Authority (2020). Co-creating Singapore's Energy Future through R&D – Digitalisation. [online] YouTube. Available at: https://www.youtube.com/watch?v=8GeVbacC9X8 [Accessed 29 October 2020].

59 Partly sourced from: Sandner, P. (2020). Blockchain, IoT and AI – a Perfect Fit. [online] Medium. Available at: https://medium.com/@philippsandner/blockchain-iot-and-ai-a-perfect-fit-c863c0761b6 [Accessed 30 October 2020]. Images from: Publicdomainvectors.org.

60 Global Blockchain Business Council (2020). Building a Blockchain-Enabled Energy Marketplace with Power Ledger. [online] YouTube. Available at: https://www.youtube.com/watch?v=SPEaXUpVu8c&feature=emb_logo [Accessed 1 November 2020].

61 Information sourced from various media including the following sources. Power Ledger (2020). Power Ledger Signed an Exclusive Partnership with TDED to Accelerate Blockchain-based Digital Energy Business in Thailand. [online] Power Ledger. Available at: https://www.powerledger.io/announcement/power-ledger-to-accelerate-thailand-renewables-with-partnership-for-digital-energy-trading/ [Accessed 1 November 2020]. BCPG Public Company Limited (2019). PEA ENCOM – BCPG Jointly Establish 'Thai Digital Energy Development'. [online] BCPG Public Company Limited. Available at: https://www.bcpggroup.com/en/news-medias/news/255/pea-encom-bcpg-jointly-establish-thai-digital-energy-development [Accessed 1 November 2020]. BCPG Public Company Limited (2018). Sansiri and BCPG Join Force to Pioneer the World's Largest Realtime Blockchain-based, Peer-to-Peer Electricity Trading Pilot Project at T77. [online] BCPG Public Company Limited. Available at: https://www.bcpggroup.com/en/news-medias/news/196/sansiri-and-bcpg-join-force-to-pioneer-the-worlds-largest-realtime-blockchain-based-peer-to-peer-electricity-trading-pilot-project-at-t77 [Accessed 1 November 2020]. Praiwan, Y. (2019). BCPG Seeks P2P Power Trade. [online] https://www.bangkokpost.com. Available at: https://www.bangkokpost.com/business/1819219/bcpg-seeks-p2p-power-trade [Accessed 1 November 2020].

62 Information sourced from various media including the following sources. Electrify (2017). Marketplace – Electrify – Asia's Electric Marketplace. [online] Electrify.asia. Available at: https://electrify.asia/marketplace/ [Accessed 1 November 2020]. Eco-Business (2020). Electrify to Tap US$60M Solar Energy Opportunity in Southeast Asia with Launch of Singapore's First Peer-to-peer Energy Trading Pilot. [online] Eco-Business. Available at: https://www.eco-business.com/press-releases/electrify-to-tap-us60m-solar-energy-opportunity-in-southeast-asia-with-launch-of-singapores-first-peer-to-peer-energy-trading-pilot/ [Accessed 1 November 2020]. Tech Collective Southeast Asia (2019). Top Blockchain Startups in Southeast Asia. [online] Tech Collective. Available at: https://techcollectivesea.com/2019/11/21/top-blockchain-startups-in-southeast-asia/ [Accessed 1 November 2020].

63 Ant Group Co., Ltd. (2020). Ant Group Co., Ltd. H Share IPO Prospectus. [online] Ant Group Co., Ltd., Hong Kong S.A.R.: Ant Group Co., Ltd., pp. 9, 171. Available at: https://www1.hkexnews.hk/listedco/listconews/sehk/2020/1026/2020102600165.pdf [Accessed 2 November 2020]. Sourced from the Hong Kong Exchanges and Clearing Limited.

64 Ant Group Co., Ltd. (2020). Ant Group Co., Ltd. H Share IPO Prospectus. [online] Ant Group Co., Ltd., Hong Kong S.A.R.: Ant Group Co., Ltd., pp. 170–171. Available at: https://www1.hkexnews.hk/listedco/listconews/sehk/2020/1026/2020102600165.pdf [Accessed 2 November 2020]. Sourced from the Hong Kong Exchanges and Clearing Limited.

65 Ant Group Co., Ltd. (2020). Ant Group Co., Ltd. H Share IPO Prospectus. [online] Ant Group Co., Ltd., Hong Kong S.A.R.: Ant Group Co., Ltd., p. 223. Available at: https://www1.hkexnews.hk/listedco/listconews/sehk/2020/1026/2020102600165.pdf [Accessed 2 November 2020]. Sourced from the Hong Kong Exchanges and Clearing Limited.

66 EY (2019). What China Can Teach the World about Digital Transformation. [online] www.ey.com. Available at: https://www.ey.com/en_gl/digital/what-china-can-teach-the-world-about-digital-transformation [Accessed 2 November 2020].

67 Segal, A. (2020). The Coming Tech Cold War with China. [online] www.foreignaffairs.com. Available at: https://www.foreignaffairs.com/articles/north-america/2020-09-09/coming-tech-cold-war-china [Accessed 2 November 2020].

68 CGTN (2020a). China to Include Quantum Technology in Its 14th Five-Year Plan. [online] news.cgtn.com. Available at: https://news.cgtn.com/news/2020-10-21/China-to-include-quantum-technology-in-its-14th-Five-Year-Plan-UM1KUlk80M/index.html [Accessed 2 November 2020].

69 Dezan Shira & Associates (2020). How Can Foreign Technology Investors Benefit from China's New Infrastructure Plan? 7 August. [online] China Briefing. Available at: https://www.china-briefing.com/news/how-foreign-technology-investors-benefit-from-chinas-new-infrastructure-plan/ [Accessed 2 November 2020].

70 China Science Daily, Science net, Science News Magazine (2020). SADI Think Tank Released the First White Paper on 'New Infrastructure' Development in the Industry. [online] news.sciencenet.cn. Available at: http://news.sciencenet.cn/htmlnews/2020/3/437386.shtm?id=437386 [Accessed 4 November 2020]. Estimates from China Electronic Information Industry Development Research Institute. (Original in Chinese.)

71 Lingyun Capital Research Center and Goldman Sachs (2020). China's New Infrastructure: Prospects for the Next Five Years. [online] www.sohu.com. Available at: https://www.sohu.com/a/425191576_733114 [Accessed 4 November 2020]. Article quotes: Linyunhui (issue 889) | China's New Infrastructure: Prospects for the Next Five Years from Lingyun Capital Research Center and Goldman Sachs. (Original in Chinese.)

72 Xinhua News (2020). Xinhua Headlines: Key CPC Session Draws 15-year Roadmap for China's Modernization. [online] www.xinhuanet.com. Available at: http://www.xinhuanet.com/english/2020-10/30/c_139476984.htm [Accessed 2 November 2020].

73 National Development and Reform Commission (2020). The National Development and Reform Commission Held a Press Conference in April to Introduce the Macroeconomic Situation and Respond to Hot Issues. [online] Ndrc.gov.cn. Available at: https://www.ndrc.gov.cn/xwdt/xwfb/202004/t20200420_1226031.html [Accessed 2 November 2020]. (Translated from the Chinese original.)

74 National Development and Reform Commission (2020a). The Development and Reform Commission Answers Reporters' Questions on 'Guiding Opinions on Expanding Investment in Strategic Emerging Industries, Cultivating and Strengthening New Growth Points and Growth Poles'. [online] www.gov.cn. Available at: http://www.gov.cn/zhengce/2020-09/24/content_5546618.htm [Accessed 3 November 2020]. (Translated; original in Chinese.)

75 National Development and Reform Commission (2020a). The Development and Reform Commission Answers Reporters' Questions on 'Guiding Opinions on Expanding Investment in Strategic Emerging Industries, Cultivating and Strengthening New Growth Points and Growth Poles'. [online] www.gov.cn. Available at: http://www.gov.cn/zhengce/2020-09/24/content_5546618.htm [Accessed 3 November 2020]. (Translated; original in Chinese.)

76 Atha, K., Callahan, J., Chen, J., Drun, J., Green, K., Lafferty, B., Mcreynolds, J., Mulvenon, J., Rosen, B., and Walz, E. (2020). China's Smart Cities Development. [online] USA: SOS International LLC, p. 38. Available at: https://www.uscc.gov/sites/default/files/2020-04/China_Smart_Cities_Development.pdf [Accessed 4 November 2020].

77 State Grid Corporation of China (2020). White Paper 2019 Internet of Things. [online] sgcc.com.cn. Available at: http://www.sgcc.com.cn/html/files/2019-10/14/20191014235609307380194.pdf [Accessed 4 November 2020]. (Chinese.)

78 Yu, Y. (2019). Chinese Grids' Transformation to Benefit Digital & Tech Companies Globally. [online] Energy Iceberg. Available at: https://energyiceberg.com/chinese-grids-transformation-to-benefit-digital-tech-companies-globally/ [Accessed 4 November 2020].

79 ZTE Corp. (2020). ZTE, China Mobile and China Southern Power Grid Complete the Industry's First 5G R16 End-to-End High-Precision Timing Power Distribution Service. [online] Zte.com.cn. Available at: https://www.zte.com.cn/global/about/news/20201029e1.html [Accessed 4 November 2020].

80 National Energy Administration (2020b). National Energy Administration Releases National Power Industry Statistics in 2019. [online] www.nea.gov.cn. Available at: http://www.nea.gov.cn/2020-01/20/c_138720881.htm [Accessed 5 November 2020].

81 Huawei Technologies Co., Ltd. (2020). Huawei Investment and Holding Co., Ltd. 2019 Sustainability Report. [online] Huawei Technologies Co., Ltd. Available at: https://www.huawei.com/en/sustainability [Accessed 5 November 2020].

82 Huawei Technologies Co., Ltd. (2020). Huawei Investment and Holding Co., Ltd. 2019 Sustainability Report. [online] Huawei Technologies Co., Ltd. Available at: https://www.huawei.com/en/sustainability [Accessed 5 November 2020].

83 Huawei Technologies (2020a). Global Power Summit: Cloud, Big Data, 5G, and AI Accelerate Digital Transformation in the Power Industry. [online] Huawei Enterprise. Available at: https://e.huawei.com/en/material/enterprise/bc9d06ee35ac4e089c5d7ec358551e2f [Accessed 5 November 2020].

Chapter 4
Financing the Growth

The development of Asia's electric power and related infrastructure as well as the digitalisation transformation will require capital investments in the trillions of dollars in the next 30 years. But, where will the trillions of dollars come from? Will the digital tech and the energy sector easily attract the required amount of equity, credit, and lending? The simple answer is undoubtedly yes. The capital raising process, however, will be more straightforward for the digital tech sector than for the energy one in the short term. The digital tech sector funding is well established, has well-proven channels, and has a strong track record. For the energy sector, funding environmentally sustainable projects is something that is somewhat less well established. This is because relative to the digital tech sector, green and sustainable energy funding is still at a relatively early stage of development.

This section of the book will first address tech financing. Following that it will assess green and sustainable energy finance. It will then dig deeper into two areas. The first is an analysis as to what is behind the surprisingly huge exponential growth of capital dedicated to sustainable development, including clean energy projects, factors that have been driven both by governments and by corporations. The second area is an evaluation of existing mainstream, or plain-vanilla, investment and financing products turning green as well as some of the new instruments, including fintech or digital financing approaches.

4.1 Financing Energy Digitalisation Tech: Rich and Well-Proven Channels

The tech sector, which includes energy digitalisation technologies, is a well-established and a well-understood one by financiers and investors. It enjoys a long track record of attracting investment and funding through a great variety of channels. It is an important, rich, and deep market. The highest profile manifestation of this in capital markets globally is perhaps the role of tech in stock markets. The tech sector takes up a high percentage of the world's largest stock markets in terms of the listed tech companies' market value. Many stock markets around the world have tech sector indices that comprise listed companies in the information and communication technology (ICT) sector. One example is the Hang Seng Tech Index in Hong Kong. Some financial centres even have a dedicated platform chiefly dedicated to these companies. Some examples are the NASDAQ in New York and the Kosdaq in Seoul, both of which were originally designed to house innovative and tech listed companies.

Many listed ICT companies are household names: corporations such as Alibaba, Alphabet (the owner of Google), Amazon, Apple, Facebook, Microsoft, and Tencent. These seven corporations were worth more than $7 trillion as of June 2020 (Figure 4.1). Given their size they can easily fund tech investments directly with their own equity, or fund them through equity and borrowing, or by co-investing with other companies.

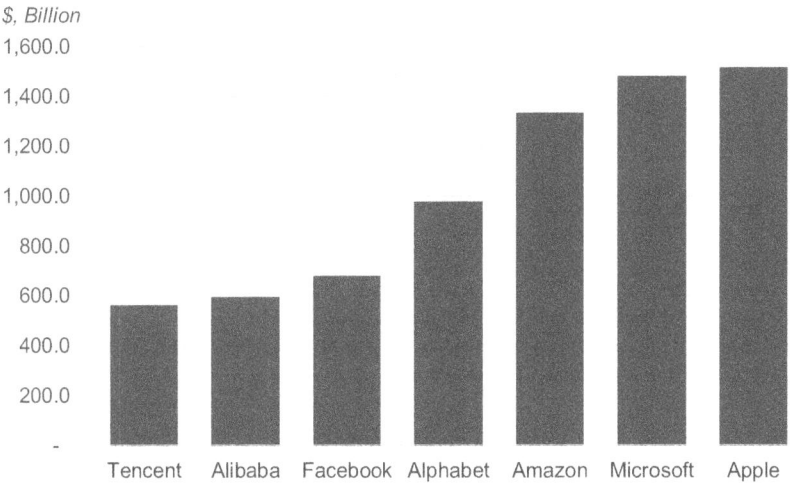

Figure 4.1: Market Capitalisation of the Largest Tech Companies in the World.
Source: Author, November 2020. Data source: Yahoo Finance.

The seven corporations and their peers invest trillions of dollars annually in ICT R&D, including ICT-related R&D directly or indirectly related to energy. They spent more than $118.6 billion on a combined basis on ICT-related R&D in 2019, which represented an increase of more than 23 per cent from the previous year (Figure 4.2).

The amount is surely set to continue to rise over the next few years. The companies have not provided long-term guidance as to their R&D spending. A scenario analysis was created to gauge their potential future spending. I found that over the next ten years, the seven are likely to pour several hundreds of billions of dollars into R&D – assuming of course that all seven are still around over the next ten years; after all we must remember that ICT companies can disappear or mutate into something else in a very short period of time. This scenario analysis assumed the lowest year-on-year growth rate in the past five years as the low case scenario and the median year-on-year growth rate as the median scenario. My simple scenario analysis shows that on a combined basis the companies will spend from $4.26 trillion to $5.62 trillion through 2030 from 2020 (Figure 4.3). This is of course only a theoretical exercise, but it does show that the potential amount of capital that will be poured into R&D is simply enormous.

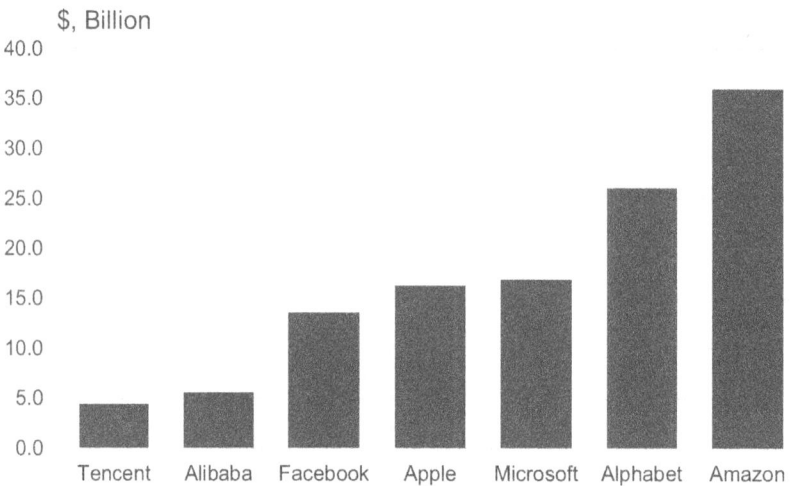

Figure 4.2: Amount Spent by Seven Digital Companies on R&D in 2019.
Source: Author, November 2020. Data source: financial statements published by Alibaba, Alphabet, Amazon, Apple, Facebook, Microsoft, and Tencent.

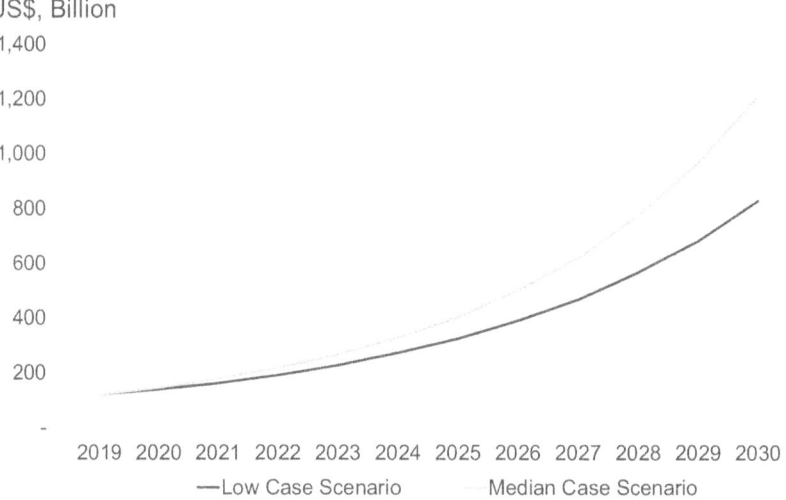

Figure 4.3: Scenario Analysis of Possible Growth in R&D Spending by Seven Tech Giants.
Source: Author, November 2020.

Market intelligence company International Data Corporation, or IDC, estimated that in 2020 global investments in ICT would still amount to about $4.3 trillion[1] despite the global economic slowdown. Gartner, Inc., another research organisation, estimated a lower amount, putting the capital required at about $3.4 trillion; this comprises data centre systems, enterprise software, devices, information technology services, and communication services.[2] Whether $3.4 trillion or $4.3 trillion, the amount is simply colossal, particularly given the global economic slowdown caused by governments' reactions to COVID-19 in 2020.

Some of the spending in ICT will be done directly by many different mainstream sectors and companies with their own balance sheets. These include owners or operators of commercial and residential buildings, hospitals, manufacturing facilities, logistics facilities such as warehouses, and others. In the green and sustainable area, investments could be related to areas such as electric power self-generation, energy efficiency or tech systems to optimise their electricity purchasing. This is more likely to happen in markets where there is no longer a generation and distribution monopoly and where the markets are open to competition such as Australia, Japan, New Zealand, or Singapore and to some extent in China, India, and the Philippines, as discussed in earlier chapters. The relationships between the sectors and the potential energy investments are quite diverse and comprehensive (Figure 4.4).

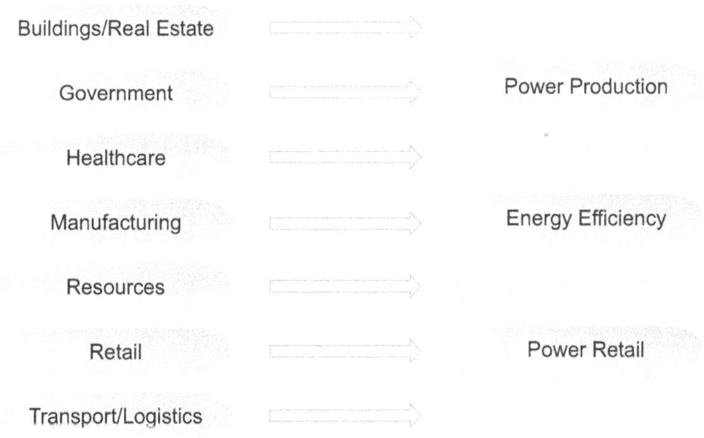

Buildings/Real Estate		
Government		Power Production
Healthcare		
Manufacturing		Energy Efficiency
Resources		
Retail		Power Retail
Transport/Logistics		

Figure 4.4: Key Sectors and Energy Technologies Relationship.
Source: Author, November 2020.

Apart from the massive tech companies, other types of investors are very active in investing in the ICT space. They range from angel investors to hedge funds to venture capital firms (Figure 4.5).

Within the ICT sector it is not an easy task to specifically identify what is and what is not an energy digitalisation technology. This is because often it is a subset of one of the many tech subgroups. It will overlap, for example, with technologies related to AI, blockchain, the cloud, chip and component making, computers and software, cybersecurity, the IoT and smartphones, as detailed in the last chapter. So, it is quite difficult to specifically identify exactly how much investment has gone into energy digitalisation tech. There is of course the amount spent by

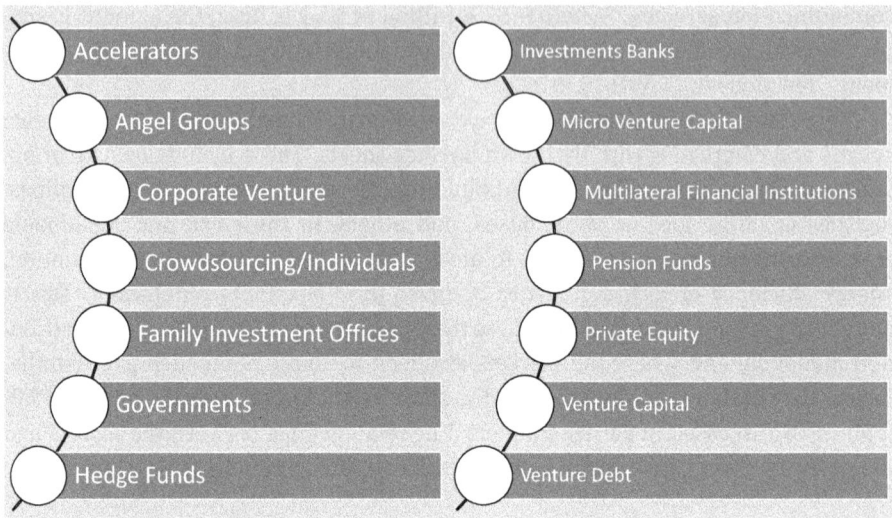

Figure 4.5: Sources of Funding for Tech Companies.
Source: Author, November 2020.

utilities. Unfortunately, not many actually disclose exactly how much budget was allocated to digital technologies. One rare example is Italy-based power company ENEL. The utility disclosed in early 2020 that it would be spending 11.8 billion Euros ($13.9 billion) in digitalisation investments in the three-year period through 2022.

4.2 Alphabet Soup and Drivers of Green Capital Growth

An energy tech related project should not find it hard to raise funds through equity, credit, or borrowing channels. However, a project that wants to be labelled green or sustainable has to take a much arduous path to funding. Such a project actually has to prove that it is what it claims, i.e., green or sustainable. Generally speaking, tech projects do not have to prove they are tech projects. The methods for verification have been rapidly multiplying in recent years. This typically will entail some form of certification from a national or supranational entity. The added process translates into more human capital having to be deployed, and, of course, more cost. While tech is tech and is generally well understood by professionals, the same cannot be said for what exactly constitutes green or sustainable and what does not. One of the many, many examples is nuclear energy. It is accepted that it is a zero carbon and emissions free source of energy. However, it is not regarded as a green and sustainable source of energy everywhere. For example, it is a clean energy in China, which is building a lot of new stations. But it is not in Germany, which is actually phasing out its nuclear power plants. In short, in the view of this author, the green and sustainable sector is still in its infancy when it comes to the capital market.

Capital markets were probably created in the 1100s in Genoa and Venice, two then powerful Italian city states. Capital markets have traditionally been most preoccupied with a business' profitability and profit growth. In the last 100 years or so, the markets have slowly but surely come to recognise that a business was not just about profitability but that its social and environmental impact also plays an important role for the investors to decide whether or not to buy the business' equities or credit and for the financier to lend to the business. But it really has not been until about three decades ago or so that corporate factors, other than the state of a corporate's profit and loss and balance sheet accounts, have gained prominence. In the past one or two decades, financial markets have progressively fine-tuned concepts such as corporate social responsibility (CSR) and environmental, social and governance (ESG). It is the advent of capital exclusively dedicated to sustainable investments that quite crucially motivated, in the view of this author, the enormous amount of work that has gone into how best to measure a corporation's CSR or ESG performance.

Unfortunately, all of the goodwill that has gone into the measuring of CSR and ESG performance has perhaps led to quite a bit of confusion given that there has been a lack of uniformity in the measures or definitions. So, before we start talking about the present and future growth of green and sustainable finance, some definitions need to be tackled.

In this discussion about financing, we will not be addressing at a granular level the standards, definitions, classifications, or taxonomies of what exactly green and sustainable financing and investing is and is not. Such discussions and opinions are readily available from a myriad of professional experts in various organisations, mentioned throughout this chapter. The primary focus lies with the medium to long-term trend of the amount of capital available for green and sustainable finance.

4.2.1 The Alphabet Soup: What's in a Definition?

Let's assume that one was to do an Internet search about issues related to the financing of investments dedicated to generating a positive impact on the environment. One is sure to find millions of results (I found about 277 million actually). They will include a colossal volume of documents and data from domestic and international institutions, financial regulators, asset management firms, and many other financial markets actors. The very broad narrative is fundamentally the same: it is about investment decision making, which at a minimum considers the environmental sustainability impact of the capital invested. For the purposes of the book no narrow definition is used. Whatever the wording used, one of earliest and still lingering terminologies in relation to green and sustainable finance is ESG – typically, with more weight on the 'E' factor.

The general concept in the context of investing first appeared in the 1960s. It attracted a lot of discussions in the 1970s and 1980s. But it was not until the 1990s that the concept of ESG investing started to appear. In the context of green and sustainable finance the key factors in ESG include carbon emissions, climate change, energy consumption, pollution, resource depletion, and waste production (Figure 4.6).

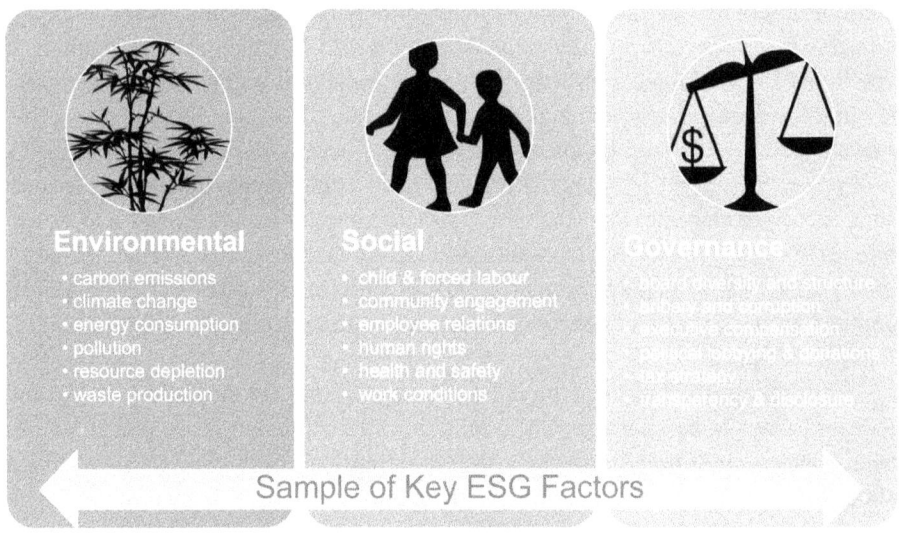

Figure 4.6: Sample List of ESG Factors (Adapted).
Source: Author, November 2020. Data source: Principle for Responsible Investment (n.d.). An Introduction to Responsible Investment | Investment tools. [online] Principle for Responsible Investment. Available at: https://www.unpri.org/investment-tools/an-introduction-to-responsible-investment [Accessed 23 June 2020]. Clipart from Clker.com.

The fact that sustainable investing and related terms are dogged with inconsistent terminology has been a very hot topic in recent years. Many leading experts have joined the debate to clarify the terms, with many highlighting the importance of reaching a uniform definition. One such expert is Barbara Novick. Novick is the co-founder of BlackRock, which is the world's largest institutional asset manager measured by funds under management and which stood at $6.47 trillion as of March 2020. BlackRock has a publicly stated strategy to broaden its sustainable investing platform and move towards making sustainability its standard for investing. Novick authored an article that addressed the definitions. She emphasised that there is a need to reach an agreed understanding as to the expectation from financial products offering sustainable investment themes.[3] She opined that such common understanding can be achieved by creating a robust system of classification or taxonomy. Such system would allow the issuer of the financial product, be it a corporate or a financial institution, to present the data regarding a product's attributes to be clear and transparent. At the same time, the creation of the classification must be done thoughtfully because too much granularity could reduce innovation and asset owner choice.[4] Novick also underlined two policy-related concerns.

One is the need for a well-regulated sustainable finance ecosystem at the global level. Another is addressing the issue that the standards must be robust and must be clear to avoid unintended 'greenwashing'. A very tight and complex definition can be misleading and may prevent the asset owner from adequately assessing the sustainability characteristic of an investment.[5] This sentiment is echoed by a great variety of other experts in the field including Fiona Reynolds,[6] the CEO of The Principles for Responsible Investment (PRI), a United Nations-backed initiative for asset owners and investment managers.

Within this decade, it is extremely likely that a broadly agreed taxonomy and frameworks will be established, in the view of this author. It is in the interest of all parties involved. Also, the momentum in the creation of sustainability related financial products has been accelerating. This has been breeding a huge push for clearer and widely accepted definitions to be established. As of 2020, the word 'sustainable' in sustainable investing and financing is often replaced with other terms, including 'climate', 'ESG-based', 'responsible', or 'impact'. This author also believes that the agreed taxonomy and frameworks will revolve around, and be tightly aligned with, the United Nations Sustainable Development Goals (UN SDGs). The 17 goals (Figure 4.7) include Affordable and Clean Energy, Industry, Innovation and Infrastructure, Sustainable Cities and Communities, Responsible Consumption and Production, and Climate Action.

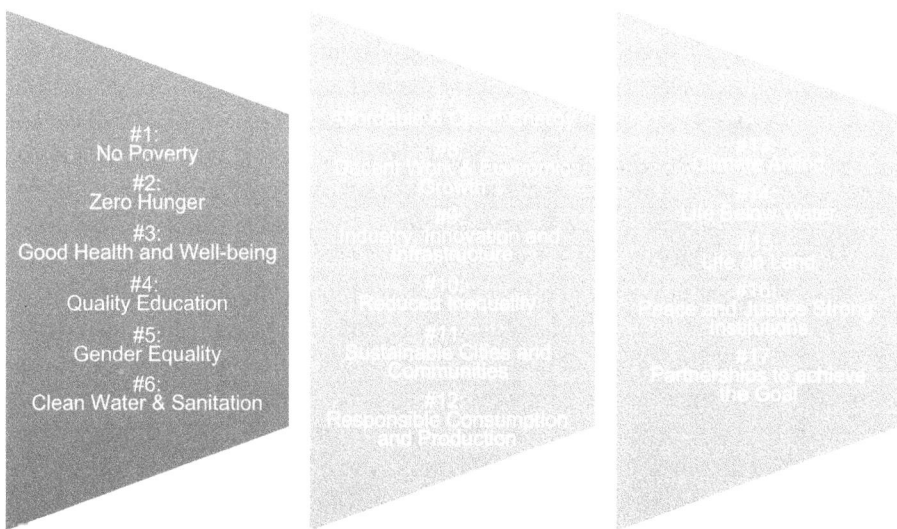

Figure 4.7: The 17 United Nations Sustainable Development Goals (SDGs).[7]

This author absolutely recognises that the UN SDGs will evolve and mutate over the coming years. Some new SDGs may be added, some may be merged, and others may be dropped. It is likely to be a gradual but progressive evolution. So, it is fairly certain that in five or ten years the wording will be different. Still the essence of the concepts will not change. They will be just fine-tuned.

4.2.2 The Drivers of Green Capital Growth

The significant momentum towards settling on a taxonomy or definition of what is sustainable investing and financing is driven by the huge growth in capital dedicated to this theme, and the momentum is driven by many different pressure points. There are bottom-up pressures, mostly from corporates and their shareholders. There are top-down pressures, chiefly from governments and regulators. And, there are pressures from financial institutions, be it asset managers or banks. The pressures are often interwoven and overlap, of course. These stress drivers are not a fad or a 'flavour of the month'. They have been mounting in the past few years. They will not dissipate anytime soon. Rather, they will gain further impetus in the 2020s. The reason for the strengthening in recent years is because of the sharp rise in shareholder, employee, and consumer activism directly or indirectly related to ESG factors in general, including sustainable investing and financing. It is crucial to note that there has been a mind shift – one away from the traditional view towards the way social and environmental issues are considered in the investment or financing process. The traditional way of thinking was that accounting for these concerns will actually cost the investor money in the form of either paying a premium for the investment or even investing at break even or at a loss. Since the mid-2010s or so, the mind shift has been that it is absolutely not necessary to 'sacrifice return when investing in solutions for sustainable development' as Dirk Schoenmaker, professor of Banking and Finance at Rotterdam School of Management,[8] adequately stated.

So, let's breakdown the pressures from governments, from within and on corporations, and from the financial actors.

Many different government authorities around the world have been pushing a move towards green and sustainable investments and financing, albeit the strength of the push has varied depending on the country. The best-known, high-level concern has been climate change. While the debate over climate change has been lively for several decades, perhaps the most important advance was the ratification of the Paris Agreement in late 2016. Of the 197 parties who are members of the United Nations Framework Convention on Climate Change (UNFCCC), 189 are Paris Agreement signatories. The Agreement's purpose is to hold the increase in global average temperature to well below 2 degrees Celsius (35.6 Fahrenheit) above pre-industrial levels to reduce the risks and impacts of climate change. It obliges the signatories to set greenhouse gas emissions reductions and to make 'finance flows consistent with a pathway towards low greenhouse gas emissions and climate-resilient development'[9] (Article 2.1c).

Some key tools available to governments to redirect capital towards low emissions and climate-resilient finance as well as increasing the supply of climate-compatible finance were summarised quite well by researchers at the World Resources Institute (Figure 4.8). They can influence behaviour in at least four ways. They can issue financial policies and regulations, such as mandating climate reporting by asset managers or owners. They can create national fiscal policy levers that direct price signals, such as structuring fee-in tariffs for renewable energy projects or lowering of cutting all together the sales tax for electric vehicles. They can use public finance to shift financial risk, as creating green banks or climate funds that can deploy capital towards decarbonisation activities. Finally, they can drive information instruments that will raise awareness about climate risks and opportunities.

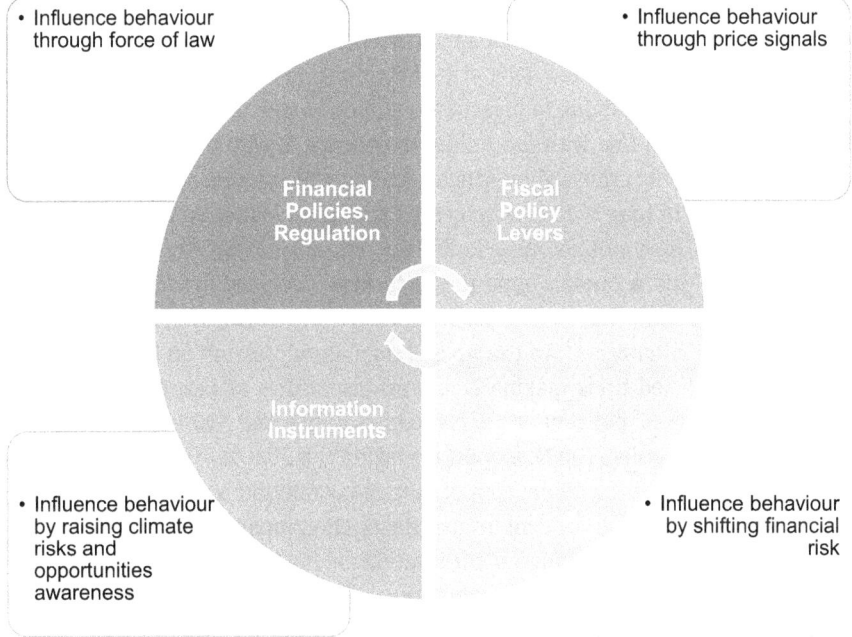

Figure 4.8: Tools Governments Can Employ to Shift Finance.
Source: World Resources Institute (2018). Aligning Finance Is the Forgotten Goal of the Paris Agreement, But It Is Vital to Successful Climate Action. [online] World Resources Institute. Available at: https://www.wri.org/blog/2018/12/aligning-finance-forgotten-goal-paris-agreement-it-vital-successful-climate-action [Accessed 1 August 2020].

The pressures on corporates towards sustainable investing is not limited to pressures that they get from governments' financial and fiscal policies and regulations. Many are also being pushed by social activism, internal staff, shareholders, or other stakeholders. Perhaps the pressure point with the highest visibility has been having to address sustainable investing as part of the acceleration in having to comply to ever growing ESG guidelines. In the past decade or so, corporations, especially those listed on stock exchanges, have increasingly realised that they must deal with ESG issues. Historically, for many this has simply been a matter of filling in forms and ticking boxes so that they may comply to the prevalent ESG rules or guidelines of the stock markets they are listed on. But in more recent years, there has been an increasing shift to taking ESG issues much more seriously. One of the many pressures driving this is internal. Improvement and change in addressing green issues for corporations have been driven by its employees, by its owners, by its shareholders, and by key stakeholders. An example of pressure from internal stakeholders and social activism was highlighted during the COVID-19 pandemic and Black Lives Matter movement in 2020, albeit the pressure was more focused on the 'Social' and the 'Governance' principles than on the 'Environment' one. Adidas, one of the largest sportswear manufacturers in the world, came under immense pressure from a group of employees to address the fact that the company lacked racial diversity. Adidas subsequently committed to allocate 30 per cent of new jobs at its US operations to African American and Latino people as well as pledging to invest $120 million in African American communities. Another example in line with the Adidas experience is with tech giant Alphabet. The company was openly criticised by African American employees for doing too little on the diversity front in June 2020 driven by the Black Lives Matter movement. This led the company's CEO to announce new initiatives and a large financial commitment worth $175 million. This 'economic opportunity package' included funding of $100 million for capital firms, start-ups, and organisations led by African American entrepreneurs, among other initiatives.[10] An example of regulatory pressure on corporations is that the number of listed firms making emissions disclosures as part of their ESG requirement has risen to 67 per cent of the companies in the S&P 500 Index versus just 53 per cent five years earlier, *The Economist* magazine calculated. In Europe, the number surged to 79 per cent from 40 per cent of Euro Stoxx 600 Index members. In Japan, it is now 46 per cent versus 13 per cent for the Nikkei 225 companies.[11]

Another dimension of pressure on corporations is having to account for climate risks. Their shareholders and their stakeholders will scrutinise their climate risk-related strategies, budgets, and business contingency plans, and the like; these shareholders and stakeholders include equity or credit fund managers of course and their role and approaches are discussed in Section 4.3. Here suffice it to say they themselves are under pressure to invest in companies with high ESG ratings. This occurs regardless of whether or not the leadership believes that the climate is changing, and this poses business risks. They are being asked to disclose their strategies to address climate risks, including credit, physical, reputational, and transition risks. They have to address questions such as how their revenues or assets may be impacted by extreme weather events. And also questions on how the company will be financially affected by the shift from fossil fuels based energy to clean energy, for example. For corporations designing blueprint plans addressing climate change related risks it is not only highly laborious but also quite complex. How tasking is it? Well, one of the many examples is a set of scenarios offered by the Task Force on Climate-related Financial Disclosures, or TCFD (Figure 4.9). The TCFD is an organisation created in 2015 to develop voluntary climate-related financial risk disclosures under the Financial Stability Board, which counts all G20 major economies and the European Commission among its members. The checklist provided by the TCFD for corporations includes items such as structuring the governance around addressing climate risk, assessing the materiality of the risks, constructing various low to high probability scenarios, evaluating the impact on the business operationally and financially, and identifying potential responses.

1. **Ensure Governance is In Place:** Integrate scenario analysis into strategic planning and/or enterprise risk management processes. Assign oversight to relevant board committees/sub-committees. Identify which internal (and external) stakeholders to involve and how

2. **Assess Materiality of Climate-Related Risks:** Market and Technology Shifts; Reputation; Policy and Legal; Physical Risks

3. **Identify and Define Range of Scenarios:** Scenarios inclusive of a range of transition and physical risks relevant to the organization

4. **Evaluate Business Impacts:** Impact on input costs, operating costs, revenues, supply chain, business interruption, timing

5. **Identify Potential Responses:** Responses might include changes to business model, changes to portfolio mix, investments in capabilities and technologies

6. **Document and Disclose:** Document the process; communicate to relevant parties; be prepared to disclose key inputs, assumptions, analytical methods, outputs, and potential management responses

Figure 4.9: Climate-Related Risks and Opportunities Scenario Analysis Application Process (Adapted).
Source: Adapted from: Task Force on Climate-Related Financial Disclosures (2017). Recommendations of the Task Force on Climate-Related Financial Disclosure: The Use of Scenario Analysis in Disclosure of Climate-Related Risks and Opportunities Technical Supplement Technical Supplement – the Use of Scenario Analysis in Disclosure of Climate-Related Risks and Opportunities. [online] Task Force on Climate-Related Financial Disclosures. Available at: https://assets.bbhub.io/company/sites/60/2020/10/FINAL-TCFD-Technical-Supplement-062917.pdf [Accessed 6 November 2020].

Part of the process of the corporation addressing the TCFD checklist will also often mean that the company will have to allocate a budget to tackle all of the risks that it identifies. Also, it may need to set aside some capital for contingent liabilities as well. An example of this is the US's Pacific Gas and Electric Company, which was forced to file for bankruptcy protection in January 2019. This was because of the extensive California wildfires that created at least more than $13 billion in claims and damages against the company as it was linked to one of the major fires when one of its power transmission lines came into contact with nearby trees. Another example is Hitachi, a Japanese conglomerate, whose many businesses include information technology and energy. In its annual report for the fiscal year through March 2020, the company highlighted that it had recognised that it needed to formulate strategies for its business operations – and those of its suppliers – from such climate-related risks of climatic events including the increased severity of droughts, floods, and typhoons as well as rising sea levels and chronic heat waves. Hitachi's management said that it would consider possible damage from floods when determining the location or equipment layout of a new plant.[12]

Yet another pressure is from another actor: banks, including commercial and multilateral ones. They have been, and will continue to be, a driving force in the exponential growth of capital dedicated to green and sustainable investment and financing. And the commitment from this actor has also been growing exponentially over the past few years. One example of the pressure from (and on) financial institutions worth looking at is the setting up of the Finance Initiative as part of the United Nations Environment Programme, or UNEP FI. The UNEP FI has recently focused on helping its lending financial institutions members to raise the amount of loans which support socially and environmentally sustainable economic activities. This, under an umbrella that they call Principles for Responsible Banking. It was launched with 130 banks holding $47 trillion in assets, about a third of the global sector's, in September 2019. Almost 220 such institutions will follow these principles, as the membership has been growing steadily.[13]

Unfortunately for now, members from the Asia region are still lacking. Looking through the 218 members as of June 2020, there were only 29 from Asia, accounting for just 13 per cent of the total (Figure 4.10).

Many large financial institutions in the region have yet to join, apart from those from Australia, Japan, and South Korea. It should be expected that there will be more members from the region's largest economies, including China, India, and Indonesia. In the case of China, there are only six financial institutions who are members but among the country's largest financial institutions only one is represented, the Industrial and Commercial Bank of China (ICBC), which is currently the largest bank in the world in terms of Tier 1 capital. It is highly likely that the other large Chinese banks will also join the UNEP FI, including China Construction Bank (world's top two), Agricultural Bank of China (world's top three), and the Bank of China (world's top four). However, while these Chinese banks have not sought membership as of mid-2020, they have their own sustainability-related lending standards; this is discussed in the Green Bank Lending section below.

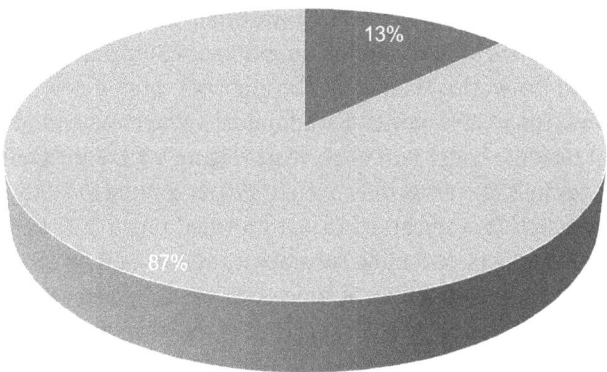

■ Members From Asia Pacific   ○ Members From Rest of the World

Figure 4.10: United Nations Environment Programme – Finance Initiative Banks' Membership by Region (Adapted).
Source: United Nations Environment Programme Finance Initiative (2019). About United Nations Environment Programme Finance Initiative – United Nations Environment – Finance Initiative. [online] Unepfi.org. Available at: https://www.unepfi.org/about/ [Accessed 6 November 2020].

4.3 Plain Vanilla Financing Turns Green

Despite the increasingly large amount of capital dedicated to green and sustainable investment and finance, there has been relatively limited financial products innovation in the past few decades, in the view of this author. The financial instruments on offer are chiefly equity-related, or credit, or loans and their derivatives. So, one could say that just what has been happening with green and sustainable investment and finance is simply turning plain vanilla products (or traditional, existing products) into green products. This has actually recently been changing in the past three to five years, and the outlook for innovation in the coming years is looking very bright.

4.3.1 Corporates Building Green Equity Credentials

We will evaluate on three levels how corporations have been, and will continue to, build their green credentials to meet financial institutions, societal, and regulatory demands. We will highlight some examples from three electricity markets in Asia, namely from China, Hong Kong, and Australia. We will then turn to examining how corporations in general are structuring their green and sustainability strategies, and how they are actually getting a lot of help and support from institutional investors as well as some international organisations. Finally, we will examine the intersection between the corporates and the equity capital markets.

At the equity finance level, change has been slow. First, there was the snail pace recognition of the importance of ESG and the adoption of ESG principles from corporations from all around the world. Then came the alphabet soup of new standards and new organisations. Today, ESG principles adoption is much easier and has been sharply rising; still it took the better part of 30 years (Figure 4.11). For example, there were 1,500 ESG indexes in 2020 versus only one in 1990, according to MSCI ESG Research LLC. Also, the research firm remarked that it provided broad ESG Ratings coverage for more than 14,000 issuers, including subsidiaries and more than 680,000 equity and fixed-income securities globally as of June 2020 versus coverage of less than 1,000 in 1990 by KLD Research & Analytics, one of the ESG ratings pioneers founded in 1988 which was subsequently acquired by MSCI. The analysis in the following sections will predominantly focus on the 'E' portion of ESG. This is because our overall discussion in this section of the book is about financing green and sustainable growth. Of course, this does not in any way whatsoever undermine the importance of the 'S' and of the 'G'.

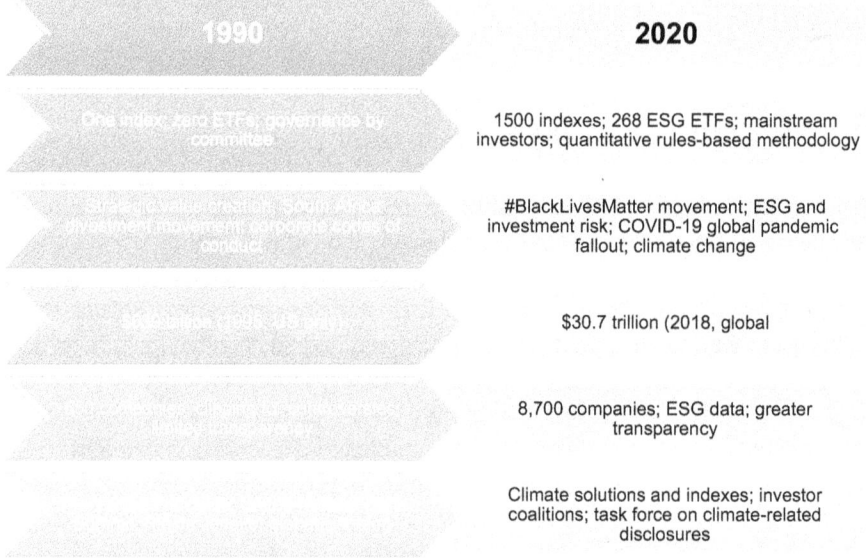

Figure 4.11: Thirty Years' Evolution of MSCI ESG Indexes.
Source: MSCI (n.d.). Explore 30 Years of ESG. [online] www.msci.com. Available at: https://www.msci.com/esg/30-years-of-esg [Accessed 6 November 2020].

The asset managers use ESG criteria to determine the score for a company. If the company's rating is below the prescribed rating, then the company may not even be investable from the fund managers' perspective. We will not delve too much on the ESG scoring system. But, it is worth noting that currently the asset managers deploy internal and external scoring systems. The problem is that there is no uniform approach in the systems used. It is almost certain, however, that some kind of uniform methodology will be adopted in the coming years given that various organisations are striving towards that goal. We can use some examples that show why it is important for a company to think about ESG in general. We will look at three companies: China's China Longyuan Power Group Corp. Ltd., Hong Kong's CLP Holdings Ltd., and Australia's AGL Energy Ltd.

Chinese electric power generation utility Longyuan Power was listed on the Hong Kong Stock Exchange on 10 December 2009. The company is principally a renewable energy generator and is the largest wind power producer in China and one of the largest in the world. At the time of its listing its clean energy asset base was tiny. Its then parent company decided to inject a large thermal coal-fired power asset into the company. The rationale behind this, in the view of this author, was that a larger company would be able to raise more funds as well as being able to attract a larger number of investors, given that not all investors invest in small market capitalisation companies.

Longyuan Power had 6,407 megawatts in operational installed capacity as at the end of 2009. This included 4,504 megawatts in wind power and 1,875 megawatts in coal generation and 29 megawatts in other renewable energies. However, because the output from coal plants is higher than from renewable energy plants, wind produced 3.41 terawatt-hours, coal produced 18.9 terawatt-hours, and other renewables produced 0.02 terawatt-hours in 2009. The output profile at the end of 2019 was sharply different. The capacity breakdown was 20,032 megawatts for wind, 1,875 megawatts for coal (unchanged), and 250 megawatts for other renewables. The generation breakdown for wind, coal, and other renewable energies was 39.3, 8.9, and 0.5 terawatt-hours, respectively (Figure 4.12).

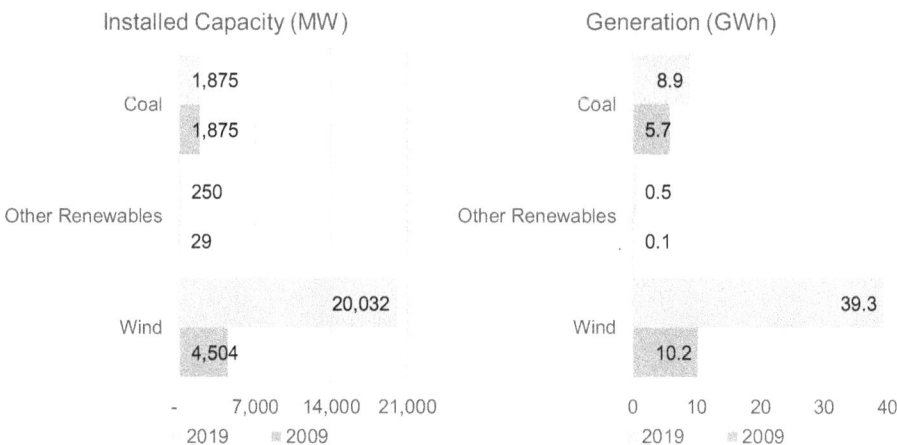

Figure 4.12: China Longyuan Power Capacity and Output Change, 2009–2019.
Source: China Longyuan Power Group Corp. Ltd. (2010, 2019). China Longyuan Power Group Corp. Ltd. 2010 Interim Result Presentation. China Longyuan Power Group Corp. Ltd. 2019 Final Result Presentation.

In percentage terms, wind capacity rose to 90 per cent of the total in 2019 from 70 per cent in 2009 while output changed to 81 per cent of the total from 64 per cent. Coal capacity dropped to 9 per cent from 29 per cent while generation fell to 18 per cent from 36 per cent. So, Longyuan Power made tremendous efforts to sharply increase its clean energy footprint. However, because Longyuan Power still owns and operates polluting thermal coal-fired generation, for some investors the company's ESG score at the environmental-end may be very low while for some the stock is not investable at all. There are, of course, many other companies in similar situations. The management will have to carefully think about their asset portfolio from a sustainability angle. In the case of Longyuan Power, the company is highly likely in the next few years to either completely shut down its coal generation facilities or possibly it could sell off these facilities, in the view of this author.

Longyuan Power of course is a little bit of an anomaly. It is an electric power utility dedicated to clean energy, but it got stuck with a polluting fossil fuel asset. For many other listed utilities in the region, the problem is actually the other way around. They historically mostly had fossil fuel power generation assets and it is only in recent years that they have tried to shift away from a brown asset portfolio to a green one.

The electric power supplier CLP Holdings Ltd., formerly known as China Light and Power, was born about 120 years ago in Hong Kong. It is today one of two vertically integrated (comprising generation, transmission, and the sale of power) companies in the special administrative region; the other one is Hongkong Electric. The group started to earnestly invest outside Hong Kong in the 2000s – if we ignore the passive investment in a nuclear power plant across the border in mainland China in the 1990s. Just like many other utilities operating in developed economies in Asia and elsewhere, it decided to invest in new geographies because a maturing local market meant reduced earnings growth prospects. It constructed and acquired power assets, including coal-fired ones, in Australia, China, India, and Thailand. The utility released 'CLP's Climate Vision 2050', a carbon reduction programme in 2007. The Climate Vision 2050 had an objective to cut the group-wide carbon intensity of its generating portfolio by 75 per cent to 0.2 kilos per kilowatt-hour by 2050 from about 0.8 in 2007; carbon intensity means the amount of carbon dioxide released per unit of electricity produced. CLP increased its target reduction to 0.15, or 80 per cent less than its 2007 baseline, in 2018 (Figure 4.13).

To some, the cut may not necessarily sound like a particularly large amount but in the world of energy it is a very ambitious objective indeed. This is because the group has a large amount of coal-fired generation in its portfolio. So, CLP had to embark on a journey of assets divestment and fresh investments in clean and renewable energy. In fact, when the company refreshed its Climate Vision in 2018, it also committed to not making any new investments in coal-fired power plants, continuing to find ways to phase out its existing coal-fired footprint, and increasing its investments in clean and renewable energy as well as in transmission and decentralised smart energy solutions. Specifically, the company actually committed to clear targets

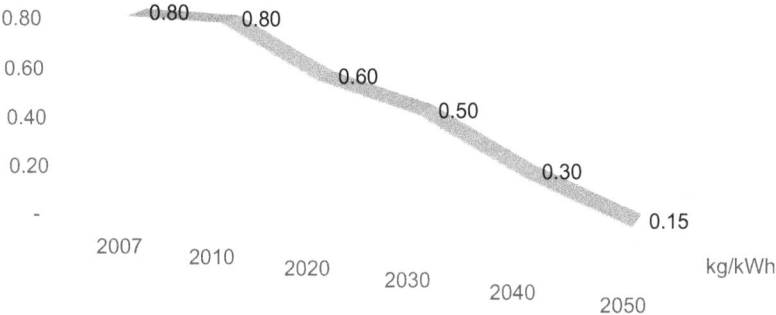

Figure 4.13: CLP Holdings' Climate Vision 2050 CO_2 Intensity Reduction Target.
Source: CLP Holdings Limited (2019). Climate Vision 2050 | CLP Group. [online] www.clpgroup.com. Available at: https://www.clpgroup.com/en/sustainability/our-approach/frameworks-strategies/climate-vision-2050 [Accessed 6 November 2020].

by 2030. It wants clean and renewable energy to take up 30 per cent by 2030, from 20 per cent in 2020, of its generation portfolio and for non-carbon generation to account for 40 per cent by 2030 from 30 per cent in 2020 of the total. It may be worth noting that the targets, first set in 2007, were voluntary. They were certainly influenced by internal and external stakeholders, but were voluntary none the less. As such, CLP was a bit of a pioneer in Asia back in the late 2000s. Elsewhere in the region, other utilities have also made some efforts since the late 2000s. However, most of them are state owned or state controlled so their decarbonisation strategies were mostly forced on them by their majority shareholder.

AGL Energy was born in 1837 in Sydney as a gas supplier known as Australian Gas Light Company. AGL Energy now serves about 3.7 users and has the largest electricity generation portfolio in the country. The company's capacity is quite heavily reliant on coal-fired generation plants. It owns and operates three such plants, totalling 6,875 megawatts out of a total portfolio of 11,330 megawatts, as of December 2019, amounting to 60.7 per cent of the total. The dominance of coal is even more evident when looking at its output, with black and brown coal being responsible for 18.9 terawatt-hours out of the company's 22.7 terawatt-hours total, or 83.3 per cent of the total production, as of December 2019 (Table 4.1).

Similar to CLP, AGL Energy decided to decarbonise its generation portfolio. It first released a new policy, the Greenhouse Gas Policy, in 2015. The company engaged itself to shut down its coal-fired power plants once they reached the end of their technical life. It subsequently updated this plan and stated that it would actually work towards closing them altogether so that the generating portfolio may be emissions free in line with Australia moving towards full decarbonisation by 2050. AGL Energy also made several climate-related commitments. These include offering its users optional carbon neutral prices products, supporting Australia's evolution in voluntary carbon markets, investing in new electricity supply sources, and responsibly

Table 4.1: AGL Energy Generation Portfolio Performance (December 2019).

Asset	Capacity (MW)	Output (TWh)	Carbon Intensity (tCO₂e/MWh)
Coal	6,875	18.9	
Oil and Gas	1,952	1.4	
Renewables	2,503	2.4	
Total	11,330	22.7	0.92

Source: AGL Energy (2020). AGL Energy Half Year Results, six months ended 31 December 2019, 13 February 2020, p. 34.

transitioning its energy portfolio. Its aim to achieve net zero emissions from generation by 2050, or earlier, includes a cut of 20 per cent in emissions by 2023 thanks to the closure of the one of its three coal-fired power plants, Liddell. This will be followed by the closure of the Bayswater and Loy Yang in 2035 and 2048, respectively.[14] In practical terms this means that AGL Energy will have time to redeploy its capital investments towards such areas as clean and renewable projects and solutions, grid scale storage, and behind-the-meter technologies, and power produced on-site and not generated on the side of the grid, such as roof-top solar power systems.

At the end of the day, in the view of this author, investors and lenders alike may look at two aspects: the long-term strategies of the utilities on the one hand and their decarbonisation track record, on the other.

But what kind of help can the utilities and other companies get to formulate and structure their sustainable investments strategies? Of course, there are consultants or research agencies who can help to support such a mission. However, green and sustainable investing is a very young field. Not only that, it is an area that is rapidly developing and mutating in a very short period of time. So even the crème de la crème consultants may at times struggle with the changing national and international frameworks and guidelines as well as changing environments. Other challenges include the evaluation of physical, reputational, and other risks. These could be the reasons why now regulators and securities exchanges operators have decided to come in more actively to offer some help. One such example is the Sustainable Stock Exchanges (SSE) initiative.

The SSE initiative was set up through a variety of UN agencies and related organisations, including the UN Global Compact and the PRI in 2009. Its stated mission is to create 'a global platform for exploring how exchanges, in collaboration with investors, companies (issuers), regulators, policymakers and relevant international organisations, can enhance performance on ESG . . . issues and encourage sustainable investment, including the financing of the UN Sustainable Development

Goals'.[15] The SSE initiative membership started to grow exponentially from the mid-2010s or so (Figure 4.14).

The SSE initiative breaks down its members, which reached 96 exchanges as of June 2020 (Figure 4.14), looking at six different criteria, including whether it produces sustainability reports (47), whether it provides written guidance on ESG reporting (54), whether it offers training on ESG topics (53), whether its markets are covered by an ESG index (42), whether it makes ESG listing a mandatory requirement (25), or whether it has ESG bond segments (31).[16]

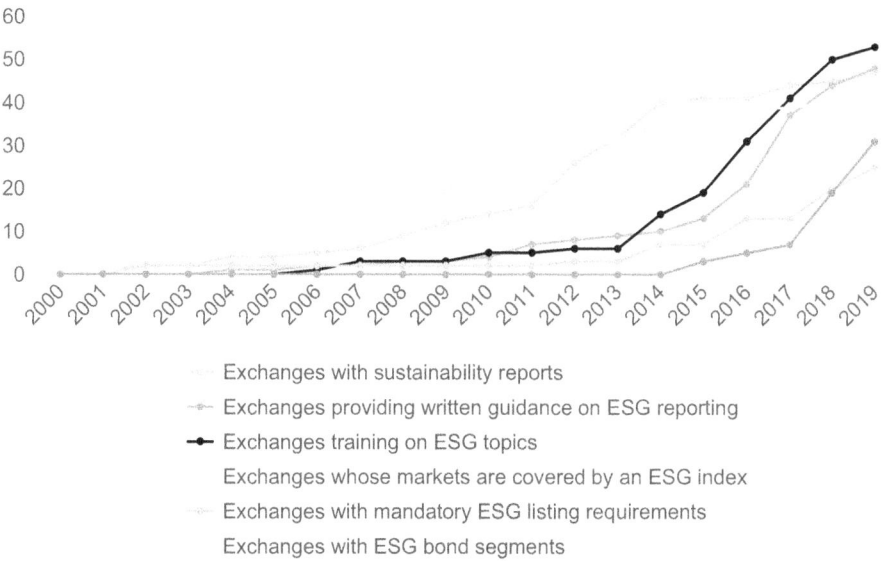

Exchanges with sustainability reports
Exchanges providing written guidance on ESG reporting
Exchanges training on ESG topics
Exchanges whose markets are covered by an ESG index
Exchanges with mandatory ESG listing requirements
Exchanges with ESG bond segments

Figure 4.14: Growth of Stock Exchange Sustainability Activities.
Source: Sustainable Stock Exchanges Initiative (2020b). Sustainable Stock Exchanges Initiative – Data. [online] sseinitiative.org. Available at: https://sseinitiative.org/data/ [Accessed 6 November 2020].

Hong Kong Exchanges and Clearing Limited (HKEX) is only one of five exchange members that meet all seven criteria. Recently, HKEX decided to go one step further. It founded a platform dedicated to sustainable finance in the middle of 2020, the first in Asia to do so. It is called the Sustainable and Green Exchange, or STAGE for short. As a first phase, STAGE is creating an online repository of green, social, and sustainable bonds and ESG-related exchange traded products by the end of 2020. In the next phases, STAGE will cover other products such as futures linked to ESG indices and will launch other new products. STAGE should prove very value added for companies that are green or have sustainable business practices, whether they are listed, or plan to list, on HKEX, as STAGE will help increase their visibility and awareness as well as those looking at issuing green financial products such as green bonds through HKEX. Needless to say, the information will also be of great help to advisors, asset managers, issuers, and market participants who are already active, or plan to be active, in the green and sustainable finance space in Hong Kong. HKEX also said that in future it could consider the introduction of other asset classes and product types such as ESG-linked derivative products.[17] STAGE is also highly likely to be extremely welcome by finance professionals given that it is a platform solely dedicated to green and sustainable financial products. But what will be STAGE's unique features? What is the overall long-term vision for this platform? This author approached the head of green and sustainable finance at HKEX, Grace Hui, and asked her these two questions. Grace said that STAGE is a platform for asset owners, investment managers, investors, issuers, advisors, and other market participants who wish to play a role in impact investing. As a responsible exchange, the goal of HKEX is to lead the market for, first, ESG information flow between issuers and investors; second, the development of a variety of impact investment products that cater for both institutional and retail investors; and, third, facilitating investors to learn and seize impact investment opportunities.

Equity investors have many avenues if they want to invest in companies in the energy space focused on green and sustainable projects. Retail and high net worth individuals can buy one of the dozens of investment funds whose mandate is to invest in such companies. Some examples, among the many, include the BlackRock Sustainable Energy Fund ($1.8 billion as of 27 July 2020), the Vanguard FTSE Social Index Fund Admiral ($7.9 billion as of 27 July 2020) or the Pictet-Quest Europe Sustainable Equities R (0.62 billion Euros, about $0.8 billion, as of 30 June 2020). Financial investors in general can choose to invest passively by buying one of the many dedicated sector ETFs like the iShares Global Clean Energy issued by BlackRock ($0.92 billion as of 26 July 2020) or the Invesco Solar ETF issued by Invesco ($0.88 billion as of 26 July 2020). A third option is for the investor to buy a whole index or one of more of its constituents. The NASDAQ alone, for example, has several indices including the NASDAQ Clean Edge Green Energy Index, the NASDAQ Clean Edge ISE Global Wind Energy Index, and the NASDAQ Clean Edge Smart Grid Infrastructure Index. So, in short, choices are not lacking. One issue with all of these funds, ETFs, and indices is that the criteria that makes up each of these financial instruments is different, so investors will still have to do a little bit of homework. However, in the view of this author, uniformity in the definitions and the criteria will likely become a reality within the next few years, as I have emphasised several times throughout the book.

It is worth noting that ESG-focused equity funds have been enjoying tremendous capital inflows. This, not just over the past few years but also during the first half of 2020 at a time when the COVID-19 pandemic was causing much havoc in the equity capital markets. Global funds network Calastone, for example, found that for the UK, ESG fund inflows rose 37-fold in the three-year period through April 2020.[18] It also estimated that the amount during the four months period through July 2020 was more than the total flow of the previous five years. Globally, the funds inflows in

2019 of about $20.6 billion were up about four times the 2018 level[19] and for the past ten years they were only two years of net outflows, in 2011 and 2012 based on research by information provider Morningstar.[20] The head for Southeast Asia of fund management giant BlackRock, Deborah Ho, noted that the firm has seen an inflow of $15.5 billion in funds set for sustainable strategies just in the first quarter of 2020, despite the COVID-19 pandemic. Ho added that when investors rebalanced their portfolio during the market turmoil, they showed a preference for sustainable funds over more traditional ones. In the first quarter, sustainable open-ended funds inflow was a record $40.5 billion in new assets, up 41 per cent against the same period the previous year. Ho also mentioned that Blackrock aims to have all active portfolios and advisory strategies to be fully ESG integrated. It has adopted three initiatives: first, building sustainable, resilient, and transparent portfolios and integrating this in all aspects of portfolio construction; second, increasing access to sustainable investing; and, third, enhancing engagement voting and transparency investment stewardship.[21]

Another two areas of equity investments and finance are investments from a corporate's own balance sheet and from private equity funds. On the investment by corporates, we have already discussed the examples of Longyuan Power, CLP, and AGL Energy. Such mode of investment is relatively straightforward to understand. It is important to note though that utilities are not the only types of corporates taking such action. They are joined by companies in a great variety of different industries, including gigantic tech companies like Amazon, Apple, and Microsoft, and even (amazingly so) oil and gas companies such as BP and Shell (Table 4.2).

Table 4.2: Examples of Companies Committed to Net Zero and Renewables Investing.

Company	Pledge
Amazon	Net zero carbon by 2040; 100% renewable energy by 2030; orders 100,000 fully-electric delivery vehicles; orders 100,000 fully-electric delivery vehicles
Apple	Supply chain and products carbon neutral by 2030; 100% renewable energy for Apple production from >70 suppliers; to invest $100 million in accelerated energy efficiency projects for suppliers
Microsoft	Carbon negative by 2030; by 2025, 100% renewable energy; electrify global campus operations vehicle fleet by 2030
BP	Net zero across operations including oil and gas production on an absolute basis by 2050 or sooner; 50% cut in carbon intensity of products sold by 2050 or sooner; increase proportion of investment into non-oil and gas businesses
Shell	Greenhouse gases emitted on average with each unit of energy sold to be cut by 30% by 2035 and by 65% by 2050

Source: Corporate press releases by Amazon, Apple, Microsoft, BP, and Shell.

In fact, many other corporates have committed to using renewable energy. This commitment was either through investing in renewable energy generation or committing to buy electricity from renewable energy sources under long-term agreements (known as Corporate Power Purchasing Agreements), which sometimes can be 20 years or longer. Total accumulated commitment between 2009 and 2019 was about 50 gigawatts, two-thirds of which – or 33.1 gigawatts – was committed to in 2018 and 2019, based on data from research firm BloombergNEF.[22] Interestingly such commitments in Asia only accounted for 3.3 gigawatts in the same period. This may sound negative but actually it is not. It is one more piece of evidence that the region has huge scope for growth when it comes to green and sustainable energy.

On the private equity side of things, there is a lot of private equity funds' capital running after green and sustainable companies for projects, too. From the global new investments in clean energy, venture capital and private equity firms broadly speaking on average invested approximately $1 billion to $2 billion per quarter from January 2006 through December 2019 based on data collected by research firm BloombergNEF.[23] This capital flow includes their investments not just in renewable energy but also on clean energy technologies, such as energy storage, albeit the bulk was on solar and wind related investments. However, the strong movement towards decarbonisation, following the ratification of the Paris Agreement, and a great variety of countries adopting renewable energy targets, the participation from venture capital and private equity firms is most likely to accelerate in the coming years. There are many examples of such funds operating in Asia. Some are country-specific while others have a more geographic scope. One of the many examples is O2 Power, which was created in January 2020. It has a total commitment of $500 million in equity to be invested in in solar and wind projects in India. O2 Power was jointly established by Temasek, the Singapore government-owned conglomerate, and EQT Infrastructure, a global investment group with 41 billion Euros ($48.3 billion) in assets under management as of January 2020. Another example is the Actis Energy 5 LP, Actis' fifth fund, started raising money in January 2020 and by July had already locked in $2.9 billion out of the $5 billion it seeks. Funds will chiefly be invested in wind and solar projects in Africa, Asia, and Latin America. A third example, South East Asia Clean Energy Facility (SEACEF), which was funded by philanthropic donors and plans to deploy its $2.5 billion climate fund in wind and solar power, grid infrastructure, energy efficiency in buildings, and electronic mobility and storage in the Southeast Asian region, includes the philanthropic organisations Sea Change Foundation International, Wellspring Climate Initiative, High Tide Foundation, Grantham Foundation, Bloomberg Philanthropies, Packard Foundation, and Children's Investment Fund Foundation.

4.3.2 Bond Issuers Learning from Chameleons

Fixed-income securities may well currently be the highest profile sustainable finance asset type. Back in November 2008, the World Bank issued a bond with a specific focus to finance projects that could help the climate. This type of bond became known as a green bond. So, green bonds have a history of more than ten years, making them high-profile financial instruments in the world of green and sustainable finance. In recent years green bonds have been joined by other types of bonds, such as blue bonds and thematic bonds. So just like some species of chameleons, sustainability-related bonds have the capability to change colour.

The World Bank thus created a blueprint for what is today one of the fastest growing areas within bond markets. How green bonds came about is itself an interesting story. The first one actually was not called a green bond. It was issued in 2007 by a multilateral financial institution, the European Investment Bank. The bank branded it 'Climate Awareness Bond' at the time, with proceeds equivalent to about $807 million dedicated to renewable energy and energy-efficiency related projects.[24] The investor-led first green bond came about because a Swedish financial group approached its clients, such as AP, the national pension fund of Sweden, proposing to find ways to directly invest in companies and projects that would have a positive impact on the environment. The CEO of AP3, one of the four national pension AP funds, Kerstin Hessius, said that at the time the investment approach was just to exclude companies that do not invest in green and sustainable projects, which was the typically negative and passive investment approach at the time that still lingers with some asset managers to this day, in the view of this author. The Swedish investors decided that they did not want to just exclude investing in companies with poor environmental governance, a rather passive approach, but wanted rather to find good companies and also investment in green and sustainable projects, although they had to be liquid, easy to understand, and credible. The pension fund-backed green bond was launched by the World Bank in 2008. The six-year 3.5 per cent-coupon bond raised 2,325 million Swedish krona ($231 million) and in addition to the Swedish AP funds also included two other Swedish financial institutions, Skandia and Länsförsäkringar.[25] Since 2008, there have been a great variety of green bonds issuances, not just multilateral financial institutions such as the World Bank, including national governments, local governments, local commercial banks, and corporations. Albeit, the first corporate issuance had to wait five years. It was another first for Sweden. It was issued by Swedish property company Vasakronan, which raised 125 million Euros (about $141 million) in 2013.

Green bonds are the more established form of fixed-income instruments given the longer history. For this reason, the total amount raised is quite staggering. Between 2009 and June 2020 a total of about $745 billion was raised by corporations based on data compiled by financial information provider Bloomberg (Figure 4.15). Looking at the country of incorporation of the issuer, the largest one has been China. It accounted for 14 per cent of the total or $104.4 billion. It was closely followed by

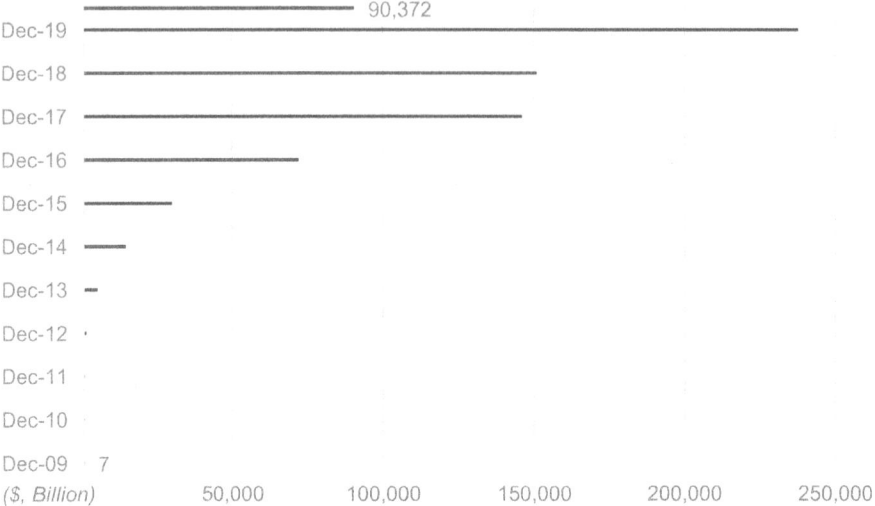

Figure 4.15: Annual Green Bonds Issuance through June 2020 Compiled by Bloomberg.
Source: Bloomberg.

French-incorporated entities that were responsible for 13.8 per cent of the total or $ 98.4 billion. Other significant issuers were the Netherlands and Germany, which made up 9.2 per cent and 9.1 per cent of the total, respectively. Interestingly, the Chinese entities issuance have slowed while those from the US, Germany, and France have accelerated. What these numbers hide is the fact that corporates have started to use new sustainable finance fixed-income instruments.

We mentioned earlier that the objective here is to give a sense of where things are now and where things are going with green and sustainable finance. As such, we have not really delved into definitions of what is and what is not green and sustainable finance and investment. It is worth, however, briefly examining some terminology approaches related to green bonds.

The Climate Bonds Initiative is a non-government organisation that is funded from a wide variety of entities including Switzerland and other governments, Bloomberg and other corporations, HSBC and other commercial banks, as well as the Inter-American Development Bank, and other multilateral organisations. The Climate Bonds Initiative also offers a certification service. Instruments such as bonds as well as loans are verified to conform with the Climate Bonds Standard. Accordingly, the Climate Bonds Initiative definition of what is and what is not a green bond is quite important and relevant. If no such standards existed anyone could issue a bond and call it 'green'. This is known as greenwash, the occurrence of which has not been uncommon in the past. We took a snapshot of some of the basic definitions, although the methodology behind the certification is vastly more complex than the short summary presented here. The organisation provides some clarity as to the categories of the financial instruments. It emphasises that it is imperative for the issuer to qualify and quantify the purpose of the proceeds from the issuance regardless of whether it is a green bond, a revenue bond, an asset-backed security, a project bond, a securitisation bond, a covered bond, a loan, or another debt instrument[26] (Table 4.3).

Table 4.3: Types of Green Bonds by the Climate Bonds Initiative.

Type	Proceeds raised by bond sale are	Debt recourse	Example
'Use of Proceeds' Bond	Earmarked for green projects	Recourse to the issuer: same credit rating applies as issuer's other bonds	EIB 'Climate Awareness Bond' (backed by EIB); Barclays Green Bond
'Use of Proceeds' Revenue Bond or ABS	Earmarked for or refinances green projects	Revenue streams from the issuers though fees, taxes, etc. are collateral for the debt	Hawaii State (backed by fee on electricity bills of the state utilities)
Project Bond	Ring-fenced for the specific underlying green project(s)	Recourse is only to the project's assets and balance sheet	Invenergy Wind Farm (backed by Invenergy Campo Palomas wind farm)
Securitisation (ABS) Bond	Refinance portfolios of green projects or proceeds are earmarked for green projects	Recourse is to a group of projects that have been grouped together (e.g., solar leases or green mortgages)	Tesla Energy (backed by residential solar leases); Obvion (backed by green mortgages)
Covered Bond	Earmarked for eligible projects included in the covered pool	Recourse to the issuer and, if the issuer is unable to repay the bond, to the covered pool	Berlin Hyp green Pfandbrief; Sparebank 1 Bolligkredit green covered bond
Loan	Earmarked for eligible projects or secured on eligible assets	Full recourse to the borrower(s) in the case of unsecured loans. Recourse to the collateral in the case of secured loans, but may also feature limited recourse to the borrower(s)	MEP Werke, Ivanhoe Cambridge and Natixis Assurances (DUO), OVG
Other debt instruments	Earmarked for eligible projects		Convertible Bonds or Notes, Schuldschein, Commercial Paper, Sukuk, Debentures

Source: Climate Bonds Initiative (2019). Explaining Green Bonds. [online] climatebonds.net. Available at: https://www.climatebonds.net/market/explaining-green-bonds [Accessed 27 April 2019].

Since the late 2010s, sustainable finance has moved beyond just green bonds. To further offer investors tailor-made financial instruments, theme-based bonds have been created. In late 2018, the Seychelles launched the first blue bond assisted by the World Bank. A blue bond is a thematic bond dedicated to supporting sustainable use of ocean resources, including projects industries including fishing, tourism, or offshore energy. Thematic bonds are simply conventional fixed-income instruments permitting investors to deploy capital to more narrow investment themes, apart from climate change, including health, food, education, financial services access, and social housing.

An interesting example of a thematic bond is what is thought to be the first SDG-related targets linked bond. Italy's ENEL is one of the largest electric power utilities in the world with a market capitalisation of more than 80 billion Euros ($90.1 billion), operating 84,742 megawatts chiefly in in Italy, Iberia, and Latin America. About 41 per cent of its earnings before interest taxes depreciation and amortisation came from operating transmission and distribution networks, 15 per cent from conventional generation, 20 per cent from retail, and 24 per cent from its renewable energy subsidiary, Enel Green Power, in the first quarter 2020 through March 2020.[27] In late 2019, Enel issued the world's first general purpose bond linked to the UN SDGs worth $1.5 billion. It offered a unique feature. If ENEL did not meet its renewables target of 55 per cent of total capacity by the end of 2021 it would have to raise the bond's coupon by +25 basis points. Soon thereafter it issued Euro-denominated bonds in four tranches worth about $2.81 billion, which was almost four times oversubscribed, according to US banking giant Citi, which was a joint active bookrunner. Citi said that the 'short-dated tranches of four and seven years followed a similar structure as the USD bond; however, the coupon on the long-dated 15 year was linked to Enel achieving greenhouse gas emissions by 2030 equal to or less than 125 grams of carbon dioxide (CO_2) per kilowatt hour, also incorporating a +25 basis point step-up mechanism if the target is not reached'. The ENEL bonds linkage to SDGs related key performance indicators were a market first.[28] This example also clearly shows that the world of sustainable credit issuance is still at a relatively embryonic stage of development and then many different iterations and transformations can be expected in the coming years.

Several interesting take-aways on the green bond market in 2019 and some key trends were identified by *Climate Bond Initiative's Green Bonds Global State of the Market*, an annual report produced by the organisation.[29]

Take-aways-wise, this author identified two. The first is from breaking down the industry categories where the proceeds from the green bonds issuance are directed. Energy-related use accounted for 32 per cent of the total, buildings 30 per cent, transport 20 per cent, water 9 per cent, land use and marine resources as well as waste accounted for 3 per cent each, and finally manufacturing and the ICT sectors accounted for 1 per cent each. Of note, most of the proceeds, about 97 per cent of the total, to be invested in these sectors – except for water as well as for land use and marine resources – had some sort of energy-related investment. Examples include energy conservation systems for buildings, electric vehicles-related spending for transport, waste-to-energy projects for the waste sector, as well as telecommunications network energy efficiency investments for the ICT sector. Second, there were still several criteria for subsectors that were under development or not yet developed as of June 2020. Examples include hydropower in the energy sector, urban development in the buildings sector, and aviation in the transport sector. These remaining criteria are, arguably, some of the tricky ones. And perhaps some may not necessarily be resolved in the short term as they are quite controversial. Examples include nuclear power in the energy sector and highly radioactive waste management in the waste sector (Table 4.4).

In terms of key trends for the global green bond market over the next one to five years, there are at least four. First, green bonds are no longer exotic, nice-to-have but not must-have products. They have become a mainstream financial instrument. Second, the increasingly diversified usage of labels will continue so as to more clearly identify the usage of the proceeds given that the decarbonisation and other transitions in key sectors will differ. This more varied labelling is simply an extension of what has been seen over the past few years with a creation of blue bonds, ESG-linked credit, social bonds, sustainability bonds, and even COVID bonds. Third, the Climate Bonds Initiative identifies taxonomies harmonisation as another key trend. One last short to mid-term trend relates to the issuers. They will be more and more from the Asia region, a continuation of the trend seen in 2018 and 2019. Also related to issuance is the trend of an increasing number of sovereign issuances, something also likely to continue and perhaps accelerate in the short to mid-term.

One of the financial actors that is increasingly active in the green and sustainable financial instruments space is the Asian Infrastructure Investment Bank (AIIB). AIIB was founded in 2015, commenced operations in 2016, and now has grown to 103 members. AIIB has increasingly been an ardent proponent of sustainable infrastructure investments. It launched its first 'global bond' in 2019. It was a sustainability-themed bond worth $2.5 billion with a coupon of 2.25 per cent to finance environmentally sustainable infrastructure investments and promote ESG investing across the Asia region. Demand for the bond was so strong that it had upsized the issue from the original $2 billion.[30]

It held its Annual General Meeting in July 2020, virtually due to the COVID-19 pandemic, and one of the sessions was on 'Investing in Climate Action'.[31] AIIB staff and some outside experts chiefly discussed some of the things that different actors could do in order to support the involvement of private capital in climate action. One of the experts was Professor Lord Nicholas Stern, a luminary figure in the world of sustainable finance. Lord Stern is a professor of economics and government at the London School of Economics and member of the AIIB International Advisory Panel among many current titles. He became particularly famous for producing the Stern Review, a 700-page report that dealt with the economics of climate change back in 2006. At the AIIB session, Lord Stern highlighted what he perceived was the main challenge to sustainable finance, the fact that more than two-thirds of emissions are from infrastructure. Lile many other

Table 4.4: Industries Issuing Green Bonds (Adapted).

Category / Industry	2019 Issuance Percentage	Overlap with Energy	Criteria Under Development	Criteria Not Developed
Energy	32%	✓	Hydro; Transmission and Distribution	Storage; Nuclear
Buildings	30%	✓	Products and Systems for Efficiency; Urban Development	
Transport	20%	✓	Water-borne	Aviation
Water	9%	✗		
Land, Marine Use	3%	✗		Fisheries and Aquaculture; Supply Chain Management
Waste	3%	✓	Radioactive Waste Management	
Industry	1%	✓		Glass, Chemical, and Fuel Production
ICT	1%	✓		Broadband Networks; Telecommuting Software and Service; Power >Management

Source: Climate Bonds Initiative (2020). Green Bonds Global State of the Market 2019. [online] Climate Bonds Initiative. Available at: https://www.climatebonds.net/files/reports/cbi_sotm_2019_vol1_04d.pdf [Accessed 6 November 2020].

experts he urged all actors to address and encourage building retrofitting, promoting electric vehicles, building more bicycle paths, restoring degraded land, and managing water systems. He believes that these areas have tremendous economic multipliers. He also emphasised that it is institutions like AIIB who play a key role in advancing work in these areas and at the same time managing early stage risk. These areas listed by Lord Stern are some of the areas of focus of the bank. At the same session a senior staff member of AIIB, Stefen Shin, introduced the investment approach and the ESG framework that will be used for its AIIB Asia Climate Bond Portfolio. The bank is keen to develop climate change related investment, adopt a framework that will mobilise private investment, and tackle the underdevelopment of the climate bond market. To this end it is launching a project that will involve a managed fixed-income portfolio of $500 million. It envisages that a further $500 million will be rallied from climate change-focused institutional investors. A portion of the funds will be directed for market education, engagement, and bonds issuer support. Explaining the investment approach, Mr Shin said that one of the attractions for other institutional investors is that AIIB will provide a user-friendly, clear ESG framework that it actually plans to provide to others at no charge before the end of 2020, which will be a significant cost and time saving for these investors in the view of this author.

How will the investment approach work in practice? AIIB will classify issuers as an A-list Climate Bond issuer or B-list issuer. The A-list ones are those issuers whose use of proceeds from the bond issuance meet all three Paris Agreement principles. The B-list ones do not quite meet the principles but meet some of them and are working towards meeting all of the principles and their ability to cope with climate change. The principles comprise: the portion of green business activities, climate mitigation, and resilience to climate change. Specifically, 'A-List issuers are companies that are leaders in aligning business practices with the three variables. B-List issuers are companies that are on the transition path in aligning themselves with the three variables (i.e. companies who are on a trajectory to enter the A-List and currently transitioning to a low-carbon and climate-resilient business model).'[32] Adhering to these principles should make the business of these companies relatively 'future proof'. The first $500 million anchor investment should be launched sometime in the third quarter of 2020 and will be managed by Amundi, a huge French global financial institution. Amundi, together with the Climate Bonds Initiative, helped to structure the framework.

One cannot talk about green bonds without talking about the Chinese green bond market. It is the second largest in the world in terms of amount issued and is also the fastest growing among the leading markets. Apart from size and growth, the Chinese green bond market has a set of very interesting features. The nation will need to annually invest between 3 and 4 trillion yuan ($448–599 billion) in green and sustainable projects so as to meet its Paris commitments, which makes the role of green bonds particularly important.

Two organisations published state of the market reports for China for 2019 identifying some key features, important issues, as well as trends. One is the China-based Climate Policy Initiative[33] (CPI) and the other, which was referred to earlier, is the Climate Bonds Initiative[34] (CBI). We should differentiate that CPI studied the whole green bond market in China, amounting to issuance of about $120 in 2019, looking at the bonds issued under the domestic taxonomy. The approach by CBI was narrower taxonomy-wise and the amount studied was about half, about $56 billion. As such, understanding of both studies gives a more comprehensive idea of the Chinese green bond market. CBI remarked that there are discrepancies between the Chinese and the international green bonds guidelines and that these revolve around projects' eligibility and disclosure. Needless to say, the Chinese issuers of offshore green bonds comply with international guidelines. These, though, relatively were few in recent years; there have only been about 36 since 2015. But what's with a definition? Actually, the taxonomy is particularly important not just because of the potential greenwashing element but more importantly because certain investors do have to adhere to strict definition parameters regarding the projects type. For example, under Chinese categorisation, efficiency-related improvements projects for coal-fired power plants are included; however, this is something that is not acceptable under an international definition. Nuclear power generation is included under the standards of one regulatory entity, China's central bank the Peoples Bank of China, but not accepted by another Chinese regulatory body, the National Reform and Development Commission (NDRC), as well as the European Investment Bank. In fact, CPI does highlight that one of the development hurdles is the number of regulating authorities. Specifically, the PBOC regulates financial bond issuers and non-financial corporate issuers. The Chinese Securities Regulatory commission regulates corporate bond issuers through the Shanghai and Shenzhen Stock Exchanges. And the Ministry of Finance regulates municipal bond issuers. CPI correctly remarks that this generates inconsistent standards, raises bond issuance and monitoring expenses, and reduces the bond issuance efficiency.

Apart from the taxonomy challenge that China will surely address in the coming years, other areas of development include the gradual increase in green bond issuance by non-financial corporations as well as from local government financing vehicles. Other key trends include the increase of green asset-backed securities as collateral, that is, where the investors' returns are based on the cash flows of an underlying asset, such as a solar power generation plant. Yet another trend is that the use of proceeds is likely to become more diversified away from clean transport, railroads, and subway lines. This category accounted for about 30 per cent to 37 per cent of issuance in 2018 and 2019, whether looking at the China or international taxonomy approach.

Since the creation of green bonds, more green credit products have been introduced. Vasakronan, the Swedish property company, mentioned at the beginning of this section, in 2018 had another first – the first issuance in the world of green commercial paper, short-term financing instruments with capital raised assigned for green assets. Vasakronan raised 610 million Swedish krona (about $66 million).[35]

4.3.3 Green Bank Borrowing

Green loans and sustainability-linked loans are yet another area whereby corporations and others can finance green projects. But, like many other areas in green and sustainable finance, green loans are still at an embryonic stage.

The issuance principles behind these loans are currently mostly still voluntary in nature. A variety of associations have come up with some parameters, highlights Linklaters LLP, a UK-based multinational law firm. Similar to green fixed-income instruments, for green loans the parameters revolve around the use of proceeds, the project selection process, the actual management of the proceeds, and, of course, the reporting processes, says Linklaters. The law firm also stressed that while green loans are often used for both commercial loans and sustainability-linked loans, the two are actually somewhat different. The later has the pricing of the loan linked to the borrower's performance and not a set of prearranged criteria.[36]

Data related to green borrowing is not so easily attained. One of the many approaches to assessing the monetary volume of this particular segment is examining sustainability-linked loans by corporates. This type of borrowing amounted to $234.2 billion cumulatively between 2010 and June 2020, based on data compiled by financial information provider Bloomberg, albeit this is just a tiny fraction of the global lending market. The sharpest growth seen with corporate borrowing actually was only since 2018. European countries have historically dominated the issuance. Six nations accounted for 68 per cent of the total. They include France (17.5 per cent of the $234.2 billion total), Spain (13.2 per cent), Great Britain (12 per cent), Italy (9.5 per cent), Germany (8.5 per cent), and the Netherlands (7.3 per cent). Just like the growth momentum in green equity-based and credit-based issuance, the total amount of green lending will continue to grow at a rapid pace. Also, the increase will be more geographically diversified, with Asian corporates progressively becoming a main actor. Singaporean corporates only entered the sustainability-linked loans league in 2017, Australian corporates in 2018, Hong Kong corporates in 2019, and those incorporated in China just in 2020, for example (Figure 4.16).

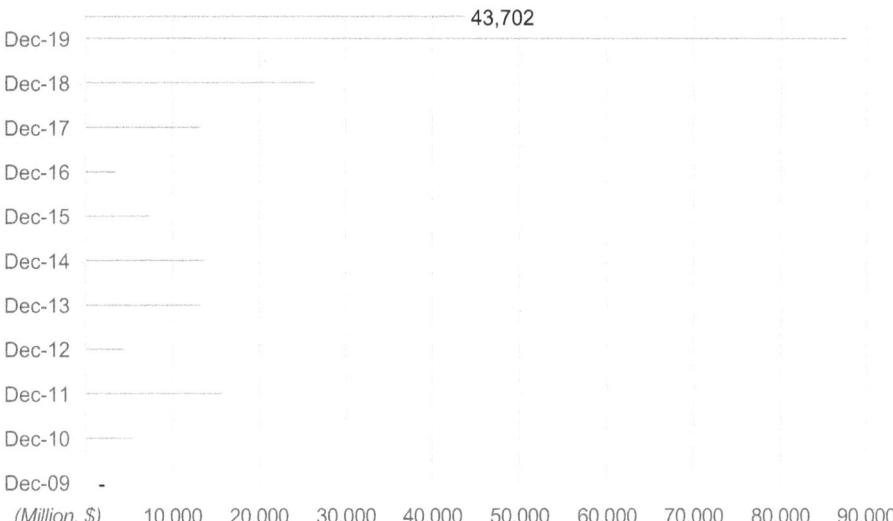

Figure 4.16: Annual Sustainability-Linked Loans Compiled by Bloomberg (Adapted).
Source: Bloomberg.

In fact, not even the COVID-19 pandemic and ensuing economic slowdown halted the growth. Nor did it stop some high-profile announcements from some commercial and investment banks in the first six months of 2020. More clear evidence that green and sustainable financing and investment is attracting an increasing amount of capital. The examples (Table 4.5) include Germany's Deutsche Bank, US's Goldman Sachs, UK's HSBC, Singapore's OCBC Bank, and UK's Standard Chartered Bank. They range from S$25 billion ($17.9 billion) to $750 billion. Of course, this is just a small sample to provide a sense of the momentum.

Table 4.5: Recent Asian Banks Sustainable Finance Commitments.

Name	Commitment	Notes	Date
Deutsche Bank	EUR200 billion	Sustainable investment by 2025	13-May-20
Goldman Sachs	US$750 billion	Investing, financing and advisory activities by 2030 to help clients with climate transition	13-Apr-20
HSBC	US$100 billion	Sustainable financing and investment by 2025	17-Feb-20
OCBC Bank	S$25 billion	Sustainable finance portfolio by 2025 from S$10 in 2019	22-Jun-20
Standard Chartered Bank	US$75 billion	Sustainable development (US$40 billion) and renewables project (US$35 billion) financing services	18-Feb-20
JPMorgan Chase	US$200 billion	Sustainable financing in 2020	25-Feb-20

Source: Shiao, V. (2020). OCBC, Riding Growth Wave, Targets S$25b Sustainable Finance Portfolio by 2025. [online] The Business Times. Available at: https://www.businesstimes.com.sg/banking-finance/ocbc-riding-growth-wave-targets-s25b-sustainable-finance-portfolio-by-2025 [Accessed 6 November 2020]. Staff, R. (2020). Deutsche Bank Targets 200 Billion Euros of Sustainable Investment by 2025. 13 May. [online] Reuters. Available at: https://uk.reuters.com/article/uk-deutsche-bank-sustainability/deutsche-bank-targets-200-billion-euros-of-sustainable-investment-by-2025-idUKKBN22P13J [Accessed 6 November 2020]. West London Business (2019). HSBC UK Launches Green Finance Proposition to Support UK Businesses. [online] West London Business. Available at: https://www.westlondon.com/hsbc-green-finance/ [Accessed 6 November 2020]. Goldman Sachs (2019). Goldman Sachs: What We Do. [online] Goldman Sachs. Available at: https://www.goldmansachs.com/what-we-do/sustainable-finance/index.html [Accessed 18 January 2020]. Standard Chartered (2020). Standard Chartered Commits USD75bn toward Sustainable Development Goals. 18 February. [online] sc.com. Available at: https://www.sc.com/en/media/press-release/standard-chartered-commits-usd75bn-towards-sustainable-development-goals/ [Accessed 6 November 2020]. Available at: https://www.ai-cio.com/news/jp-morgan-chase-commits-200-billion-sustainability-financing/.

Another factor that raises confidence in the growth momentum of green and sustainability related loans is the tremendous trend in China. First, is the amount that has been deployed by the nations' six largest state-owned banks (Figure 4.17), a total of 13,293.6 billion yuan ($1,987.1 billion) in the three years through December 2019. The six had deployed 5,026.9 billion yuan ($751.4 billion) in green loans in 2019. This was

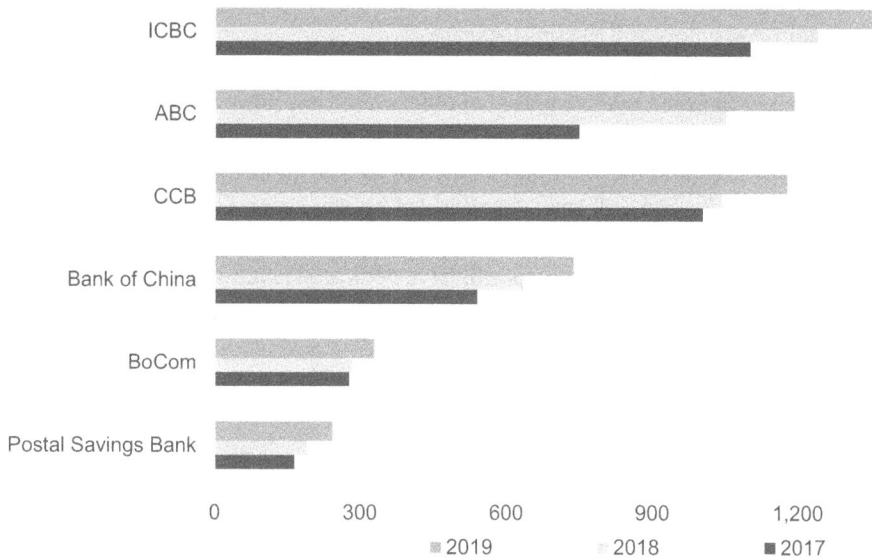

Figure 4.17: China's Largest Banks Green Loans (Billion Yuan).
Source: Sina Finance and Economics (2020). Exploration of Green Finance of Large State-owned Banks: ICBC's Largest Investment and Postal Savings Bank's Fastest Growth. [online] finance.sina.com.cn. Available at: http://finance.sina.com.cn/china/gncj/2020-04-02/doc-iimxyqwa4762190.shtml [Accessed 6 November 2020].

up 13.3 per cent from the 2018 total of 4,436.5 billion yuan ($663.2 billion), which itself had increased 15.8 per cent from 3,830.2 billion yuan in 2017 ($572.5 billion), based on the respective annual reports that they released. The top three banks were responsible for almost 75 per cent of the total, or 9,897.4 billion yuan ($1,479.4 billion), during the period, namely the Industrial and Commercial Bank of China (ICBC), the Agricultural Bank of China (ABC), and the Construction Bank of China (CCB).[37] Still, they will continue to rise rapidly in the short term as the 'big six' green lending lags behind that of other banks. In 2016, the China Banking Regulatory Commission stated that the amount of green credit for the nation's 21 largest banks was 10 per cent of total credit. However, in 2019 it was only 7.5 per cent for the big six.

The COVID-19 pandemic did not slow down the growth. To the contrary, for the banking system as a whole green lending to clean energy, environmental protection, and related sectors rose almost 11 per cent to a total of 11.01 trillion yuan ($1.65 trillion) as of June 2020 compared to the end of 2019, based on data from the People's Bank of China (PBOC), the nation's central bank. The amount dedicated to green infrastructure projects was 5.28 trillion yuan ($789 billion).[38] To further promote this, the PBOC in July 2020 announced that from 2021 it would start to assess the country's banks' green and sustainable finance performance on a quarterly basis.

The PBOC will give the banks a score based on the green lending share of the total, the annual growth rate, and the risk exposure.[39]

There are a lot of discussions in the green finance markets as to the green loans' taxonomy used in China relative to that of other countries. However, there has been a meeting of minds in recent years on that front, too. An example comes from the London branch of ICBC and the Singapore branch of ABC. Both institutions created green lending frameworks and will raise loans based on the Green Loan Principles, with French bank BNP Paribas acting as the green structuring advisor. Following the Green Loan Principles means that the banks will have dedicated lending to green infrastructure projects, in such sectors as renewable energy, energy efficiency, pollution prevention and control, and green buildings, for example.[40]

To better understand the growth potential for utilities around the region to obtain green and sustainable finance I first looked at the largest utilities in Asia listed on stock markets. I selected the top 15 by market capitalisation and these had a combined market value of only $250 billion (Table 4.6). In terms of profile, there are one each listed in Australia, Malaysia, and South Korea; two are listed in Japan; and five each in China and in Hong Kong – albeit two of the five listed in Hong Kong solely have operations in China. The list also takes into consideration whether the listed utility is included in a major index, including the MSCI AC Asia Pacific Utilities Index or the CSI 300, comprising the top 300 stocks traded on the Shanghai and the Shenzhen Stock Exchanges.

Table 4.6: Asia's Key Listed Utilities (7 August 2020).

Name of Utility Company	Stock Market Symbol	Currency	Last Price	Exchange Rate vs. US$	Market Capitalisation (US$, Billion)
China Yangtze Power	600900.SS	CNY	18.08	6.97	57.1
The Hong Kong and China Gas	0003.HK	HKD	10.94	7.75	25.1
CLP Holdings	0002.HK	HKD	74.40	7.75	24.3
CGN Power	003816.SZ	CNY	3.02	6.97	19.4
China Gas Holdings	0384.HK	HKD	22.80	7.75	15.4
Tenaga Nasional	5347.KL	MYR	11.02	4.19	14.9
ENN Energy Holdings	2688.HK	HKD	93.50	7.75	13.6
Power Assets Holdings	0006.HK	HKD	42.70	7.75	11.8
Korea Electric Power Corp.	015760.KS	KRW	19,600	1,190.01	10.6
Huaneng Power International	600011.SS	CNY	4.57	6.97	10.3

Table 4.6 (continued)

Name of Utility Company	Stock Market Symbol	Currency	Last Price	Exchange Rate vs. US$	Market Capitalisation (US$, Billion)
Huaneng Lancang River Hydropower	600025.SS	CNY	3.85	6.97	9.9
China National Nuclear Power	601985.SS	CNY	4.33	6.97	9.7
APA Group	APA.AX	AUD	11.11	1.40	9.4
Tokyo Gas	9531.T	JPY	2,160	105.93	9.0
Chubu Electric Power Co.	9502.T	JPY	1,237	105.93	8.8
Kansai Electric Power Co.	9503.T	JPY	1,016	105.93	8.6
SDIC Power Holdings	600886.SS	CNY	8.54	6.67	8.3
Osaka Gas Co., Ltd.	9532.T	JPY	1,988	105.93	7.8
AGL Energy Limited	AGL.AX	AUD	16.80	1.40	7.7

Source: Author, August 2020. Data source: Yahoo Finance.

I scanned through the sustainability, ESG, and related documents of each of the 15 listed companies. Very little evidence was found that these companies have been active in seeking green and sustainable finance. One reason is that given their large asset base or their pivotal role in the local economy, such as for KEPCO, which is the key utility in the country, these large companies can easily raise funding at attractive rates. Another reason is that they may find it quite difficult given their high pollution asset base, such as in the case of Huaneng Power International given it is chiefly a coal-fired power generation company. Among the 15 only three directly mentioned green loans, namely Japan's Chubu Electric Power and Tokyo Gas as well as Malaysia's Tenaga. For all of the others, this author did not find any readily available materials on green and sustainable borrowing as of July 2020. For Chinese utilities listed in China Yangtze Power, CGN Power and Huaneng Lancang River Hydro Power mentioned green financing but seemingly only readily mentioned fixed-income instruments, including green bonds and green short-term notes (Table 4.7).

On a combined basis, the 15 utilities had a combined total long-term debt, including debt securities, of about $176 billion (Table 4.8), calculated at the end of their latest semi-annual or annual reports as of July 2020. Should just 20 per cent or 30 per cent of the total be refinanced into green borrowing then this would amount to a massive $35 billion to $53 billion, and this is just for these top listed utilities. For the Asian utilities as a whole the total amount would likely be least ten times or more. Many of them are actually not listed and thus have low financial visibility. Still, given the new national

Table 4.7: Key Listed Utilities Asia Green Loans.

Name of Utility Company	Green Borrowing	Name of Utility Company	Green Borrowing
China Yangtze Power	?	Korea Electric Power Corp.	n.a.
Hong Kong and China Gas	n.a.	Huaneng Power International	n.a.
CLP Holdings	n.a.	Huaneng Lancang River Hydropower	?
CGN Power	?	China National Nuclear Power	n.a.
China Gas Holdings	n.a.	APA Group	n.a.
Tenaga Nasional	✓	Tokyo Gas	✓
ENN Energy Holdings	✗	Chubu Electric Power Co.	✓
Power Assets Holdings	n.a.		

Source: Author, August 2020. Analysis of published financial information of the companies.

Table 4.8: Asia Key Listed Utilities' Long-Term Debt.

Name of Utility Company	Stock Market Symbol	Exchange	Currency	Last Price	Long Term Debt (US$, billion)	Exchange Rate vs. US$
China Yangtze Power	600900.SS	SHH	CNY	18.08	9.82	6.97
The Hong Kong and China Gas	0003.HK	HKG	HKD	10.94	3.71	7.75
CLP Holdings	0002.HK	HKG	HKD	74.40	5.01	7.75
CGN Power	003816.SZ	SHZ	CNY	3.02	5.27	6.97
China Gas Holdings	0384.HK	HKG	HKD	22.80	2.01	7.75
Tenaga Nasional	5347.KL	KLS	MYR	11.02	10.15	4.19
ENN Energy Holdings	2688.HK	HKG	HKD	93.50	1.50	7.75
Power Assets Holdings	0006.HK	HKG	HKD	42.70	0.43	7.75
Korea Electric Power Corp.	015760.KS	KSC	KRW	19,600	46.75	1,190.01
Huaneng Power International	600011.SS	SHH	CNY	4.57	20.65	6.97

Table 4.8 (continued)

Name of Utility Company	Stock Market Symbol	Exchange	Currency	Last Price	Long Term Debt (US$, billion)	Exchange Rate vs. US$
Huaneng Lancang River Hydropower	600025.SS	SHH	CNY	3.85	11.87	6.97
China National Nuclear Power	601985.SS	SHH	CNY	4.33	26.97	6.97
APA Group	APA.AX	ASX	AUD	11.11	7.06	1.40
Tokyo Gas	9531.T	JPX	JPY	2,160	7.88	105.93
Chubu Electric Power Co.	9502.T	JPX	JPY	1,237	16.61	105.93

Source: Author, August 2020. Data source: Yahoo Finance.

decarbonisation requirements many will have to ramp up their green financing. This would include such companies as China's giant generation groups with a combined massive installed capacity of more than 875 gigawatts; they include China Datang Corp., China Energy Investment Corp., China Huadian Corp. China Huaneng Group Co., and State Power Investment Corp. The nation's two grid companies operate most of the nation's 2,000 gigawatts transmission and distribution networks. Some other giant corporations are completely state-owned in the rest of Asia, including Indonesia's PT PerU-Shaan Listrik Negara and Thailand's Electricity Generation Authority of Thailand.

Apart from green and sustainable loans to the energy sector, mostly comprising energy generation, another sector is the buildings sector where such loans are mostly directed to green buildings, namely energy efficiency and clean generation projects, as discussed in an earlier chapter (see 2.1.3.3 Energy Mega-Users: Buildings). And Asia is very much at the centre of this. The green buildings sector will involve almost $25 trillion in investment globally in the ten years through 2030, according to a study by the International Finance Corporation (IFC).[41] More than 72 per cent of this investment potential is in East and South Asia the IFC believes. It calculated that green buildings may involve as much as 12 per cent more costs, but the operational costs decrease could be as high as 37 per cent and may also lead to faster sales and potentially some sales price premium.

Given that the buildings sector is heavily reliant on bank financing because of the heavily upfront cost nature of the building sector, just like energy infrastructure, it is an area ripe for attracting an increasing number of sustainability-linked loans or revolving credit facilities.

During a public webinar, the IFC highlighted that in recent years it is principally only the top property developers that are investing in green buildings. Possibly, because there is a perception that green measures for buildings would involve significantly higher construction expenses. Actually, argues the IFC, it is not the construction costs per se that are higher; it is actually the compliance that is expensive.[42] This is because green buildings have to get certified, and this involves a substantial amount of time and expense. The IFC mentioned that the common green building certification used in East Asia is very comprehensive but the paperwork is also very comprehensive, and some buildings cannot afford the compliance and certification expenses. As such certification is not common today for the huge, and rapidly growing, residential markets. Green certification from the IFC is better suited for emerging markets. The developer of the real estate would have to get one or more internationally recognised green building certification systems such as GRESB or Leadership in Energy and Environmental Design (LEED). To this end the IFC launched the Excellence in Design for Greater Efficiencies (EDGE), a green building certification system for emerging markets. EDGE (edgebuildings.com) is a free, internationally recognised, modelling software for green buildings. It is quite certain, in the view of this author, that in the coming five to ten years, more uniformity will also come to the green buildings sector, making it easier to access and use as well as making the process more affordable, just like in the case of green and sustainable bonds.

To put some colour on what are green and sustainability-linked loans in the property market we briefly highlight some case studies in the Asia region. The first is in Hong Kong with Swire Properties, a leader in the region in real estate sustainability; we discussed some of its energy efficiency and conservation projects in an earlier chapter (see 2.1.3 Energy Efficiency Could be Growth Containment Factor). Green borrowing wise the property company had secured two as of August 2020. The first sustainability-linked loan was for HK$500 million ($64.5 million) and was locked in in July 2019 with French global banking giant Crédit Agricole Corporate and Investment Bank. The company converted a five-year revolving credit facility with a loan with a different financing mechanism, namely having the interest rate indexed to improvement in Swire's ESG performance year on year. If Swire meets two criteria it would result in a lower interest rate. The two were to retain its listing on the Dow Jones Sustainability World Index and meet the reduction targets for energy use intensity for its Hong Kong portfolio. The borrowing would be dedicated to the adoption of more advanced energy-saving technologies as well as funding other green building developments.[43] About a year later, in August 2020, Swire Properties secured a five-year green loan of HK$1 billion ($129 million) with Singapore's OCBC Bank. This loan would chiefly go towards the green features of an office tower to be completed in 2022, Two Taikoo Place.[44]

The second mini case study is in Singapore with Keppel Infrastructure Trust and Keppel Energy, a leading local company with diversified investments including energy, infrastructure, and real estate. The loan secured was for S$700 million ($ 510.2 million) from Singapore's DBS Bank and OCBC Bank in June 2020 for seven years aimed at lowering emissions at its jointly owned gas-fired co-generation plant, Keppel Merlimau Cogen Plant. Just like Swire's first sustainability-linked loan, it contains a mechanism that if pre-set objectives are met, the interest rate would be cut on a tiered basis. These targets comprise benchmarking the Merlimau Plant's carbon emissions intensity against national indices, as well as producing a continuous improvement in its carbon emissions intensity.[45]

A third mini case study is in Australia with AGL Energy, one of the country's top three electric power generators and retailers. The company took on a syndicated five-year sustainability-linked loan of A$600 million ($429.6 million) from ANZ and BNP Paribas in September 2019. At the time it was a first for an energy company in the Asia region. The loan had two key performance indicators comprising its emissions intensity and renewable energy and storage capacity targets.[46]

Overall, in the 30 months through 17 June 2020, commercial banks in the Asia Pacific region, apart from Japan, lent about $22.96 billion in green and sustainability-linked loans across 58 deals. The largest ten lenders were responsible for about 54 per cent of the total, or about $12.35 billion. The leading five were OCBC, DBS, HSBC, Bank of China, and ANZ Bank (Figure 4.18). As mentioned previously, this is still the very early stages, and the amount will certainly grow exponentially in the coming years in the opinion of this author.

Figure 4.18: Asia-Pacific (ex-Japan) Green and Sustainability-Linked Loans League Table.
Source: Huang-Jones, J. (2020). APAC Chart of the Week: OCBC Tops APAC (ex-Japan) Green/Sustainability Loans League Table in 2018–2020 YTD. [online] Debtwire Events. Available at: https://events.debtwire.com/apac-chart-of-the-week-ocbc-tops-apac-green-loans-league-table-in-2018-2020-ytd [Accessed 3 September 2020]. Note: Loan value is on a pro rata basis if there is more than one MLA in deal, include deals that meet LSTA/LMA/APLMA Green Loan Principles Liked Loan Principles or related independent agencies.

4.4 Future Greening of Finance and Financial Instruments

There is little doubt in my mind that green and sustainable finance principles will dominate the vast majority of capital flows globally, including in Asia. In this last section of the book, I will explain two reasons for this bullish view. One is the increase in adoption on the part of all actors. Another is avenues to increase liquidity. I will then address the potential of valuations overheating. Finally, I will introduce briefly the views of some other experts so as to present an even broader perspective on the long-term future of green and sustainable finance and investments.

4.4.1 Higher Adoption and Capital Flows

The amount of capital dedicated to green and sustainable finance will surge around the world in general, and in Asia in particular, in the next 30 years. Globally, all forms of finance and financial instruments will have to meet, at a minimum, one or more of the UN SDGs. Or, at the very least, the core concepts behind each of them; they are declared to be met by 2030, but they are likely to be redefined and fine-tuned unless all of the world's problems are all resolved by then, which is a highly unlikely event. Relative to some regions, especially Europe, Asia is coming from behind probably because the bulk of the population still lives in developing or frontier economies where the adoption of green and sustainable principles has lagged behind that of developed economies. But, one may ask, why am I so confident of the continued adoption of the core concepts behind the UN SDGs?

Relative to other disciplines in the financial field, the green and sustainable area is young. It is still at an embryonic stage and will see an enormous amount of evolution, and probably revolution, in the coming years. There are many reasons to be bullish on the acceleration of adoption, including increasing commitment, process streamlining, and perception change. The acceleration will happen because the growth momentum in the commitment to the principles has been steadily accelerating in the past few years. The number of actors participating – whether they are governments, financial institutions, or corporations – has been progressively increasing. This is especially evident with Asian corporates. The highly respected and closely followed international non-profit CDP (better known as the Carbon Disclosure Project) annually assesses the environmental data disclosure of corporates globally, now more than 8,400. Referring to the 2019 survey, the latest annual evaluation, CDP said that corporates in the Asia Pacific region accounted for a quarter of corporates globally that adopted science-based targets under the Science-based Targets initiative (SBTi). This represented an increase of 400 basis points over the 2018 result of 21 per cent. CDP also added that it expected the number of companies from within the region formally adopting science-based targets to rise.[47]

Another reason for the growth is the streamlining of the processes as they will continue to get gradually less confusing and simpler. This will allow an issuer, a borrower, and other entities to invest less resources in trying to meet the set requirements. Also, the overall support system for compliance has also been strengthening. Recent improvements have already enabled corporates, for example, to increasingly better understand sustainability in general and align their disclosures to the guidelines specifically. The confusing and overlapping terminology will also be fine-tuned. Finally, for all market participants, including investors, there is a significant improvement in understanding that aligning goals with the core concepts behind the UN SDGs does not necessarily have a negative impact, financial or other. The lopsided perception that ESG or SDG investment or financing will increase their operating costs, raise capital expenditures, or lower their returns on investments is slowly dissipating. This can be better explained through two hypothetical examples, that of a manufacturer of jeans and that of a coal-fired power producer.

The garment and textile industry is one of the heaviest users of water. The production process of a pair of jeans can use thousands of litres of water. Various estimates put the amount at between 3,700 litres to more than 10,000 litres, depending on the actual process. Let us assume that a jeans manufacturer wants to raise equity or debt. Today's sustainability-focused investors are likely to closely examine the water resources management efficiency of this manufacturer. If it is able to raise its water efficiency in a cost-effective way, it will manage to reduce its operating expenses and thus raise profits. This would also meet several UN SDGs, especially the one about safeguarding availability and sustainable management of water and sanitation for all. In other words, the evaluation of the jeans manufacturer's business by investors concerned about whether their ESG or sustainability benchmarks are being met would not simply be from the perspective of whether or not the company is doing something for the good of the world.

Another hypothetical example is that of a coal-fired power producer. Should this type of power producer wish to issue equity, raise debt, or other type of financing, it likely would face several challenges from the perspective of ESG or sustainability-focused investors and lenders. Some investors will simply deem coal-related capital markets investments a no-go area because of carbon dioxide and other emissions. For other investors, the reluctance to invest in coal-generation companies goes well beyond pollution as coal-power faces several present and future business risks. One is that a producer with one or more such plants may find it very difficult to source project finance or be able to refinance existing loans, given that an increasing number of financial institutions, including commercial and development banks, have decided that they will no longer undertake coal-related lending. In fact, coal-fired power plants investments have fallen in recent years according to a study by the International Energy Agency (Figure 4.19).[48] Also, capital dedicated to new projects has also tumbled, financing terms have turned harsher, and more than 100 financial institutions around the world have formulated limitations on financing coal.

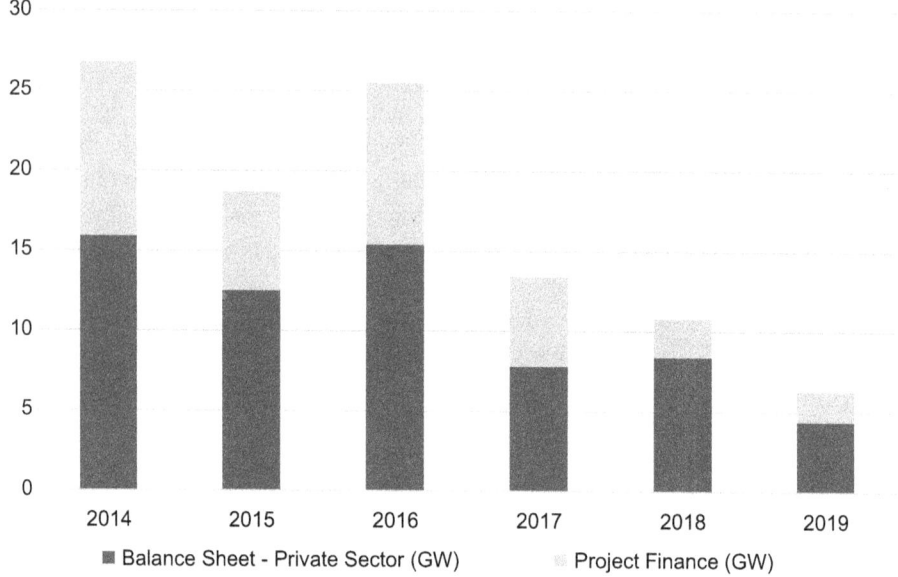

Figure 4.19: Coal Power FIDs by Financing Mechanism, 2014–2019.
Source: International Energy Agency (2020). Coal Power FIDs by Financing Mechanism, 2014–2019 – Charts – Data & Statistics. [online] IEA. Available at: https://www.iea.org/data-and-statistics/charts/coalcoal-power-fids-by-financing-mechanism-2014-2019 [Accessed 18 September 2020].

Recent examples of large financial institutions that have decided to avoid such investments include US fund management giant BlackRock and the Government Pension Fund of Norway's controlled Sovereign Wealth Fund. Financial institutions that have announced that they will no longer lend to coal-fired projects in Asia alone include Japan's Sumitomo Mitsui Banking Corp., Mizuho Bank, and the Japan Bank for International Cooperation, as well as Singapore's DBS Bank, the Overseas Chinese Banking Corp., and the United Overseas Bank.[49]

Another risk is that if the country in which the producer operates in launches a carbon or emissions trading system. Then the company will suddenly have high carbon liabilities, which will hurt its profitability or balance sheet over the long term, the latter in the case the producer decides to retire some of its plants so as to reduce its exposure though producers would not be immediately impacted as such a scheme is unlikely to be immediately implemented. Most regulators are highly likely to take a phasing-in approach to such a scheme, giving markets plenty of warning. In the region, currently only New Zealand has such a scheme operating, while Singapore and China plan to launch one. Japan, on the other hand, decided to impose a carbon tax system instead.

Another dimension of the growth is the momentum of capital dedicated to green and sustainable finance and investments will mean that trillions of dollars will be deployed with some parameters over the next one or two decades. Potentially, this may involve 60 per cent or more of total capital based on a poll straw I ran on social media platform LinkedIn in August 2020. I asked what percentage of the flow of equities, credit, or loans globally would be aligned to one or more of the UN SDGs (or their successors) by 2050. Among the 34 straw votes received, 38.2 per cent indicated the percentage would be less than 50 per cent while about 61.8 per cent of the voters believe it will be more than 50 per cent. Specifically, 26.5 per cent voted '50 per cent to 75 per cent', 20.6 per cent voted '75 per cent to 95 per cent', and 14.7 per cent voted over '95 per cent' (Figure 4.20).

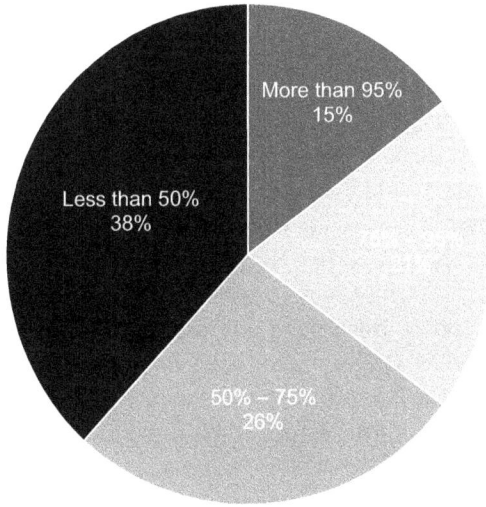

Figure 4.20: Global Future Fund Flows Poll Straw Summary.
Source: Author, September 2020.

4.4.2 New Tools and Paths to Green Liquidity

Green and sustainable capital and investments surging in the next three decades does not automatically mean that the right amount of liquidity will be created, something very fundamental for the capital markets. Better liquidity will be created by the creation of additional financial instruments, especially the digital kind.

The bulk of the world's dominant commercial, investment, multilateral, and other financial institutions have been a little slow in my opinion in getting on the green and sustainable finance band wagon. Thankfully, this is slowly changing thanks to the push they are getting from governments, corporations, and others.

One recent example can be found in New Zealand. The nation's Minister for Climate Change in September 2020 proposed that all financial institutions, whether they are private or state-controlled, in the country with more than NZ$1 billion ($675 million) will have to reveal 'governance arrangements, risk management and strategies for mitigating any climate change impact'[50] on an annual basis. If approved by the nation's parliament, New Zealand may become the first country in the world with such a requirement. I believe that this would have two effects. One is the raising of green and sustainable awareness among the local financial institutions. Another effect, whether this requirement is adopted in New Zealand or not, is that other nations are likely to follow suit, given the global acceleration in green and sustainable capital flows, and adoption of its principles, as I discussed in Section 4.4.1. The more focus that governments and financial institutions will have on green and sustainability issues, the more aggressive they are likely to be in developing new financial instruments.

One example of innovation as well as potential liquidity enhancement is Germany's first green bond issue. The ten-year zero per cent fixed coupon Green Federal bond, which was worth 6.5 billion Euros ($7.7 billion), attracted 33 billion Euros ($39.2 billion) in demand from investors, or an oversubscription rate of 5.1 times. Not only was the absolute amount quite large but also the German government targets to sell more green bonds with different maturities with the objective of establishing a yield curve that other issuers, be it governments, corporations, or other entities, can use as a benchmark to issue their own green bonds.[51]

Apart from this kind of green bond innovation, there are other new products that are being launched or will be issued in the coming years, including green syndication, green asset-backed securities, transition bonds, green derivative products, and many other instruments. The advent of these products is key to not only building up liquidity in the various asset classes, but also will be important in allowing investors to further diversify their portfolios. Green bonds are an example of a new product where demand outstrips supply. Not only will these new financial products, together with higher capital flow levels, help with the liquidity issue but it also may lead to a sharp increase in the amount of more specialised capital, such as, hypothetically speaking, the financing or an investment specifically dedicated to offshore wind projects globally or perhaps just dedicated to a specific cluster of offshore wind projects in, say, Japan. This of course also applies to financial institutions as many lack expertise in the green field. Also, it would allow for the introduction of more thematic instruments such as financial products solely related to a specific type of carbon footprint reduction, such as a project where solar cookers will be installed in a village so as to reduce the use of coal or wood for cooking. It is important to note that the above-mentioned products will be new because of their green and sustainable investment and financing parameters. However, most of the new financial instruments mentioned above are really just a tweaking of existing instruments in today's capital markets. For example, a green bond at the end of the day is a bond. So, the route for the entry of new green and sustainable products is well established.

A relatively new route that is still at an infant stage and in the process of establishing itself is the digital route. Here I am referring to corporates issuing digital coins and tokens. I personally am quite bullish on the growth of this type of green and sustainable products given they are opening the financing door to many green and sustainable projects that have found it challenging to raise funds.

In Chapter 3 various digital technologies applied to the energy industry were discussed. Technologies such as blockchain, decentralised energy, transactive energy, and others were explained. Turning to the finance aspect of digital technologies, a couple of concepts need to be first explained, namely Fintech and Defi. Fintech is a term that gained a lot of ground in the late 2010s. The broad definition is that it is at the intersection of financial services and technology. It automates the delivery and use of financial services using specialised software or algorithms.[52] Defi is the abbreviation for 'decentralised finance' applications. These applications comprise digital assets, including coins or tokens, derivatives, and exchanges. They also include financial smart contracts and protocols built on Ethereum, an open source for decentralised applications.[53] In layman's terms, it is a 'financial software built on the blockchain that can be pieced together like Money Legos',[54] a simple and clean definition I very much like.

The Fintech or Defi side of financing revolves around what some commentators call cryptographic assets (Table 4.9). They come in the form of cryptocurrencies, asset-backed tokens, utility tokens, and security tokens. They are all are 'transferable digital representations that are designed in a way that prohibits their copying or duplication'.[55] In the world of green and sustainable energy there are no examples I know of cryptocurrencies per se but there are many examples of companies issuing tokens. For asset-backed tokens, a company can sell a share of the assets. For example, a solar farm developer building a 10,000 kilowatt facility could sell 25 per cent of the underlying asset to investors by issuing 2,500 asset-backed tokens representing one kilowatt each. This developer could also sell a percentage of the output. Let's assume that the solar farm has an annual electric power output of 14,016,000 kilowatt-hours; the developer could issue 3,504,000 tokens representing one kilowatt-hour of output each. If this same developer sold utility tokens, these could represent the kilowatt-hours produced, albeit these utility tokens would have to be exchanged in the same ecosystem – very much in the same way that one would buy a ticket to a highly sought after music concert, and if the demand for the tickets is high, one may be able to resell the ticket at a higher price, albeit the number of the tickets does not change, it is the demand that increases. A security token is similar to the individual shares of a company. The buyer would be able to get a share in the company's future profits. The price of the security token would rise or fall in line with the company's fortunes as well as demand and supply.

Table 4.9: Definition of Four Subsets of Cryptographic Assets.

Subset	Digital Token Based on Blockchain Technology to:	Inherent Value
Cryptocurrency	Digital tokens or coins that currently operate independently of a central bank and are intended to function as a medium of exchange	None – derives its value based on supply and demand
Asset-backed Token	Token that signifies and derives its value from something that does not exist on the blockchain but instead is a representation of ownership of a physical asset	Derives its value based on the underlying asset
Utility token	Tokens that provide users with access to a product or service, and they derive their value from that right. They give holders no ownership in a company's platform or assets and, although they might be traded between holders, they are not primarily used as a medium of exchange	Value is derived from the demand for the issuer's service or product
Security token	Tokens that are similar in nature to traditional securities. They can provide an economic stake in a legal entity: sometimes a right to receive cash or another financial asset, which might be discretionary or mandatory; sometimes the ability to vote in company decisions and/or a residual interest in the entity	Value is derived from the success of the entity, since the holder of the token shares in future profits or receives cash or another financial asset

Source: PwC (2019). In Depth: A Look at Current Financial Reporting Issues – Cryptographic Assets and Related Transactions: Accounting Considerations under IFRS at a Glance. [online] p. 4. Available at: https://www.pwc.com/gx/en/audit-services/ifrs/publications/ifrs-16/cryptographic-assets-related-transactions-accounting-considerations-ifrs-pwc-in-depth.pdf [Accessed 19 September 2020].

Accounting firm EY put together a simple and clear, high-level comparison of the respective benefits and risks of initial public offerings (IPOs) on a stock exchange versus a security token offering (STO) or an initial coin offering (ICO); for the purposes of comparison, it treated coins and utility tokens under the same category. The key take-away is that STOs and ICOs, are riskier investments than IPOs. To this there is no denial. But the risk is due to the fact that this is a young and brave new world (Table 4.10). Regulators in individual countries are still catching up so we will see a fine-tuning of STOs and ICOs related regulations, and see these instruments mature gradually and risk elements decline progressively over the next few years.

In a study, the Asian Development Bank Institute looked at STOs concluding that an STO is a viable form of climate finance to help with the global transition to a

Table 4.10: IPOs, STOs, and ICOs Comparison.

	Initial Public Offering	Secuty Token Offering	Initial Coin Offering
Risk	Low	Medium	High
Costs	High	Medium	Low
Issued	Shares	Token representing securities	Utility token
Issuer	Public company	Start-up, public company, SMEs, large companies	Start-up, public company, SMEs
Platform	Regulated Stock Exchange	Digital (e.g. website of issuing company or on crypto exchange if IEO)	Digital (e.g. website of issuing company or on crypto exchange if IEO)
Participation	Generally, via Broker (e.g. Bank)	Directly	Directly
Accepted Funds	Generally, Fiat only	Fiat and/or crypto-assets	Fiat and / or crypto-assets
Initiated	Generally, Investment bank to underwrite the IPO	Generally, direct launch to the public with-out a centralized third party (except if IEO	Generally, direct launch to the public with-out a centralized third party (except if IEO
Documentation Requirements	Prospectus, Filings, Registration with the regulator	Prospectus, Filings, Registration with the regulator, website	Whitepaper, website
Investor Rights	Generally, Voting rights, dividends	Generally, Voting rights, dividends (if structured similar to e.g., shares)	Generally, Limited to digital access to service / application
Controlling Authority	Regulator	Regulator	None
Underlying	Asset	Asset	None
Dividends	Yes	Depending on Token structure	None
Credibility	High	Medium	Low

Source: Ernst & Young (2020). Tokenization of Assets Decentralized Finance (DeFi) Volume 1Spot on: Fundraising & StableCoins in Switzerland. Decentralized Finance (DeFi) Spot on: Fundraising & StableCoins in Switzerland. [online] Volume 1, p. 35. Available at: https://assets.ey.com/content/dam/ey-sites/ey-com/en_ch/topics/blockchain/ey-tokenization-of-assets-broschure-final.pdf [Accessed 19 September 2020].

sustainable economy. However, the institute recognised that tokenised securities currently face an immature infrastructure and regulatory uncertainty. Still, it does agree to the potential benefits of STOs including their potential to democratise

> green finance for both private (retail) investors and SME issuers, resulting in more efficient allocations of capital in developing economies and ultimately accelerating global climate finance. Tokenized securities can address both supply- and demand-side issues: a global investor base can increase the demand for green finance projects, safeguarded through enhanced transparency and auditability. Reduced minimum investment sizes and reduced transaction costs make SMEs bankable, which is particularly relevant for developing countries.[56]

The institute included in the study a comprehensive yet concise, excellent table describing STOs' strengths, weaknesses, opportunities, and threats (Figure 4.21).[57]

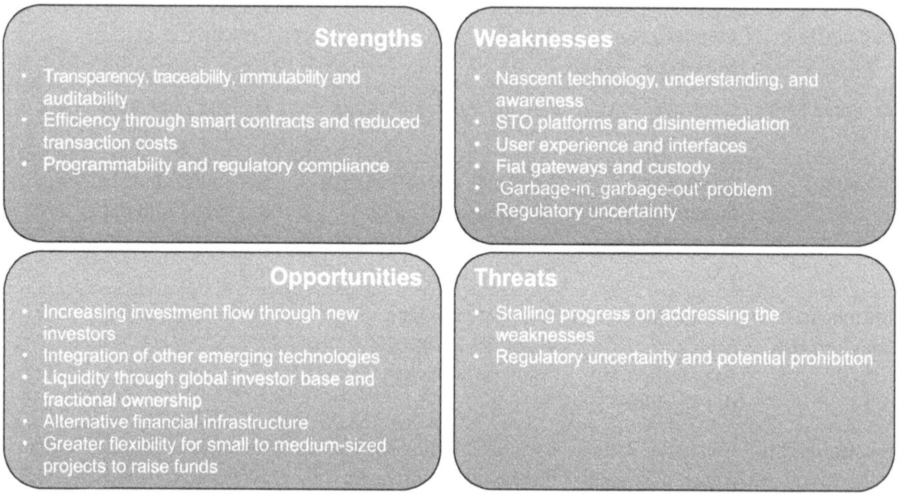

Strengths	Weaknesses
• Transparency, traceability, immutability and auditability • Efficiency through smart contracts and reduced transaction costs • Programmability and regulatory compliance	• Nascent technology, understanding, and awareness • STO platforms and disintermediation • User experience and interfaces • Fiat gateways and custody • 'Garbage-in, garbage-out' problem • Regulatory uncertainty
Opportunities	Threats
• Increasing investment flow through new investors • Integration of other emerging technologies • Liquidity through global investor base and fractional ownership • Alternative financial infrastructure • Greater flexibility for small to medium-sized projects to raise funds	• Stalling progress on addressing the weaknesses • Regulatory uncertainty and potential prohibition

Figure 4.21: Asian Development Bank Institute Securities Token Offerings SWOT Analysis. Source: Schletz, M., Nassiry, D., and Lee, M.-K. (2020). Blockchain and Tokenized Securities: The Potential for Green Finance. [online] Asian Development Bank, Tokyo, Japan: Asian Development Bank Institute, p. 4. Available at: https://www.adb.org/sites/default/files/publication/566271/adbi-wp1079.pdf [Accessed 20 September 2020]. ADBI Working Paper Series No. 1079.

Importantly, STOs, ICOs, and other similar offerings can be processed more quickly and at a lower cost than IPOs. This is particularly important in the world of green energy start-ups. Their common challenge is investment size. A venture or project that needs to raise just a few million US dollars in funding would certainly be regarded as too small for most commercial or development banks to lend to or asset managers to invest in. Also, the fund-raising processing fees, as a percentage of the total amount raised, may be too steep as well. The British Blockchain & Frontier Technologies Association (BBFTA)[58] in an article identified a survey on the subject. This survey used a hypothetical fund-raising exercise in the US of $50 million. A traditional private placement would cost at least 40 per cent less if it were done through an ICO or STO. BBFTA also highlighted the costs incurred by INX, the first STO approved by the US Securities and Exchange Commission; INX provides trading platforms for cryptocurrencies, security tokens, and their derivatives. As of August 2020, the company targeted to issue 130 million INX token coins at about $0.90 each in its STO, resulting in gross proceeds of $117 million and a net amount at $111 million, after underwriting and consulting fees, and other expenses.[59] The expected listing costs of about $6 million is equivalent to 5.1 per cent of the funds INX wants to raise. This compares to about 9 per cent to 24 per cent of proceeds, according to an analysis by PWC, for IPOs (Table 4.11).[60] It is worth pointing out that this is for a listing in the US. The cost will differ country by country but the STO versus IPO expense differential will be high nonetheless. Also, once the number of STOs on a particular exchange rise, then the average expense incurred comes down as well.

Table 4.11: Cost of a STO versus Traditional Private Placement Over Five Years.

Cost Savings Analysis – Traditional Private Placement versus Digital Securities				
	Example US$50 million Capital (Private Equity)	Traditional 5 Years ($)	Token/Digital 5 Years ($)	Percentage Difference
Offering	Legal	130,000	150,000	15%
	Accounting	40,000	40,000	0%
	Legal Filings	75,000	75,000	0%
	Travel/Roadshows	150,000	125,000	−17%
	Automation/Platform	–	50,000	N/A
	Senior Management Time	440,000	250,000	−43%
Ongoing	Regulatory Compliance	200,000	50,000	−75%
	Tax Compliance	150,000	100,000	−33%
	Ownership Transfers	250,000	–	N/A
	Listing Fees	–	10,000	N/A
	Technology Updates	–	50,000	N/A
	Admin	100,000	50,000	−50%
	Rule 144 Opinions (costs to investors based on 10 trades per year)	37,500	–	N/A
		1,572,500	950,000	−40%

Source: Survey of Entoro Capital, STO legal counsel, S&P and Pitchbook, found at BBFIA News Editor (2020). News: Will IPOs Continue to Decline as STOs Gather Momentum? [online] bbfta.org. Available at: https://bbfta.org/news/1599048573-4099c72e12/Will-IPOs-continue-to-decline-as-STOs-gather-momentum [Accessed 18 September 2020].

There are many real-world examples in the area of digital energy financing. Currently, this type of financing revolves around SMEs, most commonly start-ups, raising capital to fund some type of energy-related activity. Broadly speaking, these activities revolve around three areas: software platforms and solutions, products creation, and value or assets securitisation. Often these areas will somewhat overlap. The commoditising of the energy output or asset through marketplace software applications is particularly exciting for the future of green and sustainable finance. Just as exciting as Uber commoditising the vehicles for hire and transport services industry, and what Airbnb did for the vacation rentals market. Many of these companies successfully listed their tokens, generally through ICOs, in the second half of the 2010s. The recent slowdown in the number of such offerings is highly temporary in my view. Perhaps the relative boom in the number of offerings was a first phase of the market, and we are soon to enter a second phase where regulators and investors demand more transparency and accountability from these companies and their managements.

My bullish stance on the future of tokens for energy companies is echoed, unsurprisingly, by some of the companies. Dr Ana Trbovich, the chief operating officer and co-founder of blockchain-based energy exchange engine developer Grid Singularity, while discussing the late 2010s boom in an interview in May 2020, said that she believes that the boom was more related to financial speculation around the energy-related ICOs, and around cryptocurrencies in general. She thinks that actually if one looks beyond the noise, one can recognise that significant progress has notably been made in the application of the technology in the real energy industry. Dr Trbovich said that there are many real-world projects that are being rolled in 2020 and that while blockchain for energy is not 'the single solution to the energy complexity, it is definitely an important part of it, with the challenges that are brought about by the distributed energy resources management'.[61] I have addressed and analysed this trend in Chapter 3 where I highlighted many of the blockchain technologies for energy applications. Examples of such blockchain companies include Energy Web, Grid Plus, LO3 Energy, Power Ledger, Sun Exchange, Veridium Labs, WePower, and many others, as well as examples from more traditional energy companies such as Italian transmission system operator Terna S.p.A. and Japanese electric power utility TEPCO. Grid Singularity's COO is quite positive on the technology, highlighting that the advent of the COVID-19 pandemic underlined the real 'need for distributed management where blockchain is certainly one of the most important solutions'.[62]

4.4.3 Revaluating New Valuations

Another perspective in the discussion about the liquidity issue is the potential that the green market will overheat. If I and other like-minded individuals are correct in thinking that there will be plenty of capital available for green and sustainable equity and debt financing, could this realistically cause an increase in the valuations of the sector's equity and debt instrument? There is, potentially, the possibility that clean and renewable energy companies listed on stock markets could see their valuations rise steeply if there were a flood of new capital focused on this segment. Bonds could see a similar fate with yields falling dramatically. During my working career of over 30 years, I witnessed valuations going crazy twice. Both times it was with technology stocks. The first was roughly between 1995 and the early 2000s. It was driven by the rapid growth and quick adoption of the Internet. It is known by a variety of names probably because it was so new – the tech, dot.com, Internet, or TMT (Tech Media, and Telecom) bubble. At the time, the massive increase of the tech-heavy NASDAQ Composite Index masks the fact that there were some companies whose share price skyrocketed despite the fact they had little or no revenues or assets. As an equity analyst covering the not so sexy utilities sector, where assets and revenue generation are reasonably more transparent, I could not really understand at the time the tech-focused market exuberance. But as the NASDAQ Composite rallied from less than 1,000 in 1995 to more than 5,000 in 2000, a lot of retail and institutional investors made massive amounts of money. Then it all blew up and a global economic slowdown ensued (Figure 4.22).

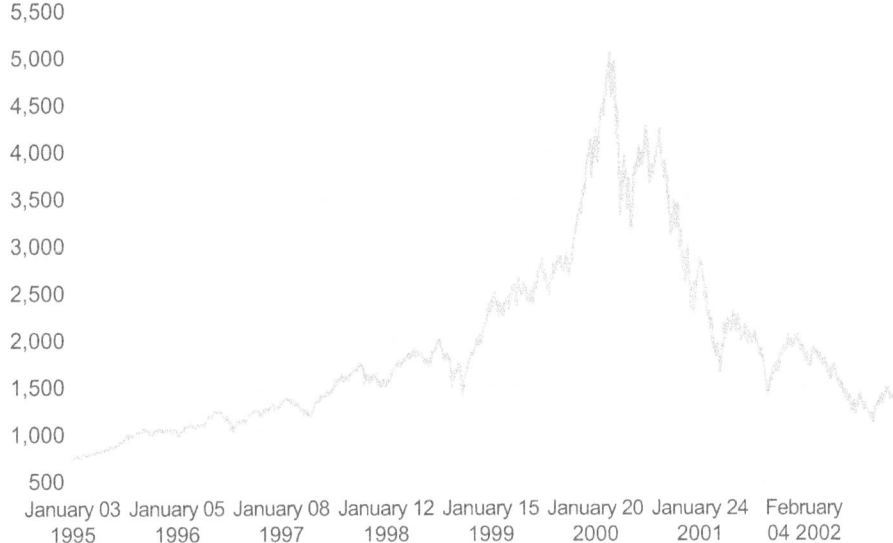

Figure 4.22: NASDAQ Composite 1990s dot-com Bubble (1995–2002).
Source: Author, January 2021. Data source: investing.com. Available at: https://uk.investing.com/indices/nasdaq-composite-historical-data.

Less than 20 years after the dot.com bubble burst, another tech-led market rally was born. A second boom saw the NASDAQ Composite rally to 12,000 in 2020 from about 1,000 in 2008. In this latest rally, some well-known names rose more than the average. Looking at the NASDAQ Composite's members weight, the largest ten companies accounted for about 47 per cent of the index' weight, and the top three, Apple, Amazon, and Microsoft, were responsible for a lofty 31.8 per cent as of 3 September 2020. The three actually rallied 156 per cent, 97 per cent, and 70 per cent respectively over the previous 12 months and now command rich price to earnings ratios; about 40x for Apple and Microsoft, and 136x for Amazon, also as of 3 September 2020 (Figure 4.23).

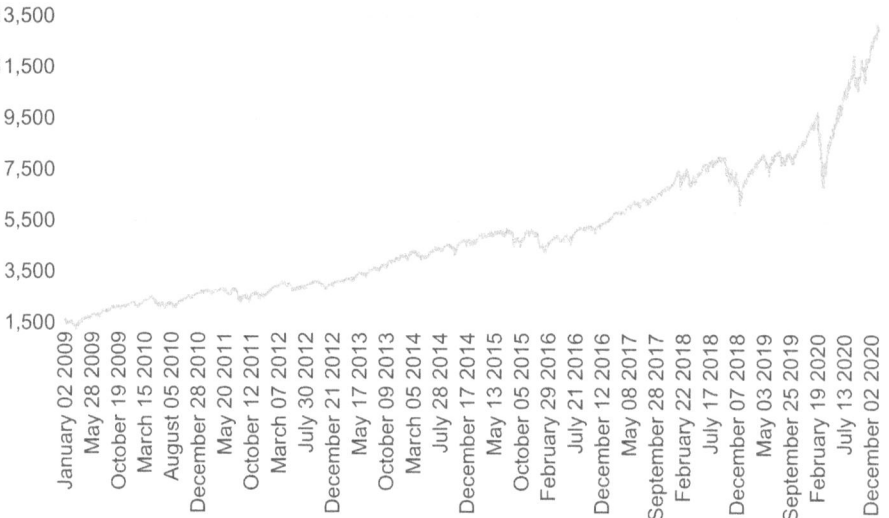

Figure 4.23: NASDAQ Composite 2010s Tech Bubble.
Source: Author, January 2021. Data source: investing.com. Available at: https://uk.investing.com/indices/nasdaq-composite-historical-data.

In the top ten, EV manufacturer Tesla is worth a quick note. It rallied 894 per cent and was at a price to earnings ratios of 959x. But this is not the only reason to mention Tesla. The company is the only representative in the top ten of the new green and sustainable economy. Its EV business brought in $8 billion in revenue and $4.7 billion in gross profit in the first three quarters of 2020. It also had a small contribution from a growing business in energy generation and storage, which earned the company $1.2 billion in revenue and $53 million in gross profit. This later business included the huge utility scale batteries Powerwall and Megapack as well as solar power generation systems (Table 4.12).

So, could we see sometime over the next few years green and sustainable companies reach stratospheric valuations similar to Tesla's current equity levels? Is it a possible benchmark? The conservative equity analyst side of me has some doubts that we will see, for example, a wind generation company attain similar heights.

Table 4.12: Tesla, Inc. 2Q 2020 Financial Results.

Nine Months to 30 September 2020	Revenue	Costs	Gross Profit
Total ($, Million)	20,792	16,228	4,564
Automotive	17,922	13,189	4,733
Energy generation and storage	1,242	1,189	53
Services and other	1,628	1,850	(222)

Source: Tesla, Inc. (2020). Q3 2020 Update. [online] Tesla Investors Events & Presentations. Available at: https://ir.tesla.com/static-files/f41f4254-f1cc-4929-a0b6-6623b00475a6 [Accessed 9 September 2020].

Still, given the flood of capital being attracted to the green and sustainable arena, it would not be farfetched to imagine that a group or several groups of companies could also see some sort of investors' euphoria. Also, I need to remind myself and likeminded analysts to be careful to apply old economy valuation methodologies when valuing new economy companies.

As discussed in Chapter 3, Asia will add more clean energy generation then the rest of the world combined in the next 30 years. So, one could draw the logical extension that some of the future clean energy giants in Asia could at some point see now hard to believe valuations.

4.4.4 Crystal Gazing into the Green Future

Forecasting financial markets developments on a 30-year view is more of an art than science. I reached out to some other experts with the aim of offering an even broader perspective on the future evolution and exponential growth of green and sustainable finance and investments in the world in general and in Asia in particular. I posed some short questions to them. Given the huge momentum in ESG, sustainable and related finance and investments witnessed in the past few years, how do you see the capital markets evolving and growing in the Asia region in the next 30 years? Will all capital flows be green or near-green or just a portion? Will there be a key landmark development, or will we see a relatively smooth progressive evolution?

Before turning to some direct quotes, I wanted to share some take-aways from having discussed the question with some friends. They are all highly seasoned professionals in the finance industry and are quite familiar with the green and sustainable side of investments and financing.

The consensus from these experts is that we should continue to see an acceleration of capital flows having to be green. Ultimately it will become the norm. All capital flows will have to have green and sustainable features. It is of course difficult to ascertain how long the conversion will take, but it is likely that it will not take another two or three decades. It is probable that it will happen within this decade. This is for several reasons. One is that the world has witnessed an acceleration and momentum from investors and commercial lenders in looking for companies to meet green and sustainable standards in the past few years. Second, is some of the investment's repercussions of COVID-19. Examples include the European Commission's post-pandemic Recovery Plan aimed at investing 1,824.3 billion Euros ($2,124.5 billion) in 2021–2027 on the green and digital transitions, the US president Joe Biden's[63] $2 trillion climate-related spending proposed when he was running his campaign, and the United Nations' call to all nations for six climate-related actions to shape the recovery, mostly revolving around green spending. Third, there is an increasing 'Fear of Missing Out' (FOMO) among corporates and financial institutions. Specifically, FOMO that if they do not carefully consider in their businesses' green and sustainable parameters, their earnings, loans, or investments could be negatively affected by new green and sustainability-related rules and regulations. For instance, a company investing in a new coal-fired generation plant may have future hurdles including carbon liabilities. Finally, all of the experts think that adoption of green and sustainable investment and financing principles among Asian corporates, governments, or institutions will sharply accelerate over the next few years.[64]

Leading Asian region tech lawyer Paul Haswell does see the sustainability factor as increasingly dominating financing and investment conversations in the coming three decades and also believes that the fintech option will gradually evolve and mature into a realistic new mainstream financing channel.

> We are already seeing a growing interest in green investment, and this will only continue as the existing generations of investors gives way to a market which is more socially, environmentally, and technologically savvy. This strongly suggests a continued shift to sustainable finance will percolate through markets globally since even institutional investors will be influenced by the move towards social and environmental awareness. With the shift to ESG being pushed by a new generation of investors comes a push to use new technologies to facilitate such investment. Thus, there will be a push towards the tokenization of energy and infrastructure assets and projects, allowing the digitalization of investment in such assets and projects and opening them up to a new pool (and generation) of investors. This evolution will not just impact green assets, but green assets are perhaps the most attractive class for modernization and may lead what will be a steady change. Already we are seeing tokenization of real estate through security token offerings; this will continue into other industries and asset classes.

Climate finance analyst June Choi at the Climate Policy Initiative concurs with my and Paul Haswell's opinion on sustainability and fintech.

> Sustainable capital markets in Asia will continue to expand rapidly, in part driven by China's ambitions for green financial reform, as well as increasing channels for small and medium enterprises, retail and foreign investor participation. The key to ensuring these finance flows contribute to green transition are clear standards, transparency, and accountability. In China, high-level political buy-in and top-down policy frameworks have enabled the rapid adoption

of green standards in the last five years. In the coming years, digitization and fintech has the potential to accelerate the market by delivering increased transparency and accountability.

Strategist, lecturer, author and fintech specialist, Paul Schulte, contributed some thoughts. He said:

> Green initiatives will be an important element of investment going forward. But I think that medium-term 'green' investment will be aimed at combating the damage for a lack of attention to the environment. For instance, Singapore will need to spend tens of billions in 'green' infrastructure to combat rising sea levels. This is a legitimate expense in the area of the environment. Other countries are in the same boat. Most humans live in coastal cities and this will only intensify going forward. So, massive investments will be needed to combat environmental degradation due to rising sea levels which wreak havoc via flooding and salivated drinking underground water. Massive funds in the ESG area will be needed to combat environmental damage before we can look forward to reducing greenhouse gases.

Last but not least another friend provided some concurring opinions. Head of Environmental Finance, Bankable Nature Solutions Asia, WWF–Hong Kong, Jean-Marc Champagne, thinks:

> We are moving in a direction where sustainable finance will become a fully integrated part of finance. Like most things that challenge the hundreds of years of status quo, change will be gradual and we will look back 30 years from now and wonder why sustainability and the environment were never fully part of finance and economics to begin with. As far as Asia goes, as long as the US continues to put money and politics ahead of science, I see Asia and Europe leading.

Notes

1 IDC (2020). Worldwide ICT Spending to Reach $4.3 Trillion in 2020 Led by Investments in Devices, Applications, and IT Services, According to a New IDC Spending Guide. [online] IDC. Available at: https://www.idc.com/getdoc.jsp?containerId=prUS46047320 [Accessed 12 July 2020].
2 Gartner (2020). Gartner Says Global IT Spending to Decline 8% in 2020 Due to Impact of COVID-19. [online] Gartner. Available at: https://www.gartner.com/en/newsroom/press-releases/2020-05-13-gartner-says-global-it-spending-to-decline-8-percent-in-2020-due-to-impact-of-covid19 [Accessed 5 November 2020].
3 Novick, B. (2020). Towards a Common Language for Sustainable Investing. [online] The Harvard Law School Forum on Corporate Governance. Available at: https://corpgov.law.harvard.edu/2020/01/22/towards-a-common-language-for-sustainable-investing/ [Accessed 5 November 2020].
4 Novick, B. (2020). Towards a Common Language for Sustainable Investing. [online] The Harvard Law School Forum on Corporate Governance. Available at: https://corpgov.law.harvard.edu/2020/01/22/towards-a-common-language-for-sustainable-investing/ [Accessed 5 November 2020].
5 Novick, B. (2020). Towards a Common Language for Sustainable Investing. [online] The Harvard Law School Forum on Corporate Governance. Available at: https://corpgov.law.harvard.edu/2020/01/22/towards-a-common-language-for-sustainable-investing/ [Accessed 5 November 2020].
6 Environmental Finance (2019). Different Paths Lead to Sustainable Investing – Environmental Finance. [online] www.environmental-finance.com. Available at: https://www.environmental-finance.com/content/analysis/different-paths-lead-to-sustainable-investing.html [Accessed 22 June 2020].

7 United Nations Department of Economic and Social Affairs (n.d.). #Envision2030: 17 Goals to Transform the World for Persons with Disabilities. [online] www.un.org. Available at: https://www.un.org/development/desa/disabilities/envision2030.html#menu-header-menu [Accessed 22 June 2020].

8 Coursera (n.d.). Coursera: Online Courses & Credentials from Top Educators. [online] Coursera. Available at: https://www.coursera.org/learn/sustainable-finance/home/week/1 [Accessed 4 July 2020].

9 United Nations (2015). PARIS AGREEMENT (UNITED NATIONS 2015). [online] unfccc.int, p. 5. Available at: https://unfccc.int/sites/default/files/english_paris_agreement.pdf#page=5 [Accessed 1 August 2020].

10 van Romburgh, M. (2020). What Google's CEO Plans to Do to Improve Leadership Diversity, Help Black Businesses. [online] Bizjournals.com. Available at: https://www.bizjournals.com/sanjose/news/2020/06/17/google-diversity-representation-commitment.html [Accessed 2 August 2020].

11 The Economist Magazine (2020). 'Hotting Up': How Much Can Financiers Do about Climate Change? [online] The Economist. Available at: https://www.economist.com/briefing/2020/06/20/how-much-can-financiers-do-about-climate-change [Accessed 6 November 2020].

12 Hitachi (2019). Hitachi Integrated Report 2019 (Year Ended March 31, 2019): Investor Relations – Hitachi Global. [online] www.hitachi.com, Hitachi, Ltd., p. 76. Available at: https://www.hitachi.com/IR-e/library/integrated/2019/index.html [Accessed 6 November 2020].

13 United Nations Environment Programme – Finance Initiative (2019). 130 Banks Holding USD 47 Trillion in Assets Commit to Climate Action and Sustainability – United Nations Environment – Finance Initiative. [online] United Nations Environment Programme – Finance Initiative. Available at: https://www.unepfi.org/news/industries/banking/130-banks-holding-usd-47-trillion-in-assets-commit-to-climate-action-and-sustainability/ [Accessed 6 November 2020].

14 Task Force on Climate-Related Financial Disclosures (2017). Recommendations of the Task Force on Climate-Related Financial Disclosure: The Use of Scenario Analysis in Disclosure of Climate-Related Risks and Opportunities Technical Supplement Technical Supplement – the Use of Scenario Analysis in Disclosure of Climate-Related Risks and Opportunities. [online] Task Force on Climate-Related Financial Disclosures. Available at: https://assets.bbhub.io/company/sites/60/2020/10/FINAL-TCFD-Technical-Supplement-062917.pdf [Accessed 6 November 2020]. AGL Energy (2020). Scenario Analysis and Task Force for Climate-related Financial Disclosure. In: AGL Energy Presentation. [online] Australia: AGL Energy. Available at: https://www.agl.com.au/-/media/aglmedia/documents/about-agl/investors/webcasts-and-presentations/2020/tcfd-speech.pdf?la=en&hash=DFC1C23569116F32C96FC7545B6A5245 [Accessed 6 November 2020].

15 Sustainable Stock Exchanges Initiative (2020a). *About the SSE Initiative.* [online] sseinitiative.org. Available at: https://sseinitiative.org/about/ [Accessed 6 November 2020].

16 Sustainable Stock Exchanges Initiative (2020b). *Sustainable Stock Exchanges Initiative – Data.* [online] sseinitiative.org. Available at: https://sseinitiative.org/data/ [Accessed 6 November 2020].

17 HKEX (2020). HKEX to Launch New Sustainable and Green Exchange. [online] www.hkex.com.hk. Available at: https://www.hkex.com.hk/News/News-Release/2020/200618news?sc_lang=en [Accessed 6 November 2020].

18 Calastone (2020). Woke British Investors Drive Unprecedented Demand for ESG Funds. [online] calastone.com. Available at: https://www.calastone.com/woke-british-investors-drive-unprecedented-demand-for-esg-funds/ [Accessed 6 November 2020].

19 Riding, S. (2020). ESG Funds Attract Record Inflows during Crisis. [online] www.ft.com. Available at: https://www.ft.com/content/27025f35-283f-4956-b6a0-0adbfd4c7a0e [Accessed 20 October 2020].

20 Iacurci, G. (2020). Money Moving into Environmental Funds Shatters Previous Record. [online] CNBC. Available at: https://www.cnbc.com/2020/01/14/esg-funds-see-record-inflows-in-2019.html [Accessed 24 February 2020].

21 Raj, A. (2020). BlackRock SEA Head Deborah Ho Talks All Things E.S.G. Citywireasia.com. Available at: https://citywireasia.com/news/blackrock-sea-head-deborah-ho-talks-all-things-e-s-g/a1385473?ref=international-asia-video-list [Accessed 6 November 2020].

22 BloombergNEF (2020). Corporate Clean Energy Buying Leapt 44% in 2019, Sets New Record. [online] BloombergNEF. Available at: https://about.bnef.com/blog/corporate-clean-energy-buying-leapt-44-in-2019-sets-new-record/ [Accessed 20 February 2020].

23 BloombergNEF (2020). Clean Energy Investment Trends, 2019. [online] Bloomberg LP. Bloomberg LP. Available at: https://data.bloomberglp.com/professional/sites/24/BloombergNEF-Clean-Energy-Investment-Trends-2019.pdf [Accessed 6 November 2020].

24 European Investment Bank (2007). EPOS II – the 'Climate Awareness Bond' EIB Promotes Climate Protection via Pan-EU Public Offering. [online] European Investment Bank. European Investment Bank. Available at: https://www.eib.org/en/investor_relations/press/2007/2007-042-epos-ii-obligation-sensible-au-climat-la-bei-oeuvre-a-la-protection-du-climat-par-le-biais-de-son-emission-a-l-echelle-de-l-ue.htm [Accessed 6 November 2020].

25 B, C. (2008). AP2 and AP3 Back World Bank Climate Bond. IPE Magazine. [online] 12 November Available at: https://www.ipe.com/ap2-and-ap3-back-world-bank-climate-bond/29672.article [Accessed 6 November 2020].

26 Climate Bonds Initiative (2019). Explaining Green Bonds. [online] climatebonds.net. Available at: https://www.climatebonds.net/market/explaining-green-bonds [Accessed 27 April 2019].

27 EnelS.p.A (2020). Q1 2020 Consolidated Results. [online] enel.com. Italy: EnelS.p.A. Available at: https://www.enel.com/content/dam/enel-com/documenti/investitori/informazioni-finanziarie/2020/trimestrali/1q-2020-risultati.pdf [Accessed 6 November 2020].

28 Citigroup (2020). Environmental, Social and Governance Report. [online] Citigroup Inc. Available at: https://www.citigroup.com/citi/about/esg/download/2019/Global-ESG-Report-2019.pdf [Accessed 6 November 2020].

29 Climate Bonds Initiative (2020). Green Bonds Global State of the Market 2019. [online] Climate Bonds Initiative. Available at: https://www.climatebonds.net/files/reports/cbi_sotm_2019_vol1_04d.pdf [Accessed 6 November 2020].

30 Barclays (2019). AIIB's Inaugural Bond – Driving Sustainable Development in Asia. [online] www.investmentbank.barclays.com. Available at: https://www.investmentbank.barclays.com/banking/financing-a-sustainable-future-with-AIIB.html [Accessed 6 November 2020].

31 Asian Infrastructure Investment Bank (AIIB) (2020) Annual General Meeting, Session: 'Investing in Climate Action' [Webinar]. Available at: http://www.slideshare.net/thetalentproject/hr-the-workplace-of-the-future [Accessed 29 July 2020].

32 Asia Infrastructure Investment Bank (2019). Project Summary Information: Asia Climate Bond Portfolio. [online] aiib.org. Asia Infrastructure Investment Bank. Available at: https://www.aiib.org/en/projects/details/2019/approved/_download/multicountry/PSI-Asia-Climate-Bond-Portfolio-20200724.pdf [Accessed 6 November 2020].

33 Escalante, D., Choi, J., Chin, N., Cui, Y., and Lund Larsen, M. (2020). The State and Effectiveness of the Green Bond Market in China. [online] climatepolicyinitiative.org. Available at: https://www.climatepolicyinitiative.org/wp-content/uploads/2020/06/The-State-and-Effectiveness-of-the-Green-Bond-Market-in-China.pdf [Accessed 6 November 2020].

34 Meng, A.X. (2020). China Green Bond Market 2019 Research Report. [online] climatebonds.net. Climate Bonds Initiative and China Central Depository & Clearing Research Centre (CCDC Research). Available at: https://www.climatebonds.net/resources/reports/china-green-bond-market-2019-research-report [Accessed 6 November 2020].

35 Vasakronan (2018). Another World First, Vasakronan Issues Green Commercial Paper to Finance Green Assets. 21 September. [online] Cision. Available at: https://news.cision.com/vasak

ronan/r/another-world-first–vasakronan-issues-green-commercial-paper-to-finance-green-assets, c2624587 [Accessed 6 November 2020].
36 Linklaters (2019). Sustainable Finance: The Rise of Green Loans and Sustainability Linked Lending. 12 June. [online] Linklaters.com, p. 5. Available at: https://www.linklaters.com/en/insights/thought-leadership/sustainable-finance/the-rise-of-green-loans-and-sustainability-linked-lending [Accessed 6 November 2020].
37 Sina Finance and Economics (2020). Exploration of Green Finance of Large State-owned Banks: ICBC's Largest Investment and Postal Savings Bank's Fastest Growth. [online] finance.sina.com.cn. Available at: http://finance.sina.com.cn/china/gncj/2020-04-02/doc-iimxyqwa4762190.shtml [Accessed 6 November 2020].
38 *China Daily* (2020). China's Green Loans See Rapid Growth in H1. [online] www.chinadaily.com.cn. Available at: https://www.chinadaily.com.cn/a/202008/03/WS5f27c699a31083481725ddca.html [Accessed 6 November 2020].
39 *The Business Times* (2020). China Plans Quarterly Look into Banks' Green Finance Performance. [online] *The Business Times*. Available at: https://www.businesstimes.com.sg/banking-finance/china-plans-quarterly-look-into-banks-green-finance-performance [Accessed 6 November 2020].
40 BNP Parisbas (2019). Seeding China's Banks with Sustainable Lending – BNP Paribas CIB. [online] cib.bnpparibas.com. Available at: https://cib.bnpparibas.com/our-news/seeding-china-s-banks-with-sustainable-lending_a-33-2964.html [Accessed 6 November 2020].
41 International Finance Corporation (2019). Green Buildings: A Financial and Policy Blueprint for Emerging Markets. [online] www.ifc.org. Available at: https://www.ifc.org/wps/wcm/connect/topics_ext_content/ifc_external_corporate_site/climate+business/resources/green+buildings+report [Accessed 5 November 2020].
42 'Green Building Webinar: Accessing a $18 Trillion Investment Opportunity', Hong Kong Green Finance Association, 29 July 2020.
43 Swire (2019). Swire Properties Announces First Sustainability-Linked Loan. [online] www.swireproperties.com. Available at: https://www.swireproperties.com/en/media/press-releases/2019/20190730_swire-properties-announces-first-sustainability-linked-loan [Accessed 14 August 2020].
44 Swire (2020). Swire Properties Secures HK$1 Billion Green Loan from OCBC Bank. [online] www.swireproperties.com. Available at: https://www.swireproperties.com/en/media/press-releases/2020/20200804_green-loan [Accessed 14 August 2020].
45 OCBC (2020). Keppel Infrastructure Trust and Keppel Energy Secure Singapore's First Sustainability-linked Loan in the Energy Sector from DBS Bank and OCBC Bank. [online] www.ocbc.com. Available at: https://www.ocbc.com/group/media/release/2020/keppel-sustainability-linked-loan-from-dbs-and-ocbc.html [Accessed 14 August 2020].
46 Sutherland, P. (2019). Herbert Smith Freehills advises AGL Energy Limited on its Innovative A$600 Million Sustainability-linked Loan. [online] Herbert Smith Freehills, Global Law Firm. Available at: https://www.herbertsmithfreehills.com/news/herbert-smith-freehills-advises-agl-energy-limited-on-its-innovative-a600-million [Accessed 14 August 2020].
47 CDP (2020). Press release: Companies in Asia Pacific Show High Awareness of Climate-related Issues, but More Actions Needed to Achieve Paris Agreement Targets – CDP. [online] www.cdp.net. Available at: https://www.cdp.net/en/articles/media/press-release-companies-in-asia-pacific-show-high-awareness-of-climate-related-issues-but-more-actions-needed-to-achieve-paris-agreement-targets [Accessed 16 September 2020].
48 International Energy Agency (2020b). Energy Financing and Funding – World Energy Investment 2020 – Analysis. [online] IEA. Available at: https://www.iea.org/reports/world-energy-investment-2020/energy-financing-and-funding [Accessed 18 September 2020].

49   Buckley, T. (2020b). IEEFA Australia: Coal Lobbyists Claim Commitment to Paris Agreement Targets, Five Years after Their Members Did. [online] Institute for Energy Economics & Financial Analysis. Available at: https://ieefa.org/ieefa-australia-coal-lobbyists-claim-commitment-to-paris-agreement-targets-five-years-after-their-members-did/ [Accessed 3 September 2020]. Buckley, T. (2020). IEEFA Asia: Asian Financial Institutions also Beginning to Exit Coal Financing. [online] Institute for Energy Economics & Financial Analysis. Available at: https://ieefa.org/ieefa-asia-asian-financial-institutions-also-beginning-to-exit-coal-financing/ [Accessed 3 September 2020].

50   Burton, M. (2020). New Zealand to Require Financial Firms Report Climate Change Risks. 17 September. [online] Reuters. Available at: https://www.reuters.com/article/climate-change-newzealand/corrected-new-zealand-to-require-financial-firms-report-climate-change-risks-idUSL4N2GB23E [Accessed 18 September 2020].

51   White & Case LLP (2020). White & Case Advises Federal Republic of Germany on First Issuance of Green Federal Bond with Principal Amount of €6.5 Billion | White & Case LLP. [online] www.whitecase.com. Available at: https://www.whitecase.com/news/press-release/white-case-advises-federal-republic-germany-first-issuance-green-federal-bond [Accessed 18 September 2020]. Bahceli, Y. (2020). Germany Raises 6.5 Billion Euros from First-ever Green Bond. 3 September. [online] Reuters. Available at: https://www.reuters.com/article/bonds-green-germany/germany-raises-6-5-billion-euros-from-first-ever-green-bond-idUSKBN25T2UP [Accessed 18 September 2020].

52   PricewaterhouseCoopers (2016). Q&A: What is FinTech? [online] PwC. Available at: https://www.pwc.com/us/en/industries/financial-services/library/qa-what-is-fintech.html [Accessed 3 September 2020].

53   Kotow, E. (2019). 'DeFi Vs. FinTech'. HedgeTrade Blog. 14 May. Available at: https://hedgetrade.com/defi-vs-fintech [Accessed 3 September 2020].

54   DeFi Pulse (2019) 'What Is DeFi?' 3 September. Available at: https://defipulse.com/blog/what-is-defi/ [Accessed 3 September 2020].

55   PwC (2019). In Depth: A Look at Current Financial Reporting Issues – Cryptographic Assets and Related Transactions: Accounting Considerations under IFRS at a Glance. [online] p. 4. Available at: https://www.pwc.com/gx/en/audit-services/ifrs/publications/ifrs-16/cryptographic-assets-related-transactions-accounting-considerations-ifrs-pwc-in-depth.pdf [Accessed 19 September 2020].

56   Schletz, M., Nassiry, D., and Lee, M.-K. (2020). Blockchain and Tokenized Securities: The Potential for Green Finance. [online] Asian Development Bank, Tokyo, Japan: Asian Development Bank Institute, p. 10. Available at: https://www.adb.org/sites/default/files/publication/566271/adbi-wp1079.pdf [Accessed 20 September 2020]. ADBI Working Paper Series No. 1079.

57   Schletz, M., Nassiry, D., and Lee, M.-K. (2020). Blockchain and Tokenized Securities: The Potential for Green Finance. [online] Asian Development Bank, Tokyo, Japan: Asian Development Bank Institute, p. 4. Available at: https://www.adb.org/sites/default/files/publication/566271/adbi-wp1079.pdf [Accessed 20 September 2020]. ADBI Working Paper Series No. 1079.

58   BBFTA News Editor (2020). News: Will IPOs Continue to Decline as STOs Gather Momentum? [online] bbfta.org. Available at: https://bbfta.org/news/1599048573-4099c72e12/Will-IPOs-continue-to-decline-as-STOs-gather-momentum [Accessed 18 September 2020].

59   Various news releases from INX Limited. Available at: https://www.inx.co/ [Accessed 20 September 2020].

60   BBFTA News Editor (2020). News: Will IPOs Continue to Decline as STOs Gather Momentum? [online] bbfta.org. Available at: https://bbfta.org/news/1599048573-4099c72e12/Will-IPOs-continue-to-decline-as-STOs-gather-momentum [Accessed 18 September 2020].

61   Covid-19 and Energy – the Use of Blockchain post-Covid 19 with Ana Trbovich, Grid Singularity (2020). 18 May. [Podcast] Montel AS. Available at: http://covid19.montelnews.com/937606/3792029-18-may-2020-the-use-of-blockchain-post-covid-19-with-ana-trbovich-grid-singularity [Accessed 15 September 2020].

62 Covid-19 and Energy – the Use of Blockchain post-Covid 19 with Ana Trbovich, Grid Singularity, (2020). 18 May. [Podcast] Montel AS. Available at: http://covid19.montelnews.com/937606/3792029-18-may-2020-the-use-of-blockchain-post-covid-19-with-ana-trbovich-grid-singularity [Accessed 15 September 2020].

63 As of 24 September 2020.

64 The following expert opinions are from email correspondence with the author in October–November 2020.

Chapter 5
Conclusions and Suggestions

Asia is already a massive energy user and will become much larger. It consumes about two-fifths of the globe's primary energy and generates almost half of the world's electricity. The consumption will continue to grow rapidly over the next three decades. I estimate that it will increase threefold from the current level at the very minimum. Most likely more than that. Probably fourfold or more. It will also dynamically and dramatically change and transition. This is giving rise to new opportunities directly and indirectly related to the region's electricity markets. The new developments are attracting hundreds of billions of dollars in public and private capital, which is pouring into new ventures and investments. The doors are being opened by the advent of two concurrent tectonic shifts. Individual power markets in the region are increasingly prioritising green over dirty and polluting generation. These markets are also eagerly adopting new digital technologies and solutions to reduce costs and raise efficiencies. While these colossal shifts have already begun to take place, they are neither widespread nor are they uniform. Yet, they are here and now. I believe that it is a perfect time for corporates and investors to identify new businesses, projects, and other investments possibilities in the sector.

The enormous size of the industry means that the dynamics are highly complex and multifaceted. To better understand this, the first stop is evaluating the overall economic outlook for the region. Economies are closely tied to the electric power sector as its production and supply is the lifeblood of any economy. No matter whether the growth is centred on the manufacturing industry or the services sector. While this can be applied to any economy anywhere around the globe, the world's centre of economic gravity has been shifting to Asia for many years. And it will continue to do so over at least the next few decades. The region will be consuming a lot more energy given the expected economic and population growth, three-quarters of a billion people for a total of more than 5.3 billion by 2050 in the next three decades.

The Asian energy landscape will dramatically change by 2050. Usage will surge, especially in emerging economies. A simple calculation measure is electricity generation per capita. Developing economies in Asia had 16 times the population of developed ones in 2018 but the quantity of electricity consumed was merely four times greater. A three-fold to four-fold rise in electric output or more will be required to bring per capita levels in developing markets to that of developed ones, I estimated. The total could be higher under several scenarios. One such scenario would be higher than expected electricity demand due to an above-expectations rise in BEVs ownership. This is because individual governments in the region are progressively adopting new targets or updating existing ones on capping ICE vehicles. For example, in 2020 Singapore announced that it would phase out ICE vehicles by 2040 and

China said that it would do the same by 2035, except for hybrid vehicles. Apart from traditional carmakers progressively shifting production to non-ICE vehicles, a new industry is rising including batteries and other equipment providers as well as new EV start-up companies. One such company is China-based WM Motor with its highly innovative smart cars. Possibly offsetting sharper than expected electric power usage would be more stringent conservation and efficiency measures, including rules around lighting, buildings' energy consumption, as well as a wider use of energy-management companies. There are many examples of successful business models in these areas. Buildings efficiency wise, Hong Kong real estate giant Swire has proven to be a leader in Asia, winning many sustainability-related international awards. In China, a foreign-owned company, Adenergy, has proven to be one of the leaders in smart energy-management systems, including smart energy-saving solutions for industrial buildings – a model that it is now successfully exporting to other countries. In the field of lighting, UK-based company Energys is winning contracts in Hong Kong and other economies in the region in LED retrofit lighting and other lighting solutions.

Another key factor helping to construct a new backdrop are regulatory changes. These are already occurring, or will evolve, in individual Asian power markets in ways similar to the sector reforms in Australia, China, Japan, New Zealand, and Singapore. In essence, the sectors' restructuring is about opening up electricity markets to retail competition, giving consumers actual choice in their energy provider. In Australia, New Zealand, and Singapore, companies can only control generation and retail or own transmission and distribution assets, but not both. This has led to brutal competition where the energy retailers focus on gaining and retaining customers. China and Japan are well on their way to a similar environment, which will lead the market incumbents to lose sizeable amounts of market share. These transformations are driving the metamorphosis of incumbent market participants, facilitating the entry of new players, and promoting business models innovation. A great many companies have had to modify their traditional business model following on the footsteps of utilities in the UK where markets opened up in the late 1980s. In New Zealand, the four biggest energy retailers control about 73 per cent of the power market while a myriad of others control the balance. Competition is fierce even though New Zealand is one of the region's smallest electricity markets. In Japan, an interesting example is Tokyo Gas, which not only supplies natural gas but also has become a significant power retailer. Another is telecom giant KDDI, which has started to retail electricity as part of its wide offering of services to its telecom customers. In Singapore, many new retail power companies were created in recent years. Some are brand new, such as iSwitch, while others are subsidiaries of already well-established diversified companies (or conglomerates). In a similar way to Tokyo Gas, in China city-gas retailer ENN, which is not a state-owned company, is also successfully pursuing new energy-related businesses including power retail and smart energy solutions. Another example is China Resources Power, a state-owned company, which has been one of the most progressive and successful companies in the area of energy retailing and smart power services.

Apart from the consumption growth and regulatory transformations, Asia is undergoing two concurrent tectonic shifts in the fuel sources it uses for electricity and in the adoption of new digital technologies and solutions. These changes are not unique to the region; they are also happening in the EU, for example. The difference is the size of the region's energy market and that the majority of these markets are still developing, thus offering greater scope for absolute growth.

Subsequent to the Paris Agreement coming into force in late 2016, an increasing number of nations worldwide have published their respective pledges to decarbonise their economies. A variety of forecasts indicate a sharp increase in non-fossil fuel generation worldwide in the next 30 years. These forecasts do not necessarily all concur as to the amount to be added and also use a different base for their calculations. Still, they all agree that the upsurge will be staggering. IRENA estimates the share of non-fossil fuels in the generation mix will rise to between 36 and 61 per cent by 2050 from less than 5 per cent in 2015. The US EIA expects a share of 49 per cent from 21 per cent in 2010. McKinsey estimates 73 per cent from 18 per cent in 2005. These forecasts were published before 2020, a year when several major Asian energy users pledged to decarbonise their economies by 2050 (Japan and South Korea) or by 2060 (China). I believe that the share in the electric power generation mix from non-fossil fuel sources will be at the high end of these forecasts. The energy mix shift from brown to green will have some tremendous repercussions to many economies in Asia including Australia, those in East Asia, and China.

In the case of Australia, the shift has a number of dimensions and complexities. These include the nation's heavy reliance on fossil fuels as a domestic source for electricity, the importance of these commodities to its exports, and the large amounts of capital expenditure it will require for renewables. The latter is particularly significant given the high reliance of the country's power generation facilities on fossil fuels, especially thermal coal. Also, the region's increased amount of clean energy build-up will affect Australia's thermal coal and natural gas export volumes. In the case of the three East Asian electricity markets of Japan, South Korea, and Taiwan, the shift to clean power also has dramatic multifaceted implications. Their drive towards clean energy is all the more urgent compared to many other markets in the region because it will reduce their energy commodities import bills and raise supply security, which is very much an ongoing central concern to them given the limited amount of energy from domestic sources. Coal, natural gas, and oil were responsible for about nine-tenths of primary energy consumed in the three markets in 2018. In the case of China, the nation has been, in my view, the undisputed global leader in clean energy additions in the past five years or so. The amount of solar and wind generation capacity that the nation added is not just large, relative to that of the rest of the world combined, it is astronomical. In September 2020, the president of the People's Republic announced that the nation had doubled down on its commitment towards clean energy by pledging that it would arrive at net-zero by 2060 at the latest. The push will help the nation to further sharply reduce harmful greenhouse gases. It is also quite positive for economic activity including job formation and adding to its exports, for example. A positive side effect for the rest of Asia and the world is that China is incredibly well placed to drive down the LCOE of newer technologies such as energy storage given the massive economies of scale it offers. It has already successfully done so for solar generation equipment, for example, an industry that it currently leads globally; all top five of the world's largest photovoltaic module manufacturers in 2020, measured by shipments, were based in China, including JinkoSolar, JA Solar, Trina Solar, LONGI Solar, and Canadian Solar.

One final theme about the impact of changing energy fuel sources is the impact on the prices of fossil fuels from Asia's energy transition to green from brown power in the next 30 years. The price per unit of energy produced from solar and wind power will continue to fall in the next several years. Today, these unit costs are already lower than those for fossil fuel-based power generation in several countries, including Australia. Over the next few years, I expect that the renewable resources' unit price advantage will widen further, and it will be ubiquitous in the region. Asian power markets will want to not only build clean energy generation facilities because of their respective decarbonisation targets, but also because it makes a lot of economic sense. Coal, natural gas, and oil will not completely disappear, but their usage will progressively dwindle at varying speeds.

The second tectonic shift lies with the opening of the digital door. It is the advent of Energy 4.0, the digital transformation in the energy sector. Digital technologies are getting better and cheaper, and prediction and computational tools are becoming faster. In the electricity industry, they are driving down costs throughout the value chain of the production and supply of energy, they optimise operations, facilitate raising environmental sustainability, and effectively increase the efficiency of energy-related financial transactions. Of particular note is the intersection, or combining, of AI, blockchain, and IoT. These solutions are estimated, for example, to be able to cut generation expenses by more than 10 per cent through process automation, digital enablement, and advanced analytics, according to McKinsey. The cost reductions are even greater for transmission and distribution (approximately 26 per cent), customer and retail expenses (about 25 per cent), and for general corporate costs (approximately 15 per cent). An energy and automation digital solutions company, Schneider Electric, estimates that cost savings can be even greater than those projected by McKinsey. It estimated that the investments in Energy 4.0 could produce a payback in as short a time as less than one year.

Energy 4.0 is attracting an enormous amount of innovation. The innovators are coming from a variety of places. They can be large corporations such as the vertically integrated utility KEPCO, they can be national governments such as Singapore's, and they can be a new breed of market entrants. In the region, I have estimated that there are AI, blockchain, and IoT-related energy pilot projects in 18 out of the 22 electricity markets. It is a very widespread phenomenon. Despite the very vibrant activity on this front throughout the region, the Energy 4.0 innovation leader is China. The nation has made enormous advances in these areas. This is being led not only by government authorities or state-owned enterprises, such as two main grid companies SGCC and CSG, but also by the non-state-owned sector. Participants include e-commerce giant Alibaba, drone maker DJI, electronics and smartphones maker Huawei, and a smorgasbord of start-ups, too, such as wind measurements equipment maker Blue Horizon. The perception that China is technologically backwards is very much misplaced and the new tech Cold War may result in the country's leading position extending even further as it seeks more technological self-sufficiency.

The development of electric power and related infrastructure as well as the region's digital transformation will require investments amounting to many trillions of dollars in the next 30 years. The capital will be sourced from both the public and private sectors. I believe that money will in no way be lacking. To the contrary, it will be quite plentiful.

I need to segregate the two different targets of the capital: digitalisation and energy infrastructure. For the former it is more straightforward than for the latter. Diversified fund raising for the tech sector has a longer history than that for clean energy infrastructure. Traditionally, infrastructure projects have more heavily relied on long-term loans from large financial institutions as well as by the issuance of bonds. Also, tech has often provided higher returns on investment, the higher-risks-higher-returns principle. In my personal experience investors' general understanding of the tech sector is relatively deeper and broader. It obviously helps that many tech companies today are household names, such as the FANGs (Facebook, Amazon, Netflix, and Google), for example. High-profile Energy 4.0 companies, such as Schneider Electric, are far from being household names, while other companies are only known for higher profile products. Huawei, for example, is better known for its smartphones and 5G tech but not so much for its Energy 4.0 innovations. So, for tech the funding channels are rich and well proven.

When it comes to the financing of green and sustainable projects, especially those in the energy field, in recent years there has been a massive acceleration in interest from the financial markets. Many new products have been created with a special focus on being green and sustainable, such as green bonds. It is highly likely that a vast percentage of the capital flows of financial institutions globally will mandate green and sustainable credentials. For example, lending institutions extending loan facilities to coal-fired power plants are fewer and fewer. Even asset managers, from giants like BlackRock and Fidelity to smaller ones, now see a risk in investing in companies with high carbon footprints. Also, some solutions have appeared for providing loans to projects that are harder to finance such as small isolated microgrids. The solutions include the issuance of coins or tokens, part of the new world of fintech. These are new and face many development challenges. But I believe that this particular solution will soon become mainstream, particularly as it becomes better regulated. I also believe that many more new solutions will be created from the green and sustainable momentum.

Asia's electric power sector will need massive amounts of capital because of a much larger energy infrastructure over the coming 30 years. This infrastructure will progressively transition to being green and sustainable from being fossil fuels-centric. And we will also witness an Energy 4.0 revolution. Who will this benefit, apart from consumers who will get cheaper, higher quality service, as well as environmentally sustainable energy? If many trillions of dollars would be invested in these two concurrent transitions in the region, who are the companies that will benefit the most. And. how can new entrants take advantage of this? Will the playing field be even, or will it be simply dominated by large corporations and institutions? What will be the impact on passive and active investing? What kind of new strategies and new business models are likely to emerge? What kind of risks and challenges will the two concurrent shifts create?

There are likely to be a great number and variety of actors who will greatly benefit from the changes. Albeit just like any industry in transition, there will be some market incumbents that will prosper and some that will disappear. And it will be the same for new players as well; some will thrive while many will die. And the opportunities are abundant. Asia has a great number of power markets, and many have stand-alone submarkets – such as Australia, China, and India. They include large and homogeneous ones as well as isolated and micro ones, that is, micro-grids and distributed energy resources.

I believe that as government planners, financial institutions, and other industry actors become more familiar with the clean energy sector, capital will be more easily deployed. Conservative corporate cultures and management practices at corporations and financial institutions will increasingly shift to a more understanding approach toward green and sustainable. Just a decade ago in my corporate experience, financing a utility-scale solar power generation plant was not easy due to a lack of familiarity on the part of financial institutions. This is today rapidly changing Asia-wide.

The common denominator is that institutional attitudes towards clean energy are rapidly changing. There is an educational process going on at an accelerated speed. There is a growing realisation that if one does not get involved now, one may miss the boat and miss out on a great number of new income-generating possibilities. Conversely, those companies and financial institutions that get on the bandwagon early are likely to have a significant strategic advantage. They are able to structure their operations in such a way as to benefit from the multitude of green and sustainable opportunities. The risk for those companies and institutions that do not change is a likely progressive loss of business and market share. The industry peers wanting to copy the model of those corporates that have instituted change will face a very severe upward slope to climb. This is because the first corporate change to be undertaken is that of corporate culture, one that in the energy industry tends to be conservative and often outdated.

I have qualified and quantified that the transition to green and sustainable energy and to digitilisation in Asia is dynamic and multifaceted. Asia's energy revolution is already gaining speed. It means big challenges and huge opportunities. It is high time to get on board and grab some of the many new business and investment possibilities.

List of Figures

1.1	Asia and World Population Forecasts —— 2	
1.2	Global GDP Share Breakdown by Region —— 3	
1.3	World Average Annual GDP Growth 2000–2018 —— 3	
1.4	Global Energy and Power Output Share in 2019 —— 6	
2.1	Potential Power Generation Multiple Growth in Developing Asia 2050 versus 2018 (x) —— 26	
2.2	Europe Brent and Cushing, OK WTI Spot Prices Freight-On-Board —— 30	
2.3	Weltmeister EX6 Plus —— 36	
2.4	Australian Energy Intensity and Energy Productivity —— 39	
2.5	Ten Years to June 2018 Electric Power Output in Australia —— 39	
2.6	Hong Kong Energy Consumption by Sector (Adapted) —— 41	
2.7	Allocation of Green Bond Proceeds (as at 30 September 2019) —— 42	
2.8	Swire Properties Taikoo Portfolio Map (Simplified maps are not to scale and for illustrative purpose only; Gross Flow Area figures are for reference only) —— 43	
2.9	Swire Properties Change in Gross Floor Area and Energy Consumption —— 44	
2.10	ISO 50001 Energy Management System Certifications by Region, 2011–2017 —— 45	
2.11	Compressors Using Optimisation: Compressor Room Before and After Adenergy's EMS and Leakage Work —— 47	
2.12	Typical Lighting System —— 50	
2.13	IntelliDim – Dynamic Lighting Control —— 51	
2.14	Australia Electricity Market Structure (Adapted) —— 53	
2.15	Profile of National Electricity Market and Western Australian Wholesale Electricity Market —— 57	
2.16	Annual Average Spot Prices in Australia's Three Largest States —— 58	
2.17	Earnings of Australia's Largest Electricity Retailers —— 59	
2.18	New South Wales Power Prices 1 January Through 14 July 2020 —— 60	
2.19	National Electricity Market One Month Annualised Retail Transfer Statistics (Churn Rate) —— 61	
2.20	Japan's Electricity Market System Reform —— 64	
2.21	Japan Electric Power Exchange (JPEX) Total Transaction Volume (GWh) —— 65	
2.22	Number Electricity Accounts Switching Applications in Japan (March 2016–June 2020) —— 66	
2.23	Singapore's Electricity Market System Reform —— 70	
2.24	Singapore Peak Demand versus Generation Capacity —— 71	
2.25	Uniform Singapore Energy Prices (USEP) 1 January Through 8 July 2020 —— 72	
2.26	Singapore Electricity Retail Licensees —— 74	
2.27	Singapore Electricity Retail Licensees' Market Share as of 31 March 2020 —— 74	
2.28	China's Electricity Market System Reform —— 76	
2.29	Percentage of Electricity Under Market-Oriented Trading —— 79	
2.30	Electricity Supply Market Shares by Company in Great Britain (2Q 2020) —— 85	
2.31	Retail Electricity and Gas Market Share for Quarter 3, 2019–2020 —— 87	
2.32	Electric Power Retail Market in New Zealand Market Share —— 90	
2.33	Tokyo Gas Electricity's New Contracts and Total Sales —— 94	
2.34	Major Electricity Retailers (ex-Major Electric Power Companies) in April 2020 —— 95	
2.35	China Resources Power Return on Equity Versus Peers (%) —— 102	
3.1	Australian Energy Production, by Fuel Type —— 113	
3.2	AEMO's Five Capacity Forecast Scenarios Definitions —— 116	

List of Figures

3.3	AEMO's Australia Central Case Scenario Capacity Forecast	116
3.4	AEMO's Australia Renewable Energy Capacity Forecast	117
3.5	Japan, South Korea, Taiwan Nuclear Power Production	120
3.6	China's CO_2 Total Emissions	125
3.7	China's CO_2 Emissions (Kilotons)	126
3.8	China's Copenhagen and Paris Environmental Commitments	126
3.9	China's Historical Greenhouse Gases Emissions	127
3.10	China Installed Capacity Additions (January 1995 to November 2020)	129
3.11	Capacity Utilisation Rates of Different Power Sources in China	131
3.12	China Solar Power Generation Capacity	132
3.13	China Wind Power Generation Capacity	133
3.14	China Wind Power Average Curtailment Rate	134
3.15	China Installed Capacity Projections (Adapted)	136
3.16	Electrochemical Batteries	139
3.17	China's Historical and Estimated Electrochemical Storage Capacity	140
3.18	IRENA's Renewable Power Generation Costs ($/MWh) in 2019	145
3.19	Lazard's Cost of Energy Comparison – Unsubsidised Analysis	146
3.20	BloombergNEF Global LCOE Ranges, Second Half 2020 Updates	147
3.21	Newcastle Freight-On-Board 6,000 Kilocalories per Kilo Net-As-Received Monthly Price – $ per Metric Tonne	149
3.22	Japan Spot LNG Price Statistics	150
3.23	Summary of Energy Industry Intersection with Energy 4.0	155
3.24	Internet of Things Applications for Energy Industry (Adapted)	156
3.25	Disruptive Forces' Potential Impact on Businesses in the Power Sector (Adapted)	160
3.26	Worldwide Cost Savings from Enhanced Digitalisation in Power Plants and Electricity Networks over 2016–2040 (Adapted)	161
3.27	KEPCO's Research and Development Expenditure	166
3.28	KEPCO's New Industries Expansion (Adapted)	169
3.29	Schematic View and Some Features of Smart Grid	170
3.30	Convergence of Blockchain, IoT, and AI	171
3.31	How Blockchain Works (Adapted)	172
3.32	Two Different Breakdowns of New Energy Infrastructure Spending Estimated for 2020–2025	180
3.33	State Grid Corp. of China Spending on Digitalisation (Billion Yuan)	186
3.34	China Power Meters Additions and Replacements, and Smart Meters Spending	187
3.35	Snapshot of Huawei's Businesses	188
3.36	Blue Aspirations ZX300M Model Floating Lidar System	190
4.1	Market Capitalisation of the Largest Tech Companies in the World	199
4.2	Amount Spent by Seven Digital Companies on R&D in 2019	200
4.3	Scenario Analysis of Possible Growth in R&D Spending by Seven Tech Giants	200
4.4	Key Sectors and Energy Technologies Relationship	201
4.5	Sources of Funding for Tech Companies	202
4.6	Sample List of ESG Factors (Adapted)	204
4.7	The 17 United Nations Sustainable Development Goals (SDGs)	205
4.8	Tools Governments Can Employ to Shift Finance	207
4.9	Climate-Related Risks and Opportunities Scenario Analysis Application Process (Adapted)	209
4.10	United Nations Environment Programme – Finance Initiative Banks' Membership by Region (Adapted)	211

4.11	Thirty Years' Evolution of MSCI ESG Indexes —— 212	
4.12	China Longyuan Power Capacity and Output Change, 2009–2019 —— 213	
4.13	CLP Holdings' Climate Vision 2050 CO_2 Intensity Reduction Target —— 215	
4.14	Growth of Stock Exchange Sustainability Activities —— 217	
4.15	Annual Green Bonds Issuance through June 2020 Compiled by Bloomberg —— 222	
4.16	Annual Sustainability-Linked Loans Compiled by Bloomberg (Adapted) —— 229	
4.17	China's Largest Banks Green Loans (Billion Yuan) —— 231	
4.18	Asia-Pacific (ex-Japan) Green and Sustainability-Linked Loans League Table —— 237	
4.19	Coal Power FIDs by Financing Mechanism, 2014–2019 —— 240	
4.20	Global Future Fund Flows Poll Straw Summary —— 241	
4.21	Asian Development Bank Institute Securities Token Offerings SWOT Analysis —— 246	
4.22	NASDAQ Composite 1990s dot-com Bubble (1995–2002) —— 249	
4.23	NASDAQ Composite 2010s Tech Bubble —— 250	

List of Tables

2.1	Average Rates of Increase in GDP, Final Energy Consumption, Electricity Sales, and Peak Demand —— 23	
2.2	Power Generation and Population Size of Key Power Markets in Asia in 2018 —— 25	
2.3	China Electricity Consumption Estimates by Various Domestic Institutions —— 28	
2.4	National Electric Car Deployment Targets in Some Key Asian Markets —— 32	
2.5	Progress of Measures to Improve Energy Efficiency —— 40	
2.6	Status of Electric Power Market Reform in Key Asian Economies —— 54	
2.7	Markets Where Domestic Utilities Invest Abroad —— 55	
2.8	Australia's Key Energy Retailers' Basic Services Profile —— 88	
2.9	New Zealand versus Australia Power Market —— 90	
2.10	ENN Energy Key Growth Indicators 2001–2019 —— 98	
3.1	IRENA's Renewable Energy Forecasts —— 111	
3.2	World's Net Electricity Generation from Renewables —— 112	
3.3	Global Power Generation and Renewables —— 112	
3.4	Japan, South Korea, Taiwan Energy Mix —— 119	
3.5	Japan's Solar and Wind Capacity Forecast by the Renewable Energy Institute —— 121	
3.6	South Korea Capacity Addition Plan —— 122	
3.7	Taiwan'S Government Solar and Wind Capacity Forecast —— 122	
3.8	China Installed Capacity Projections (Adapted) —— 136	
3.9	Hydrogen Production Pathways Comparison (World Energy Council) —— 142	
3.10	China's Hydrogen and FCEVs Industry Overall Targets —— 143	
3.11	Sample of Hydrogen-Related Projects by Chinese Energy Companies in China —— 143	
3.12	Digital Applications Categories Based on Findings in Recent Literature (Adapted) —— 153	
3.13	Illustrative Potential Worldwide Cost Savings from Enhanced Digitalisation in Power Plants and Electricity Networks over 2016–2040 (Adapted) —— 161	
3.14	Schneider Electric Survey of 12 Key Business Benefits of Digital Transformation —— 162	
3.15	KEPCO 2030 Technical Strategies and Developing Eight Core Strategic Technologies (Adapted) —— 167	
3.16	Major Technologies of KEPCO 4.0 (Adapted) —— 168	
3.17	IMD Smart City Index 2019 and 2020 —— 170	
3.18	Irena Innovation Landscape Brief: Blockchain (Adapted) —— 173	
3.19	Snapshot of AI/Blockchain/IoT/Pilot Projects in Individual Asian Power Markets —— 174	
3.20	Size and Scale Summary of Ant Group (Adapted) —— 178	
3.21	Leading Enterprises in Key Smart City Technology Domains (Adapted) —— 184	
3.22	China's Main Grids Financial Snapshot (Six Months to June 2020) —— 185	
4.1	AGL Energy Generation Portfolio Performance (December 2019) —— 216	
4.2	Examples of Companies Committed to Net Zero and Renewables Investing —— 219	
4.3	Types of Green Bonds by the Climate Bonds Initiative —— 223	
4.4	Industries Issuing Green Bonds (Adapted) —— 226	
4.5	Recent Asian Banks Sustainable Finance Commitments —— 230	
4.6	Asia's Key Listed Utilities (7 August 2020) —— 232	
4.7	Key Listed Utilities Asia Green Loans —— 234	
4.8	Asia Key Listed Utilities' Long-Term Debt —— 234	
4.9	Definition of Four Subsets of Cryptographic Assets —— 244	
4.10	IPOs, STOs, and ICOs Comparison —— 245	
4.11	Cost of a STO versus Traditional Private Placement Over Five Years —— 247	
4.12	Tesla, Inc. 2Q 2020 Financial Results —— 251	

https://doi.org/10.1515/9783110699852-007

List of Abbreviations

$	United States Dollars
4G	fourth generation of cellular data technology
5G	fifth generation of cellular data technology
AC	alternating current
ADB	Asian Development Bank
AI	artificial intelligence
AIIB	Asian Infrastructure Investment Bank
ASEAN	Association of Southeast Asian Nations
B2C	business-to-customer
BEV	battery electric vehicle or battery-powered electric vehicle
China	People's Republic of China
CO_2	carbon dioxide
CSG	China Southern Power Grid
DC	direct current
Deloitte	Deloitte Touche Tohmatsu Limited (professional service network of firms)
EBA	European Battery Alliance
EC	European Commission
EMS	Energy Management System
Energy 4.0	refers to the digitalisation of energy industry technologies and processes
ESCO	Energy Service Company
ESG	Environmental, Social, and Governance
ETS	Emissions Trading System
EU	European Union
EV	electric vehicle
EY	Ernst & Young Global Limited (professional services network of firms)
FCEV	fuel cell electric vehicle
HEV	hybrid electric vehicles
ICE	internal combustion engine
ICO	initial coin offering
IEA	International Energy Agency
IFC	International Finance Corporation
IoT	Internet of Things
IPO	initial public offering
KPI	key performance indicator
KPMG	KPMG International Limited (professional services network of firms)
LCOE	levelised cost of energy
LED	light-emitting diode
LNG	liquefied natural gas
LPG	liquefied petroleum gas
McKinsey	McKinsey and Company
METI	Ministry of Economy, Trade and Industry, Japan
NREL	National Renewable Energy Laboratory
PHEV	plug-in hybrid electric vehicle
$PM_{2.5}$	refers to atmospheric particulate matter with a diameter of less than 2.5 micrometres
Prosumer	refers to a person/entity consuming and producing energy that can be sold to others

PV	photovoltaics
PwC	PricewaterhouseCoopers (professional services network of firms)
RCEP	Regional Comprehensive Economic Partnership
SGCC	State Grid Corporation of China
STO	security token offering
UHV	ultra-high voltage
UK	United Kingdom
UN	United Nations
US	United States
yuan	Chinese Yuan or Renminbi

Note on Currency Exchange Rates

As of 1 November 2020, one United States dollar was equal to:

Australian dollars	1.43
Chinese yuan	6.69
Euros	0.86
Hong Kong dollars	7.75
Indian rupees	74.57
Indonesian rupiahs	14,671
Japanese yen	104.83
Korean won	1,137
Malaysian ringgits	4.15
New Zealand dollars	1.51
Singapore dollars	1.37
Taiwan dollars	28.63
Thai bhat	31.15

About the Author

 **Giuseppe (Joseph) Jacobelli** is a recognised expert with over 30 years of experience in Asia's energy and capital markets, and is driven by a passion for green and sustainable development. He spent over 25 years as a financial analyst at Bloomberg Intelligence (BI) and at several leading global investment banks, including Bear Stearns, HSBC, and Merrill Lynch, attaining top rankings in Asiamoney, Greenwich, and Institutional Investor. At BI, Joseph was instrumental in building its Asia operations and served as a team leader and regional research content leader, leveraging his expertise and contacts developed across Asia's markets. Joseph worked at two energy corporations. As a group director at CLP Holdings, he gained broad execution expertise in the alternative power and carbon transactions as well as clean tech investments. He leads Asia project sourcing, development and financing for green electricity start-up Cenfura and is the founder of investment and advisory firm Asia Clean Tech Energy Investments. He is a former AmCham Hong Kong Energy Committee Chair and is a Member of the Hong Kong Institute of Directors. He regularly speaks at industry conferences and appears in broadcast, digital and print media. Joseph studied Mandarin in the 1980s and received a Master's in International Law from the National Taiwan University. He speaks English, French, Italian, and Mandarin as well as some Catonese, Japanese, and Portuguese.

Index

4G 163, 273
5G 9, 163, 180–181, 185, 187, 189, 197, 263, 273

Aden Group VII, 45, 162
AGL Energy Ltd. 213
Ant Group Co., Ltd. 178, 195
artificial intelligence (AI) XIII, 13, 35, 49, 92, 151, 155, 157, 159, 164, 166, 169, 171–174, 176–177, 179–181, 183, 188–189, 195, 197, 201, 262–263, 273
Asian Development Bank (ADB) 4, 13, 20, 148, 244, 246, 257, 273
Asian Infrastructure Investment Bank (AIIB) 225, 227, 255, 273
Association of Southeast Asian Nations (ASEAN) 7, 11, 38, 118, 273
Australia 1–2, 5, 12, 21, 24, 33, 38–39, 51–54, 56–59, 61–62, 65, 69, 73, 80, 82, 84, 86, 88–91, 96, 99–100, 103–105, 109, 112–118, 123, 141, 144, 149, 158, 174, 176, 184, 190–191, 201, 210–211, 213–215, 232, 237, 254, 257, 260–262, 264
Australian Energy Market Commission (AEMO) 53, 57, 60, 115–117, 191
Australian–ASEAN Power Link (a.k.a. Sun Cable project) 117–118

battery electric vehicle (BEV) 28, 273
BCPG Public Company Limited 175, 195
blockchain XIII, 92, 151, 155–156, 163–164, 169, 171–172, 174–177, 181, 189, 195, 201, 243, 245, 257–258, 262–263
bonds
– green bonds 218, 221–222, 224–225, 227–228, 233, 242, 263
– thematic bonds 221
BP plc 5–6, 20, 24–26, 97, 107, 111, 119–120, 132–133, 148, 190–192, 219

carbon dioxide (CO_2) 31, 103, 107, 123–127, 141, 192, 214–215, 224, 239, 273
Carbon Disclosure Project (CDP) 238, 256
China (People's Republic of China) XII, XIII, 24–5, 8–9, 11–15, 17–21, 26–29, 31, 33, 35–36, 48–49, 54, 62, 75–78, 80–84, 86, 89, 97, 99, 101–102, 104–106, 109–110, 112–113, 115, 118, 123–138, 140–144, 148, 151–152, 157, 162–164, 174, 176–177, 179–181, 183–189, 192, 195–197, 201–202, 210–211, 213–214, 221, 227–232, 235, 237, 240, 253, 255–256, 260–261, 263–264, 273–274
China Southern Power Grid Company Limited (CSG) 263, 273
Climate Bond Initiative (CBI) 224, 227
Climate Policy Initiative (CPI) VII, 227, 252
CLP Holdings Ltd. 213–214
coal XII, 29, 66, 78, 81–82, 88, 101–102, 106–110, 113–115, 118, 121, 123–124, 127–128, 132, 135, 137, 146–151, 182, 191, 193, 213–215, 228, 233, 239–240, 242, 252, 257, 261, 263
COVID-19 XI, 4–5, 11–12, 18, 21, 27, 31, 33, 72–73, 103, 123, 149, 152, 201, 208, 218, 225, 229, 231, 248, 252–253

Defi 243
Digital Twins 155, 162
digitalisation VII, XII, XIII, 16–88151, 153, 157, 161, 168–169, 176–177, 183–186, 189, 194, 198, 201, 263, 273

electric vehicle (EV) 9, 21, 28–29, 31–36, 68, 88, 103–104, 167, 180, 207, 225–226, 250, 260, 273
Emissions Trading System (ETS) 273
Energy 4.0 153, 155–160, 162–164, 166–167, 174, 176, 179, 181–184, 187, 189, 262–264, 273
Energy Management System (EMS) 44–47, 273
Energy Market Authority, Singapore (EMA) 70–71, 73–74, 169, 195
Energy Service Company (ESCO) 273
EnergyAustralia 59, 86–87, 89
Energys Group Limited 50
ENN Energy Holdings Limited 98
environmental, social, and governance (ESG) 42, 203–206, 208, 212, 214, 216–218, 225–226, 233, 236, 239, 251–255, 273
European Commission (EC) 34, 124–125, 150, 209, 252, 273

European Union (EU) 34, 38, 107, 125, 141, 192, 255, 261, 273

fintech 13, 198, 252–253, 257, 263
fuel cell electric vehicle (FCEV) 142, 273
Fukushima Daiichi Nuclear Power Plant meltdown 40, 62, 110, 119, 130

gas 30, 43, 57, 61, 65–67, 69, 73, 80–81, 83–88, 91–94, 97–101, 107–110, 113–115, 118, 121, 127–128, 130, 135, 137, 146–149, 158, 168–169, 191–193, 206, 215, 219, 224, 236, 260–261, 273
– liquefied natural gas (LNG) 59, 110, 273
– natural gas 59, 72, 83–84, 91, 97, 99, 106, 113–114, 118, 146, 148, 150–151, 168, 260–262
Global Energy Interconnection Development and Cooperation Organization (GEIDCO) 135–137, 139–140

Hong Kong VII, 5–6, 20–21, 24–25, 33, 41–45, 48–49, 97, 99, 101, 103–104, 174, 176–178, 187, 195, 198, 211, 213–214, 217, 229, 232, 236, 253, 256, 260, 274
Hong Kong Exchanges and Clearing Limited (HKEX) VII, 178, 195, 217, 254
hybrid electric vehicles (HEV) 273
hydrogen 69, 118, 135, 138, 141–143, 149–150, 168, 182, 193

I Switch Pte Limited (iSwitch) 84, 96–97, 260
Indonesia XII, 2, 5, 26, 31, 54, 75, 99, 118, 168, 176, 210, 235
initial coin offering (ICO) 244, 247, 273
InterContinental Energy VII
International Energy Agency (IEA) 24, 29–30, 32, 37, 103–104, 141, 148–149, 159, 161, 193–194, 239–240, 256, 273
International Finance Corporation (IFC) 235, 256, 273
International Renewable Energy Agency (IRENA) 110–111, 144–145, 172–173, 261
Internet of Things (IoT) XIII, 9, 156, 164, 171–174, 176, 179–181, 183–184, 189, 193, 195–196, 201, 262–263, 273
IoT 151, 156–157, 166, 169, 171

Japan XII, XIII, 2–58, 9, 12, 16, 21–22, 24–27, 29–31, 33, 40, 48, 54–55, 62–69, 80–82, 84, 89, 91–93, 98–100, 109–110, 118–121, 128, 130, 148–150, 158, 174–176, 191, 201, 208, 210, 232–233, 237, 240, 242, 246, 257, 260–261, 273

levelised cost of energy (LCOE) 144, 146–147, 159, 192, 262, 273
liquefied natural gas (LNG) 59, 96–97, 100, 109, 114, 150, 190, 193, 273

Ministry of Economy, Trade and Industry, Japan (METI) 40, 150, 193, 273

National Development and Reform Commission, People's Republic of China (NDRC) 81, 182, 228
New Zealand 1, 5, 21, 33, 51, 54, 84, 89–91, 96, 174, 201, 240, 242, 257, 260, 274

oil 8–9, 28–29, 33, 67, 80–81, 83, 93, 96–97, 106–110, 118, 128, 141, 149, 151, 168, 191, 219, 261–262
Organisation for Cross-regional Coordination of Transmission Operators, Japan (OCCTO) 63, 65–66

photovoltaics (PV) 43, 66, 135, 144, 146–147, 154, 169, 192, 274
plug-in hybrid electric vehicle (PHEV) 28, 273
$PM_{2.5}$ 124, 273
Principles for Responsible Investment (PRI) 205, 216

renewable energy 38, 42–43, 66, 69, 77–78, 82, 109–112, 115, 118, 120–121, 128, 134, 137, 141, 150, 154, 163, 165, 168, 175, 182, 189, 207, 213–214, 220–221, 224, 232, 237, 249

security token offering (STO) 244, 247, 252, 274
Singapore 5–6, 8, 17, 21, 24, 32–33, 41–42, 48, 54, 62, 69–71, 73–74, 82, 84, 89, 96–97, 100, 103, 118, 168–169, 174–175, 191, 195, 201, 220, 230, 232, 236, 240, 253, 256, 259–260, 262, 274
solar power XII, 62, 106, 114, 122, 130–131, 144, 146, 168, 216, 220, 228, 250

South Korea (Republic of Korea) 9, 11, 13, 17, 21, 29, 31, 48, 54, 81, 86, 106, 109, 115, 118–121, 128, 131, 148, 164–165, 167, 174, 210, 232, 261
State Grid Corporation of China (SGCC) 77, 196, 263, 274
Swire Properties Limited 42–44

Taiwan 6, 9, 11, 17, 20, 29, 45, 54, 118–122, 176, 191, 261, 274
Task Force on Climate-related Financial Disclosures (TCFD) 209, 254
Tesla, Inc. 251
tokens 248
Tokyo Electric Power Company Holdings, Incorporated (TEPCO) 22–23, 65, 93, 95

Uniform Singapore Energy Prices (USEP) 72
United Nations (UN) VII, 1–2, 20, 25–26, 123, 127, 192, 205–206, 210–211, 216, 224, 238–239, 241, 252, 254, 274
United Nations Framework Convention on Climate Change (UNFCCC) 127, 206
United Nations Sustainable Development Goals (SDGs) 205–206, 217, 224, 230, 238–239, 241

wind power XII, 62, 121–122, 127–128, 130, 132, 134–135, 138, 144, 213, 262
WM Motor Technology Group Company Limited 36
World Bank Group 2

www.ingramcontent.com/pod-product-compliance
Lightning Source LLC
Chambersburg PA
CBHW080911170426
43201CB00017B/2288